Translated Texts for Historia

300–800 AD is the time of late antiquity and the early middle ages: the transformation of the classical world, the beginnings of Europe and of Islam, and the evolution of Byzantium. TTH makes available sources translated from Greek, Latin, Syriac, Coptic, Arabic, Georgian, Gothic and Armenian. Each volume provides an expert scholarly translation, with an introduction setting texts and authors in context, and with notes on content, interpretation and debates.

A full list of published titles in the **Translated Texts for Historians** series is available on request. The most recently published are shown below.

The Chronicle of Pseudo-Zachariah Rhetor: Church and War in Late Antiquity
Translated by GEOFFREY GREATREX, with ROBERT PHENIX and CORNELIA HORN; introductory material by SEBASTIAN BROCK and WITOLD WITAKOWSKI
Volume 55: 2010; ISBN 978-1-84631-493-3 cased, 978-1-84631-494-0 limp

Bede: On the Nature of Things and On Times
Translated with introduction and notes by CALVIN B. KENDALL and FAITH WALLIS
Volume 56: 371pp., 2010, ISBN 978-1-84631-495-7

Theophilus of Edessa's Chronicle
Translated with introduction and notes by ROBERT G. HOYLAND
Volume 57: 368pp., 2011, ISBN 978-1-84631-697-5 cased, 978-1-84631-698-2 limp

Bede: Commentary on Revelation
Translated with introduction and notes by FAITH WALLIS
Volume 58: 343pp., 2013, ISBN 978-1-84631-844-3 cased, 978-1-84631-845-0 limp

Two Early Lives of Severos, Patriarch of Antioch
Translated with an introduction and notes by SEBASTIAN BROCK and BRIAN FITZGERALD
Volume 59, 175pp., 2013, ISBN 978-1-84631-882-5 cased, 978-1-84631-883-2 limp

The Funerary Speech for John Chrysostom
Translated with an introduction and notes by TIMOTHY D. BARNES and GEORGE BEVAN
Volume 60, 193pp., ISBN 978-1-84631-887-0 cased, 978-1-84631-888-7 limp

The Acts of the Lateran Synod of 649
Translated with notes by RICHARD PRICE, with contributions by PHIL BOOTH and CATHERINE CUBITT
Volume 61, 476pp., ISBN 978-1-78138-039-0 cased

Macarius, *Apocriticus*
Translated with introduction and commentary by JEREMY M. SCHOTT and MARK J. EDWARDS
Volume 62, 476pp., ISBN 978 1 78138 129 8 cased, ISBN 978 1 78138 130 4 limp

Khalifa ibn Khayyat's *History* on the Umayyad Dynasty (660–750)
Translated with introduction and commentary by CARL WURTZEL
and prepared for publication by ROBERT G. HOYLAND
Volume 63, 332pp., ISBN 978 1 78138 174 8 cased, 978 1 78138 175 5 limp

Between City and School: Selected Orations of Libanius
RAFFAELLA CRIBIORE
Volume 65, 272pp, ISBN 978 1 78138 252 3 cased, 978 1 78138 253 0 limp

For full details of **Translated Texts for Historians**, including prices and ordering information, please write to the following: **All countries, except the USA and Canada:** Liverpool University Press, 4 Cambridge Street, Liverpool, L69 7ZU, UK (*Tel* +44-[0]151-794 2233, *Fax* +44-[0]151-794 2235, Email janmar@liv.ac.uk, http://www.liverpooluniversitypress.co.uk). **USA and Canada:** Turpin Distribution, www.turpin-distribution.com.

Translated Texts for Historians
Volume 66

Isidore of Seville

On the Nature of Things

Translated with introduction, notes, and commentary by
CALVIN B. KENDALL and FAITH WALLIS

Liverpool
University
Press

First published 2016
Liverpool University Press
4 Cambridge Street
Liverpool, L69 7ZU

British Library Cataloguing-in-Publication Data
A British Library CIP Record is available.

ISBN 978 1 78138 293 6 cased
ISBN 978 1 78138 294 3 limp

Typeset by Carnegie Book Production, Lancaster
Printed and bound by CPI Group (UK) Ltd, Croydon CR0 4YY

We dedicate this book to the memory of our teachers,
especially Leonard E. Boyle (FW)
and Charles W. Jones (CBK).

CONTENTS

List of Diagrams, Figures, and Tables   xi
Acknowledgements   xii
List of Abbreviations   xiii

Introduction   **1**

1   Isidore's Life, Times, and Writings   3
       Education   6
       Grammar as a principle of knowledge   7
       Church discipline and biblical exegesis   9

2   *On the Nature of Things* in Context   10
       Structure   14
       Occasion   16
       Purposes and preoccupations   18
       Appeal to reason   21
       Wider ends: a Christianized erudition?   24

3   A Work of Composite Construction   27
       Text and image   28
       Fontaine's theory of three recensions   30
       Single or multiple authorship?   32
       The short recension: two types   34
       The medium recension   42
       Three Spanish interpolations?   46
       The long recension: chapter 44(–) and the mystical addition   47

4   Out of Spain and into the Future   51
       Ireland and Anglo-Saxon England   55
       Traffic between Spain and Italy   57
       Gaul   59
       Germany and Switzerland: the Zofingen metamorphosis   60
       From the Carolingian period to the age of print   63

5 Inventory of Manuscripts and Editions of Isidore's *On the*
   Nature of Things   66
        Manuscripts of Isidore's *De natura rerum*   66
        Editions of Isidore's *De natura rerum*   100

6 Principles Governing this Translation   102

The Text   **103**

Isidore of Seville, *On the Nature of Things*   105
Preface: Isidore, to his Lord and Son, Sisebut   107
List of Chapters   109
 1 Days   111
 2 Night   114
 3 The Week   115
 4 The Months   117
 5 The Concordance of the Months   120
 6 The Years   120
 7 The Seasons   123
 8 The Solstice and the Equinox   126
 9 The World   127
10 The Five Circles of the World   128
11 The Parts of the World   130
12 Heaven and Its Name   133
13 The Seven Planets of Heaven and Their Revolutions   135
14 The Heavenly Waters   136
15 The Nature of the Sun   137
16 The Size of the Sun and the Moon   138
17 The Course of the Sun   139
18 The Light of the Moon   141
19 The Course of the Moon   143
20 The Eclipse of the Sun   144
21 The Eclipse of the Moon   145
22 The Course of the Stars   146
23 The Position of the Seven Wandering Stars   147
24 The Light of the Stars   150
25 The Fall of the Stars   150
26 The Names of the Stars   151
27 Whether the Stars have a Soul   155
28 Night   156

29	Thunder	157
30	Lightning	157
31	The Rainbow	159
32	Clouds	159
33	Rains	160
34	Snow	161
35	Hail	161
36	The Nature of the Winds	162
37	The Names of the Winds	163
38	Signs of Storms or Fair Weather	165
39	Pestilence	167
40	The Ocean's Tide	168
41	Why the Sea Does Not Grow in Size	169
42	Why the Sea has Bitter Waters	170
43	The River Nile	170
44	The Names of the Sea and the Rivers	170
45	The Position of the Earth	172
46	Earthquake	173
47	Mount Etna	174
48	The Parts of the Earth	175

Commentary 177

Appendices 255

1	The Verse Epistle of King Sisebut	257
2	Introductory Formulas for the Diagram of the Winds (Diagram 7) in Chapter 37	261
3	Extracts from Chapter 37 arranged within the Diagram of the Winds	263
4	*The Poem of the Winds*	265
5	Textual Insertions in Chapter 48 and T-O Map	272
6	The Zofingen and English Types of the Long Recension	275

Bibliography	281
Index of Sources	298
General Index	303

DIAGRAMS, FIGURES, AND TABLES

Diagram 1   The Months   119
Diagram 2   The Seasons   125
Diagram 3   The Circles of the World   129
Diagram 4   The Elements   131
Diagram 5   The Macrocosm and Microcosm   132
Diagram 5A   The Phases of the Moon   143
Diagram 6   The Planets   149
Diagram 7   The Winds   164

T-O Map (diagram)   The World   176

Figure 1   Calcidius' Parallelepiped   199
Figure 2   Eastwood's Reconstruction   201
Figure 3   Zone Diagram Showing 8 Angles   205
Figure 4   Zone Diagram Showing 5 Lines   206
Figure 5   Macrobian Spheres   207

Table 1   Chart of the Three Recensions of Isidore's *De natura
             rerum*   36–41
Table 2   The Names of the Planets in Isidore's *De natura
             rerum* and *Etymologiae*   222
Table 3   The Names of the Winds in Isidore's *De natura
             rerum* 37 and Diagram 7   238
Table 4   The Chapters of the Zofingen Type, with their
             Equivalents in the Regular Order of *De natura rerum*   275

ACKNOWLEDGEMENTS

We welcome this opportunity to express our thanks to the many colleagues and friends who have helped us to bring Isidore's extraordinary survey of the universe into English. Our gratitude to Mary Whitby, who bore with our delays so graciously, and who read the long-awaited translation with thoroughness and care, is of long standing and is deeply felt. Nicholas Parmley, acting as our research assistant for the academic year 2011/2012, prepared an annotated bibliography of Spanish and Portuguese scholarship on Isidore, which greatly facilitated the initial stages of our journey. John Contreni and Wesley Stevens responded generously to enquiries and supplied correctives and new information at various steps along the way. At journey's end, Michael Lapidge, the external reviewer for the Press, gave us both encouragement and highly detailed and judicious suggestions for improvement. We are grateful to the directors of the Biblioteca Nacional de Catalunya, the Real Biblioteca del Monasterio del Escorial, and the Museu Episcopal de Vic for permission to examine the MSS of Isidore of Seville's *De natura rerum* in their keeping, and to the staff of the Interlibrary Loans division of the McGill University Libraries, who ransacked the planet to find some very obscure publications. A University of Minnesota Professional Development Grant for Retirees from the Office of the Vice President for Research and the University of Minnesota Retirees Association provided timely support for this project. And finally a special word of thanks for helping to create the line drawings to represent and typify the diagrams that are a feature of Isidore's work goes to Faith's husband Kendall Wallis.

ABBREVIATIONS

Ambrose, *Hex.*	*Ambrose, Hexaemeron*
Anth. Lat.	*Anthologia Latina* (ed. Buecheler and Riese)
ASE	*Anglo-Saxon England*
ASMMF	*Anglo-Saxon Manuscripts in Microfiche Facsimile*
Augustine, *DCD*	Augustine, *De ciuitate Dei*
Augustine, *DGAL*	Augustine, *De Genesi ad litteram*
Augustine, *En. in Ps.*	Augustine, *Enarrationes in Psalmos*
Bede, *DNR*	Bede, *De natura rerum*
Bede, *DT*	Bede, *De temporibus*
Bede, *DTR*	Bede, *De temporum ratione*
Cassiodorus, *Institutions*	Cassiodorus, *Institutions of Divine and Secular Learning*
CCSL	Corpus Christianorum Series latina
Clement/Rufinus, *Recognitiones*	Rufinus, [Ps.-Clementis] *Recognitiones Rufino Interprete*
CLA	*Codices Latini Antiquiores*
CSEL	Corpus Scriptorum Ecclesiasticorum Latinorum
Fontaine, *Isidore et la culture*	Fontaine, *Isidore de Séville et la culture classique dans l'Espagne wisigothique*
Fontaine, *Isidore: Genèse*	Fontaine, *Isidore de Séville: Genèse et originalité de la culture hispanique au temps des Wisigoths*
Fontaine, *Traité*	Fontaine, *Traité de la nature*

Isidore, *DNR*	Isidore, *De natura rerum*
Isidore, *Etym.*	Isidore, *Etymologiae*
Kendall/Wallis	Kendall and Wallis, *Bede: On the Nature of Things and On Times*
Lewis and Short	Lewis and Short, *A Latin Dictionary*
MGH: AA	Monumenta Germaniae Historica: Auctores Antiquissimi
OLD	*Oxford Latin Dictionary*
PG	Patrologiae cursus completus. Series graeca
PL	Patrologiae cursus completus. Series latina
Pliny, *NH*	Pliny, *Natural History*
TTH	Translated Texts for Historians
Wallis, 'Calendar & Cloister'	Wallis, 'The Calendar and the Cloister'
Wallis, *Reckoning*	Wallis, *Bede: The Reckoning of Time*

Introduction

ISIDORE'S LIFE, TIMES, AND WRITINGS

Chapter 44 of Isidore of Seville's *On the Nature of Things* opens with an explanation of the difference between 'the Ocean' and 'the Sea': 'the external sea is the Ocean; the internal sea is the Sea which flows out of the Ocean'. The great Ocean was thought to ring the whole *orbis terrarum*, and 'our Great Sea' – the Mediterranean – was enclosed by this land-mass.[1] The distinction between Ocean and Sea distils the conviction of Graeco-Roman civilization that its exceptional character was inscribed in the order of nature itself. As chapter 44 goes on to explain, the Sea is the nucleus of human habitation; it is articulated into the Upper and Lower zones, and punctuated by straits, estuaries, shoals, and gulfs. 'The Ocean', by contrast, is literally peripheral; it also has no features, save the boundary which separates it from the Sea. Culturally as well as geographically the Ocean is 'outside'; the Sea is 'inside'.

While there is some debate whether chapter 44 was part of the plan of *On the Nature of Things*, or even composed by Isidore of Seville[2] – this juxtaposition of Ocean and Sea can still serve as a fitting emblem of Isidore's life, writings, and historical situation.[3] Isidore was born around 560 into a family from Cartagena, on the shores of 'the Sea', but at the time of his birth the family was already displaced from their lands, and the Byzantine occupation of 552 and subsequent conflicts definitively

---

1 Isidore, *DNR* 48.3. Isidore was the first to use the term 'Mediterranean', not in *DNR*, but in *Etym.* 13.16.1; see Peregrine Horden and Nicolas Purcell, *The Corrupting Sea: A Study of Mediterranean History* (Oxford: Blackwell, 2000), 12 (erroneously referenced there as *Etym.* 12.16.1).

2 This issue is discussed below, pp. 44–50.

3 Details of Isidore's biography are skimpy, and must be extracted from his own works and rare documentary sources such as the acts of the Councils of Toledo; see Jacques Fontaine, *Isidore de Séville: Genèse et originalité de la culture hispanique au temps des Wisigoths* (Turnhout: Brepols, 2000), esp. chs. 4–7, as well as summary notices by the same author in *Lexikon des Mittelalters* 5, cols. 677–80 and *Reallexikon für Antike und Christentum* 18, cols. 1002–27.

prevented their return.[4] For reasons unknown, they chose to settle in permanent exile in the province of Baetica. On the death of his father Severianus, the oldest son Leander took charge of his siblings Fulgentius, Florentina, and Isidore. All were destined for careers in the Church: Fulgentius eventually became bishop of Ecija (ancient *Astigi*) near Seville, Florentina became a nun, and Isidore succeeded Leander as bishop of Seville around 600, serving as bishop until his death on 4 April 636.[5] Thus Isidore came from a solidly Mediterranean background, but he spent his entire life in a region which arcs across the northern flank of the Straits of Hercules (Gibraltar), and thus fronts both 'the Sea' and 'the Ocean'. His episcopal see, Seville (ancient *Hispalis*), is situated on the banks of the Guadalquivir River, which empties into the Atlantic 80 kilometres downstream. The river is navigable up to Seville, and from Phoenician times this city, along with Cadiz, was a major emporium on the 'Ocean' flank of Spain.

Isidore's life likewise unfolded across an uncertain cultural and historical frontier between the end of Antiquity and the beginning of the Middle Ages. Though period labels such as these are always contested, and their usefulness is limited, a persuasive argument can be made that the end of the sixth and beginning of the seventh centuries marks a watershed.[6] The Byzantine effort to reconquer North Africa, Spain, and Italy resulted in little but destruction and disruption, and ultimately failed to achieve Justinian's goal of reconstituting the old Empire. War with a resurgent Persian Empire brought Constantinople to its knees; but, just as significantly, there was little confidence amongst even the Romanized peoples in the western provinces that a return to the imperial *status quo ante* was possible or even desirable. The eastern Emperors were not seen as saviours of civilization. Their policies seemed to oscillate between neglecting the West and bullying its Church with novel government-sponsored heresies.

---

4 Jacques Fontaine, 'Qui a chassé le carthaginois Sévérianus et les siens? Observations sur l'histoire familiale d'Isidore de Séville', in *Estudios en Homenaje a Don Claudio Sánchez Albornoz en sus 90 años, I* (Buenos Aires: Instituto de Historia de España, 1983), 349–400.

5 The four siblings were all eventually canonized by the Catholic Church.

6 Admittedly, the case for this watershed is stronger in Byzantium than in the West. The death of the emperor Maurice in 602 is the terminus of A.H.M. Jones's magisterial *The Later Roman Empire, 284–602: A Social, Economic and Administrative Survey* (Oxford: Blackwell, 1964); Jones dates 'the collapse of the East' (p. v) to this point. These views are echoed by Andrew Louth, 'The Byzantine Empire in the Seventh Century', in Paul Fouracre (ed.), *The New Cambridge Medieval History*, vol. 1, c.*500–c.700* (Cambridge: Cambridge University Press, 2005), 291–316.

Isidore's ambivalence towards the Roman Empire, and his awareness of living in a post-Roman world, can be detected in his historical writing. He composed two works of history: two versions of a *Chronicle*,[7] and the *History of the Goths* (more correctly, *The Origins of the Goths*). And while he appreciated the role that history played in the classical theory of literary genres – and hence its significance for his own activity as a 'grammarian' – as well as the moral function of history as a source of instruction (*institutio*) and examples (*Etym.* 1.43), his own histories are not classical in conception. The chronicle represents the new Christian genre of synthetic universal history. Its pioneers were Eusebius of Caesarea and his Latin translator and continuator Jerome; but Isidore innovates even further by structuring his *Chronicle* according to Augustine's schema of the Six Ages of the World, and by beginning his story at Creation. As Jacques Fontaine observes, this reveals Isidore's interest in wedding the story of the *world* with the story of *time*, both stemming from God's initial act of Creation. The quest for origins is manifest in all domains of Isidore's thought, including in *On the Nature of Things*.[8] This obsession with origins may seem conservative, even antiquarian; but it was also the key to escaping traditional forms of historical thinking. This is particularly evident in the second edition of the *History of the Goths*, composed in 624.[9] While other 'barbarian histories' dignified the new Germanic rulers with origins that placed them within the sphere of Graeco-Roman ethnography and history, Isidore gives the Goths an additional pedigree: they are not only the descendants of the Scythians described in classical sources, but also of the biblical nations of Gog and Magog. On both planes, the Goths pre-date and out-rank the Romans. The *History* ends, significantly, with the defeat of the Byzantines by Suinthila, whom Isidore characterizes as the first to hold sovereign kingship over all of Spain. The *History* is organized chronologically by the years of the Spanish Era,[10] with parallel imperial and Visigothic regnal dates. It conveys the message

---

7 The *Chronica maiora* and the *Chronica minora* or *Chronicorum epitome*. The *Chronica minora* was incorporated into the *Etymologies* as chs. 38–39 of book 5.

8 Fontaine, *Isidore: Genèse*, 222. On Isidore's conception of history-writing in relation to *grammatica*, see Fontaine, *Isidore de Séville et la culture classique dans l'Espagne wisigothique*, 2nd edn. (Paris: Études augustiniennes, 1983), 180–85.

9 *Historia Gothorum Wandalorum Sueborum*, ed. Theodor Mommsen, MGH: AA 11.2 (1894), 267–303; trans. Kenneth Baxter Wolf, *Conquerors and Chroniclers of Early Medieval Spain*, 2nd edn. TTH 9 (Liverpool: Liverpool University Press, 1999), 79–110.

10 Year 1 of the Spanish era, which remained in use in much of Spain until late in the Middle Ages, was 38 BC.

that Spain is an empire unto itself, and its Visigothic kings are successors of a defunct Roman hegemony. The Visigoths came from far away, but their divinely destined *patria* is Spain, where they will unite with the Hispano-Romans in a new Christian kingdom. This destiny is announced at the beginning of the second edition of the *History* by an encomium of Spain itself, the fertile 'mother of princes and of peoples'.[11]

EDUCATION

As a youth, Isidore received an excellent education under the tutelage of his brother Leander, though whether this took place within the cathedral, in Leander's adjacent monastery, at home, or even in a public school – or all of the above – is impossible to determine. On the other hand, we know a great deal about aspirations concerning clerical education in Visigothic Spain, thanks to the canons of the 'national' church synods. From the 530s onwards, these synods, held at Toledo, laid the groundwork for the revival of clerical learning throughout the entire peninsula after the terrible material devastations of the 400s. The council of 527 contains the oldest reference to education in canon law; it speaks of founding new schools where there are none, and restoring older ones. The acts of the 531 synod mandated formal schooling within cathedrals. These provisions were explicitly designed to ensure a steady supply of trained and orthodox clergy that would unify the Spanish church under its Catholic kings. A more uniform liturgy would require literacy, which meant proficiency in reading Latin; so likewise would the need to convey the teachings of Scripture and the Fathers in sermons that were doctrinally correct and spiritually edifying. Isidore's own conception of education therefore stressed proficiency in Latin *grammatica* – a term which encompasses not only grammar in the restricted sense, but also rhetoric, semantics, and lexicography; and, for Isidore, as for generations of Roman students, this entailed knowledge of classical literature and the secular sciences. Isidore assumed, and with good reason, that this traditional erudition would be available in monastic or cathedral schools. Between 550 and 570, the African abbot Donatus fled to Spain from the disturbances of

---

11 Isidore's Spanish patriotism and coolness towards Constantinople is affectingly paralleled by Pope Gregory I's anxiety over *terra mea*, i.e., Italy, and his growing accommodation to post-imperial realities; see R.A. Markus, *Gregory the Great and his World* (Cambridge: Cambridge University Press, 1997), esp. chs. 6–7.

the Byzantine reconquest with 70 monks and a large quantity of books, many apparently of a didactic and scholarly character. He established a monastery in Spain, and his successor Eutropius was a personal friend of Leander of Seville. Isidore's literary executor Braulio was educated in the 'secular disciplines' by his brother in the monastery of Zaragoza. Isidore could also take it for granted that lay elites would also receive this kind of traditional Roman education.[12]

GRAMMAR AS A PRINCIPLE OF KNOWLEDGE

Isidore's vision of a renovated Christian Spain entailed a programme of cultural and educational reform supporting and supported by the monarchy, but designed and executed under the leadership of the Church. His astounding literary output was dedicated to this project, and his choice of topics was dictated by the need to shape a Christian consensus around a core of biblical and Patristic erudition.[13] This required recovering and refocusing the heritage of ancient and Christian learning. To this end, Isidore composed approximately 20 works on *grammatica*, Church discipline, biblical exegesis, and history.[14] His grammatical works mirror the classical categories of grammatical analysis: differences, synonyms, etymologies. Book 1 of the *Differences* (*De differentiis*) treats distinctions

12 Fontaine, *Isidore et la culture*, 6–10 and 1021–22 and Fontaine, 'Fins et moyens de l'enseignement ecclésiastique dans l'Espagne wisigothique', *La scuola nell'occidente latino dell'alto medioevo*, Settimane di studi del Centro italiano di studi sull'alto medioevo 19 (Spoleto: Presso la sede del Centro, 1972), 1:145–202; Pierre Riché, *Education and Culture in the Barbarian West From the Sixth through the Eighth Century*, trans. John J. Contreni (Columbia: University of South Carolina Press, 1976), 262–65.

13 On the 'Isidorian renaissance', its aims and methods, see Yitzak Hen, *Roman Barbarians: The Royal Court and Culture in the Early Medieval West* (Basingstoke and New York: Palgrave Macmillan, 2007), ch. 5.

14 Isidore's basic bibliography is established by the list prepared by his disciple Braulio of Zaragoza, *Renotatio librorum domini Isidori*, in *Scripta de vita Isidori Hispalensis episcopi*, ed. José Carlos Martín, CCSL 113B (Turnhout: Brepols, 2006), 195–207; see also Fontaine, *Isidore: Genèse*, App. I, 431–35. While this list is probably in chronological order, the exact dating of many of Isidore's works is uncertain. For up-to-date discussion, see Jacques Elfassi, D. Poiret, Carmen Codoñer, José Carlos Martín, and M. Adelaida Andrés Sanz, 'Isidorus Hispalensis Ep.', in Paolo Chiesa and Lucia Castaldi (eds.), *La trasmissione* (*Te. Tra.*) 1:196–226 and Carmen Codoñer, José Carlos Martín, and M. Adelaida Andrés Sanz, 'Isidorus Hispalensis Ep.', in Paolo Chiesa and Lucia Castaldi (eds.), *La trasmissione* (*Te. Tra.*) 2:274–417.

between related words, while book 2 applies the same method to related entities.[15] *Synonyms* (*Synonyma*) illustrates synonymy through a dialogue-meditation on repentance and virtue.[16] Isidore's most famous work, the *Etymologies*,[17] is often described as an encyclopaedia, but it is in fact a work of philosophical grammar. It rests on the axiom that a word's etymological *origin* provides insight into the *cause* (and thus the essence) of the thing which the word denotes (1.29.2).[18] Its division into twenty books was the work of Isidore's friend, editor, and bibliographer Braulio (bishop of Zaragoza, d. 651), but it naturally falls into two halves, each representing a classical format for works of comprehensive erudition. Books 1–10 progress from an encyclopaedia of disciplines inspired by Varro (grammar, rhetoric, dialectic, mathematical sciences, medicine, and law: books 1–5) to its fulfilment in the doctrine and worship of the Church (books 6–8). Languages, peoples, and social groups (book 9–10) form the bridge to a survey of natural history after the manner of Pliny the Elder (books 11–14, 16), but broadened to encompass techniques of building, agriculture, spectacle, warfare, clothing, food preparation, and tool-making (books 15, 17–20).

15 See the overview of book 1 by Codoñer in *La trasmissione* (*Te.Tra.*) 2:308–12 and of book 2 by Andrés Sanz in Paolo Chiesa and Lucia Castaldi (eds.), *La trasmissione* (*Te.Tra.*) 2:318–22.

16 See Elfassi's discussion in *La trasmissione* (*Te.Tra.*) 1:218–226. This work enjoyed an important diffusion; see Claudia Di Sciacca, *Finding the Right Words: Isidore's Synonyma in Anglo-Saxon England* (Toronto: University of Toronto Press, 2008).

17 The familiar two-volume edition by W.M. Lindsay, *Isidori Hispalensis Episcopi Etymologiarum sive Originum libri XX* (Oxford: Clarendon Press, 1911), is now being replaced by book-by-book critical editions, with facing-page translations in the vernacular of the editor, published by Les Éditions Les Belles Lettres in Paris since 1983. Our translations, unless otherwise indicated, are from Stephen A. Barney, et al. (trans.), *The Etymologies of Isidore of Seville* (Cambridge: Cambridge University Press, 2006). The bibliography on the *Etymologies* is extensive; see the overview by Codoñer (*Te.Tra.*) 2:274–99.

18 On etymology and, more broadly, grammar as a kind of metaphysics that aims not to reduce the world to the dimensions of language, but expand language to the scope of the world, see Fontaine, *Isidore et la culture*, 40–44 and 202–03; Andy Merrills, 'Isidore's *Etymologies*: on Words and Things', in Jason König and Greg Woolf (eds.), *Encyclopaedism from Antiquity to the Renaissance* (Cambridge: Cambridge University Press, 2013), 301–24; and the unusual but stimulating studies by John Henderson, *The Medieval World of Isidore of Seville: Truth from Words* (Cambridge: Cambridge University Press, 2007) and 'The Creation of Isidore's *Etymologies or Origins*', in Jason König and Tim Whitmarsh (eds.), *Ordering Knowledge in the Roman Empire* (Cambridge: Cambridge University Press, 2007), 150–74.

CHURCH DISCIPLINE AND BIBLICAL EXEGESIS

Compared to his output on grammar, Isidore wrote very little in the way of biblical exegesis apart from the *Expositions* (or *Questions*) *on the Old Testament* (*Mysticorum expositiones sacramentorum* or *Quaestiones in Vetus Testamentum*). Instead, he poured his energy into composing reference works to assist the preacher or devout reader. The *Book of Numbers* (*Liber numerorum*), Isidore's manual of number symbolism, is discussed at greater length below. The *Prefaces* (*Prooemia*) furnish short introductions to the books of the Bible; this is supplemented by biographies of biblical personages in *The Lives and Deaths of the Fathers* (*De ortu et obitu patrum*) and summaries of the typological and moral meaning of various Old Testament and New Testament figures in the *Allegories* (*Allegoriae*). Concern with clerical formation and doctrinal integrity is also reflected in Isidore's works on Church discipline and pastoral care. *On the Origin of Offices* (*De origine officiorum* or *De ecclesiasticis officiis*), dedicated to his brother Fulgentius, expounds both meanings of 'office', i.e., Christian liturgy and the duties of different ranks of the clergy. While *officium* recalls the work of Ambrose, and ultimately Cicero, *origo* reveals Isidore's conviction that pristine tradition was the guarantee of unity and orthodoxy. *On Heresies* (*De haeresibus*), a work of contested authenticity, and *On the Catholic Faith against the Jews* (*De fide catholica contra Iudaeos*), dedicated to his sister Florentina, are based on established Christian genres, but their contents reflect contemporary Spanish conditions. *On Famous Men* (*De uiris illustribus*) builds on Jerome's and Gennadius' prosopographies of Christian worthies, but again with special emphasis on Spaniards. Finally, the *Sentences* (*Sententiae*), after the *Etymologies*, Isidore's most widely copied work in the Middle Ages, and the first to be printed, combines a sort of catechism with a discussion of spiritual life and Christian ethics.

Probably close to the end of his life, Isidore completed the oldest version of an influential canon law collection called the *Hispana* or *Isidoriana*, which culled prescriptions from papal decretals and the acts of Eastern, African, Gallic, and Spanish councils up to and including the Fourth Council of Toledo held in 633.[19] The *Isidoriana* is the ancestor of a number of later compilations in Spain and in the Frankish lands, and even spawned a Carolingian forgery, the *Pseudo-Isidorean Decretals*, incorporating the so-called 'Donation of Constantine'.

---

19 Fontaine, *Isidore: Genèse*, 140–41.

ON THE NATURE OF THINGS IN CONTEXT

Amidst this vast and various literary output, *On the Nature of Things* stands out as something of an anomaly. It cannot be regarded a rehearsal for the later *Etymologies*, because its scope, design, and method are quite different. It is not explicitly presented as an aid to the exegete and preacher, or as a work of *grammatica*, though it partakes of features of both these genres. It is one of the rare works by Isidore which can be precisely dated, and it seems to have been written in haste, which suggests that it might have been triggered by a specific event, possibly the unusual number of lunar and solar eclipses in 611–612.[1] Whatever the cause, the result was something remarkable: the first sustained treatment by a Latin Christian author of cosmography which does not take the form of an exegesis of the biblical creation story.

The Commentary on the present translation attempts to bring to the fore the structure and internal logic of *On the Nature of Things*, to explain what Isidore is doing with his materials, to fill in the background to his exposition, and to address some of the difficulties posed by the text and its illustrations. We think it important nonetheless to explain at the outset some of the challenges this last goal poses. *On the Nature of Things* seems to be mined with contradictions, gaps, and even errors. When it is possible to compare the accounts in *On the Nature of Things* and the *Etymologies*, it is evident that Isidore's later version improves the exposition by filling in the explanatory details or correcting mistakes; occasionally, he seems deliberately to discard problematic material found in *On the Nature of Things*. This reinforces the impression that *On the Nature of Things* was an *œuvre de circonstance*. We should always bear in mind that Isidore was a bishop with many liturgical, administrative, and political responsibilities, and not judge him as if his principal vocation were scholarship, let alone 'science'. He was often obliged to work quickly and discontinuously.[2]

---

1 Fontaine, *Traité*, 3–6 (eclipses of 611–12) and 13 (evidence of hasty composition).
2 Fontaine, *Isidore et la culture*, 332.

On the other hand, differences between *On the Nature of Things* and the *Etymologies* also reflect the different aims of the two works. The hemerology, meteorology, and cosmology of the *Etymologies* are dispersed within a structure based on the liberal arts and the divisions of philosophy; more importantly, the style and fundamental conceptual basis of the two works is quite different. The *Etymologies* is universal in scope, but diverse and particular in its internal organization; *On the Nature of Things* is restricted to time, the cosmos, and 'meteorology'. In *On the Nature of Things*, the significant recurring words are 'natures' and 'causes'; while the *Etymologies* is also concerned with causation, the dominant phrases are 'is derived from' and 'is so named', because it is about language as a metaphysical key to all knowledge.[3]

In many respects *On the Nature of Things* is closer to the *Book of Numbers* than to the *Etymologies*. The authenticity of this work has been contested,[4] but is defended by the editor of the critical edition, Jean-Yves Guillaumin.[5] Its resemblance to *On the Nature of Things* has been noted by Jacques Fontaine. Just as the *Book of Numbers* conveys a numerology quite foreign to the mathematical sections of the *Etymologies*, so *On the Nature of Things* provides a coherent map of the universe which seems implicitly, and sometimes explicitly, to serve pastoral and exegetical ends. Like *On the Nature of Things*, the *Book of Numbers* is original in its conception. Isidore is the first Latin author to compose a technical treatise on number symbolism. Just as *On the Nature of Things* was modelled on pagan cosmological treatises, the *Book of Numbers* is a Christian adaptation for monastic readers and clerical preachers of the classical numerology found in book 6 of Martianus Capella's *Marriage of Philology and Mercury*. It opens by defending number lore against the charge of being 'superfluous', in a gambit not unlike the prologue of *On the Nature of Things* where cosmology is acquitted of being 'superstitious'. Just as *On the Nature of*

---

3 The difference between the *Etymologies* and *On the Nature of Things* should not, however, be overdrawn. Merrills' thoughtful assessment of the *Etymologies* ('Isidore's *Etymologies*') points out that etymology was not the only intellectual method employed in this work, nor was it what medieval readers valued most or found most distinctive. In particular, book 3 on astronomy leans more towards causal than etymological explanations. However, the argument of Bernard Ribémont that *DNR* is merely a dry run or a 'fragment' of *Etymologies* is, in our view, untenable: *Les Origines des encyclopédies médiévales d'Isidore de Séville aux Carolingiens* (Paris: Champion, 2002), 218–39.

4 See Martín, 'Isidorus Hisp.', 407–11.

5 *Liber numerorum / Le Livre des nombres*, ed. with French translation by Jean-Yves Guillaumin (Paris: Belles Lettres, 2005), vii–xi. Guillaumin accepts the dates 612/615.

Things juxtaposes classical physics with biblical symbolism, so the *Book of Numbers* places pagan arithmology side by side with Christian symbolism, without subordinating the secular to the sacred. Isidore's expositions are always first arithmetical, and then numerological. Numbers are part of the natural world as well as Scripture; indeed, the *Book of Numbers* can be read as a sort of parallel cosmology to *On the Nature of Things*.[6]

In *On the Nature of Things* Isidore displays genuine enthusiasm for his subject, and is exceptionally hospitable to ancient pagan scientific learning, notably when compared to Augustine, who dismissed astronomy as next to useless.[7] But even if we take into account his working conditions and particular intentions, there is no evading the fact that *On the Nature of Things* seems to present an impoverished picture of the ancient scientific heritage. Its weaknesses betray the cultural rupture and decline of the age. The natural philosophy and scientific learning available even to one of Isidore's social standing and resources seem to be attenuated, insecure, and derivative, especially when compared to the knowledge base of Latin Fathers of the fourth and fifth centuries like Ambrose and Augustine. Indeed, Isidore often 'reverse engineers' ancient scientific learning by mining it out of the works of Church Fathers who benefited from richer and more vital classical educations. In the process, the Christian source and its context are sometimes occluded. At the same time, *On the Nature of Things* is the product of an intelligent, creative, and even audacious mind, whose ambitions seem to outrun his resources of knowledge and his ability to work optimally with what he has,[8] but whose breadth of vision and appetite for learning command our respect.

Isidore dedicated *On the Nature of Things* to 'his Lord and Son' King Sisebut. Sisebut reigned for a period of about ten years from 612 to 621.[9]

---

6 Fontaine, *Isidore et la culture*, 370–73; Guillaumin, *Liber numerorum / Livres des nombres*, Introduction, xviii–xx, xxiii–xxv. The opening words of the *Book of Numbers* are, 'To pay attention to the principles [*causas*] underlying the numbers found in Scripture is not superfluous' (ed. Guillaumin, 5). Note the use of the term *causas*.

7 *De doctrina christiana* 1.28.43. Eastwood argues that *DNR* intends to put Augustine's programme of learning into action: Bruce S. Eastwood, 'Early Medieval Cosmology, Astronomy and Mathematics', in David C. Lindberg and Michael H. Shank (eds.), *The Cambridge History of Science*, vol. 2, *Medieval Science* (Cambridge: Cambridge University Press, 2013), 306. However, there is no direct or indirect evidence that Isidore drew inspiration from *De doctrina christiana*.

8 This contrast between Isidore's efforts and his resources is frequently noted by Fontaine, *Isidore et la culture*, e.g., pp. 397, 399, 439.

9 His regnal dates are variously given: 612–621 (Fontaine, *Isidore: Genèse*, 132; Luis

Fontaine dates the dedication to 613; if we can accept that date Isidore may have sent a draft of the work to Sisebut in Toledo in the spring of that year.[10] The dedication was accompanied by a prefatory letter from Isidore to the king. While it follows some of the conventions of such addresses, the preface also tells us much about Isidore's conception of his work. First, he emphasizes its originality. Sisebut had developed an interest in 'the nature and causes of things', but apparently there was no suitable treatise to guide his curiosity. Isidore offered the king a work which would explain not only the physical structure and phenomena of the universe, but also 'the reckoning of the days and the months, as well as the periods of the year and the alternation of the seasons'. He claims to have compiled this work based on 'the scholars of antiquity and especially in the works of catholic authors'. Knowledge of the nature of things is 'not superstitious learning' and is particularly appropriate for wise kings, as Solomon himself declared: 'He has given me true knowledge of the things that are, that I may know the disposition of the heavens, and the virtues of the elements … the alterations of their courses, and the changes of seasons, the revolutions of the years, and the dispositions of the stars (preface 2 [Wisdom 7:17–18])'. It bestows the character of divine sanction, and even the force of revelation, on the cultivation of knowledge about the cosmos and time.[11]

Isidore's vigorous apology for secular scientific learning in a Christian *paideia* is worthy of close attention. The overwhelming majority of Church Fathers were happy to accept the classical view of the material universe (with a few adjustments) as both correct and compatible with religious doctrine. But there was a trend in sixth-century Christian thought which opposed the Graeco-Roman model of the universe to the more ancient Middle Eastern world-picture occasionally hinted at in the Old Testament. This trend is represented by the *Christian Topography* ascribed to Cosmas Indicopleustes, a fierce sixth-century polemic in favour of a biblically sanctioned flat, rectangular earth set over a watery abyss, and against the Hellenistic philosopher-scientists' spherical universe. Cosmas was far from being an isolated eccentric, and while his view was distinctly a minority one his book had a surprisingly robust career. The issue was a 'live' one

---

A. García Moreno, *Historia de España Visigoda*, 3rd edn (Madrid: Cátedra, 2008), 147 and 153); 611/12–620 (Roger Collins, *Visigothic Spain 409–711* (Oxford: Blackwell, 2004), 75); 611/12–621 (Barney, et al., *Etymologies*, 7).

10 Fontaine, *Traité*, 3.

11 Fontaine, *Isidore: Genèse*, 303.

in Isidore's day, and Isidore's point about 'superstition' may be directed at proponents of Cosmas's position.[12]

But knowledge of the world must be informed by 'wise and sound doctrine', and, for this reason, Isidore will begin his book with the day.

STRUCTURE

The preface justifies what is actually a rather innovative move on Isidore's part: joining a hemerology, or survey of units of time, to a cosmography or summary account of the universe, and to a discussion of 'meteorology' in the classical sense of the study of phenomena that involve the interaction of atmosphere, water, and earth. As Isidore explains, he elected to begin with a hemerology because, in the biblical account, the day was created before all else, even the sun and moon whose movements measure time. Day was 'the first in the order of visible things' after light; moreover, it is not just 'the presence of the sun between its rising and its setting', but a span of time that transcends light and darkness, as even the pagans attest (ch. 1.1–2). According to Augustine, day as it moves from light into darkness and darkness into light symbolizes the whole drama of sin and redemption (ch. 1.3).

The preface reveals much about how Isidore handled his materials for *On the Nature of Things*, and what his purposes were. As he himself declared, he drew both on pagan works of learning and on the writings of the Fathers. For pagan learning Isidore turned to the syntheses of cosmology and cosmography of the Roman school tradition, notably Hyginus' *On Astronomy*, but also Solinus' *Collectanea*, supplemented by miscellanies of erudition such as Suetonius' lost *Meadows* (*Prata*).[13] Isidore's debt to Lucretius, whose *De rerum natura* may have furnished him with a title for

---

12 See, in particular, Maja Kominko, *The World of Kosmas: The Byzantine Illustrated Codices of the Christian Topography* (Cambridge: Cambridge University Press, 2013) and Kominko, 'The Science of the Flat Earth', in Jeffrey C. Anderson (ed.), *The Christian Topography of Kosmas Indikopleustes (Firenze, Biblioteca Medicea Laurenziana, plut. 9.28): The Map of the Universe Redrawn in the Sixth Century, with a contribution on the Slavic Recensions* (Rome: Edizioni di Storia e Letteratura, 2013), 67–82. For further discussion, see Commentary.

13 Nineteenth-century philologists bent on exploiting Isidore's works for *testimonia* of lost ancient sources tended to exaggerate his dependence on the *Prata*: Reifferscheid, editor of *C. Suetoni Tranquilli praeter Caesarum libros reliquae*, even argued that the *Etymologies* was largely a silent reworking of Suetonius. Fontaine, *Isidore et la culture*, 16–19, by

his book, is difficult to assess. Although he borrows extensively from Book 6 for his account of lightning in chapter 30 and especially for his discussion of the cause of pestilential disease in chapter 39, these two chapters are the only ones where we can trace a direct influence.[14] But from a broader perspective he may have taken a central theme of his book from Lucretius, who conceived of his *De rerum natura* as an antidote to superstition and fear.

Apart from the hemerology (chs. 1–8), *On the Nature of Things* is structured to mirror the vertical axis of the universe itself, beginning at the top with the heavens (chs. 9–27) and then moving downwards through the atmosphere, the bodies of water, and finally the earth itself (chs. 28–48).

But while the structure of *On the Nature of Things* has deep roots in the Roman secular tradition of scientific writing, Isidore's named 'authorities' are in fact the poets, notably Vergil and Lucan. In this respect, he gestures to another classical tradition, namely of science conveyed through commentaries on works of philosophy and literature. In the Middle Ages, the most famous and durable of these were Calcidius' commentary on the *Timaeus* of Plato (*c.*AD 321), and Macrobius' exposition of the 'Dream of Scipio' episode in Cicero's *Republic* (early 5th century AD). This tradition was adapted by Christian writers, who presented expositions of the material world through exegesis of the account of the creation in the opening chapter of Genesis. Such treatises or sermon sequences are collectively known as *hexaemera*, 'the work of the Six Days'. Isidore draws very heavily on the *Hexaemeron* of Ambrose of Milan but supplements this with Pseudo-Clement's *Recognitions* – a sort of Christian *Prata* – and the *Morals on Job* (*Moralia in Iob*) of Gregory the Great, a work dedicated to

---

contrast, argues that Isidore may not even have had access to the integral text of this work, but used anthologized excerpts.

Older historiography argued that the innovation and speculative daring of the Greeks in scientific matters had little appeal for the practical-minded Romans, who were content to repeat what they could understand and felt to be useful. The transmission of Greek science to the West depended on its being incorporated into popular Latin handbooks; see, e.g., William H. Stahl, *Roman Science: Origins, Development, and Influence to the Later Middle Ages* (Madison: University of Wisconsin Press, 1962). Recent scholarship seeks to reframe Roman science in a more sophisticated cultural analysis; see, e.g., Daryn Lehoux, *What Did the Romans Know? An Inquiry into Science and Worldmaking* (Chicago: University of Chicago Press, 2012). A re-evaluation of Isidore's science in the light of this new analysis is a desideratum.

14 A more generous assessment of Lucretius' influence is proposed by Giovanni Gasparotto, *Isidoro e Lucrezio. Le fonti della meteorologia isidoriana* (Verona: Librería, 1983).

Isidore's brother Leander. He also calls upon works by Augustine, notably *The Literal Interpretation of Genesis* (*De Genesi ad litteram*), *Expositions of the Psalms* (*Enarrationes in Psalmos*), *The Quantity of the Soul* (*De quantitate animae*), *The City of God* (*De civitate Dei*), and *Against Faustus* (*Contra Faustum*), Jerome (commentaries on Isaiah, Daniel, Amos, Hosea, Zechariah, Ecclesiastes), and Hilary of Poitiers's *On the Psalms* (*Tractatus super Psalmos*).

So why did Isidore not simply send Sisebut a copy of Ambrose's *Hexaemeron*? Why did he choose to discuss 'the nature of things', following the title and template of ancient science and philosophy, rather than 'the work of the Six Days'? When we delve deeper into the text, the questions multiply. Exactly how did Isidore understand the relationship between the ancient pagan poets, or his nameless 'worldly philosophers' on the one hand, and the Bible and the Church Fathers on the other? Finally, why did he include the hemerology, when it is not justified by either the classical or the hexaemeral tradition? The key must be sought both in the immediate circumstances of the composition of *On the Nature of Things*, but also in Isidore's broader conception of his mission as a writer.

OCCASION

On the Nature of Things was composed in 612, and almost immediately sent to King Sisebut in Toledo. Isidore seems to have enjoyed particularly close relations with this monarch, to whom he dedicated not only *On the Nature of Things* but also the initial version of his masterpiece, the *Etymologies*. Sisebut responded to Isidore's gift by composing a poem of 61 hexameters. In it he laments that, unlike his clerical friend, he cannot devote himself unremittingly to the pleasures of literary activity due to the pressure of governmental affairs. Then he proceeds to set out an explanation of the mechanism of eclipses. This poem, referred to as Sisebut's *Epistle*, accompanies *On the Nature of Things* in some manuscripts[15] and is included with the present translation.

Braulio of Zaragoza states that *On the Nature of Things* was written to 'resolve certain obscure matters concerning the elements'.[16] Its aim, then,

---

15 For which, see below, pp. 34–42.

16 Braulio, *Renotatio librorum domini Isidori*, ed. Martín, CCSL 113B: 201–02. Trans. Barney, et al., *Etymologies*, 7–9.

was to provide an answer to some difficult questions or problems about the material world. What were these 'obscure matters'? The composition of *On the Nature of Things* coincided with an unusual string of solar and lunar eclipses. There were two total eclipses of the moon in 611, one on 4 March at dawn, and another on 29 August at nightfall. These were followed by two partial lunar eclipses on 22 February and 17 August 612. By then, news of a total solar eclipse, whose line of totality passed over south-west Portugal and the southern tip of Spain on 2 August 612, must have reached Seville and even Toledo.[17] Fontaine suggests that this convergence ignited displays of pagan 'superstition' among the barely Christianized rural population.[18] In his view, *On the Nature of Things* might be regarded as an exercise in *correctio rusticorum*.[19] It was a lesson slow to be learned. In the ninth century, Hrabanus Maurus recounted the fear and frenzy of the pagans around Fulda at an eclipse of the moon.[20] As late as the American Revolution, George Washington was thankful of being alerted by the Philadelphia Council of Safety to an imminent eclipse of the sun, which he knew could cause panic among his soldiers if they were not forewarned.[21] That being said, there is no evidence that anyone in Spain or elsewhere was struck with panic by this particular cluster of eclipses. And if *correctio rusticorum* was one of Isidore's motivations, it remains unspoken. Unlike King Sisebut, who describes several superstitious beliefs about eclipses in his *Epistle* (lines 18–22), Isidore never mentions popular superstitions at all.

The issue, however, may not have been either popular or superstitious. Multiple eclipses could also kindle speculation among the clergy and

17 Fontaine's claim (*Traité*, 4–5) that the solar eclipse was visible in all of Spain was based on nineteenth-century calculations, which have been superseded by more accurate calculations made by NASA. We thank Immo Warntjes for pointing this out in his review of Kendall/Wallis in *The Medieval Review* 12.08.01; see also D.J. Schove, with Alan Fletcher, *The Chronology of Eclipses and Comets, AD 1–1000* (Woodbridge: Boydell, 1984), 114–15, which records a magnitude of 0.9 over Seville and totality over Morocco.

18 Fontaine, *Traité*, 5. This view is accepted by Stephen C. McCluskey, 'Natural Knowledge in the Early Middle Ages', *Cambridge History of Science*, 2:288.

19 Marina Smyth, *Understanding the Universe in Seventh-Century Ireland* (Woodbridge: Boydell, 1996), 305–06 cites 'pagan superstitions' and 'astrology' as among Isidore's targets in *DNR*. See below (n. 33).

20 Wesley M. Stevens, 'Compotistica et Astronomica in the Fulda School', in Margot H. King and Wesley M. Stevens (eds.), *Saints, Scholars and Heroes: Studies in Medieval Culture in Honour of Charles W. Jones*, 2 vols. (Collegeville, MN: Saint John's University, 1979), 2:46.

21 Ron Chernow, *Washington: A Life* (New York: Penguin, 2010), 286.

literate laity that the cosmic terrors described by Christ himself in the 'little Apocalypse' discourses of Matt. 24:1–25 and 36, Mark 13:1–37, and Luke 21:5–38 were coming to pass, and that they announced the impending end of the world. Even Pope Gregory I placed on record his conviction that recent 'changes in the sky and terrors from the heavens, unseasonable tempests, wars, famine, pestilence, and earthquakes' proved that the Last Judgement was very near.[22] Eclipses, storms, plague, and earthquakes were standard topics in ancient treatments of meteorology, and often presented as antidotes to ignorance and panic, but some of Isidore's specific comments seem designed to avert apocalyptic speculations.

But one must not exaggerate the purely topical character of *On the Nature of Things*. First, it fits into a pattern of thinking about the natural world which emerges in Isidore's other writings; secondly, the text itself, when examined closely, does not seem to be especially intent upon *correctio*.

PURPOSES AND PREOCCUPATIONS

Isidore's interest in the natural world was manifest in his earliest book, the *Differences* (*De differentiis rerum*), composed around 600.[23] Book I discusses distinctions between words of similar meaning. It begins, interestingly, with the vocabulary of cosmology, before moving on to psychology, ethics, society, and geography. Book 2 deals with *res* rather than *verba*, and its organization mirrors the vertical sequence from heaven down to earth found in *On the Nature of Things*. In the *Differences*, this hierarchy is religious in character, beginning with God and moving down the scale to angels and demons, and finally to humanity in all its physical, psychological, and theological complexity. There is an 'encyclopaedic' feel to this second book, where difference moves from being merely a grammatical

---

22 Gregory expressed these views in his letter to King Æthelberht of Kent: Bertram Colgrave and R.A.B. Mynors (eds.), *Bede's Ecclesiastical History of the English People*, 1.32 (Oxford: Clarendon Press, 1969), 112–13.

23 This is the date proposed by Roger Gryson, et al., *Répertoire général des auteurs ecclésiastiques latins de l'Antiquité et du haut moyen âge*, vol. 2, *Répertoire des auteurs I–Z* (Freiburg: Herder, 2007), 596, correcting AD 598–615, as proposed by Aldama in *Miscellanea Isidoriana* (1936) and reproduced by Fontaine, *Isidore: Genèse*, 436. In *Isidore et la culture*, 34–35, however, Fontaine suggests that *Differences* was composed during Leander's lifetime, and was even subtly criticized by Gregory the Great in the dedicatory epistle of the *Moralia*.

category to being a way of envisioning the universe and humanity. In this respect, it is closely allied to the *Etymologies*. Nonetheless, the methods and preoccupations of *On the Nature of Things* are also there in germ.[24] For example, Isidore explains the difference between the creation of the universe and its formation, the first being the instantaneous production of matter, the second unfolding over the biblical six days as distinct parts of the world are formed in sequence (2.11). He also explains the ontological distinction between different orders of creatures: non-living, living, irrational, mortal, and immortal (2.13). In sum, the grammar of words and a grammar of the world overlap.

The *Differences* overlaps the categories of 'encyclopaedia', 'grammar', and 'theology'; but it is harder to assign *On the Nature of Things* to any of these categories, singly or in combination. A closer parallel to *On the Nature of Things* is once again afforded by the *Book of Numbers*. Each chapter begins by briefly introducing the mathematical properties of the number, and then proceeds to unfold its allegorical significance in the Bible. For example, the number four is both 'solid' and 'perfect': solid because it is the first number that can make a figure in three dimensions, and perfect because the sum of its factors is ten. It is thus a fitting number for the evangelists, the rivers of Paradise, the colours on the vestments of the ancient Jewish priests, and so forth. Four is the number of the elements that compose the world, the seasons of the year, the qualities of the human body (hot, cold, wet, and dry), the cardinal virtues, the ages of life, the classes of the animal kingdom, the colours of the rainbow, and the days that cumulate into the leap-year day.[25]

A case can be made that *On the Nature of Things* was likewise designed as a resource for preachers and exegetes, because it pairs natural explanations with allegorical interpretations.[26] It has even been characterized as 'a technical treatise [which serves] as the vehicle for theological

24 'Les réalités (*res*) de ce second livre sont d'ordre immatériel et religieux; elles proposent un abrégé encyclopédique, de contenu strictement intellectuel, théologique, spirituel, et dont les notices sont nettement plus développées que celles du livre premier. L'idée et le genre même d'une encyclopédie y mûrissent, mais dans des secteurs encore restreints'. Fontaine, *Isidore: Genèse*, 169–70. See also Fontaine, *Isidore et la culture*, 38–40.

25 *Liber numerorum* 5 (Guillaumin, 26–31); cf. Fontaine, *Isidore: Genèse*, 195–96, 337. On the symbolism of the number four, see Heinz Meyer, *Die Zahlenallegorese im Mittelalter. Methode und Gebrauch* (Munich: Fink, 1975), 123–27 and Meyer's *Lexikon der mittelalter-lichen Zahlenbedeutung* (Munich: Fink, 1987), cols. 332–402.

26 No work has been done on this potential use of *DNR* since Angel Benito Dúran, 'Valor catequético de la obra "De natura rerum", de San Isidoro de Sevilla', *Atenas* 9 (1938): 41–51.

and epistemological teaching' like the *Synonyms*.[27] However, closer examination reveals a work which is less evidently instrumental. First, not every topic in the book calls forth a spiritual commentary: Isidore has nothing to say about the possible meaning of the month (chs. 4–5), the solstice and equinox (ch. 8), the climates and elements of the world (chs. 10–11), the comparative size of the sun and moon (ch. 16), the courses of the moon and stars (chs. 19, 22), falling stars (ch. 25), why the sea is salt and does not increase in size (chs. 41–42), the Nile (ch. 43), or the divisions of the earth (ch. 48). Secondly, when Isidore does invoke religion, he does so in a variety of registers. Sometimes he shows how his information aids in understanding biblical texts, particularly those which are expressed in an obscure or figurative manner. For example, he points out that the book of Daniel uses the word 'week' to denote a period of seven years (ch. 3), and that St John's statement 'the world knew him not' refers to worldly people, not the cosmos (ch. 9.2). On other occasions, he offers a sort of theological reflection: the seasonal course of the sun illustrates God's Providence in moderating heat and cold (ch. 17); the position of the sun in the middle of the hierarchy of planets illustrates 'divine reason', because the sun is the noblest of the planets (ch. 23); there are many plausible explanations for tides, but only God, who created them, knows which is actually correct (ch. 40), nor can anyone definitively grasp the 'perfection of the divine art' that holds the earth upright and stable in the middle of the universe (ch. 45). However, by far the majority of Isidore's religious comments take the form of allegories, that is, similitudes or analogies. Everything in the natural world is imbued with meaning; so, snow signifies unbelievers (ch. 35); rain signifies the eloquence of the apostles (ch. 33). Things exist both in themselves and for the sake of their meanings. The world becomes an allegory of the moral truths that lie behind it. At times, Isidore blurs the line between 'signifying' and 'being', as in his discussion of the winds (ch. 36.3), where it would be easy to read his words as meaning that the winds *are* the good or evil spirits.

Even unpredictable and catastrophic phenomena like eclipses, thunder and lightning, hail, or earthquakes are expounded as *figurae* of moral and theological realities; by implication, they are not to be regarded as portents

---

27 Vivien Law, *Wisdom, Authority and Grammar in the Seventh Century: Decoding Virgilius Maro Grammaticus* (Cambridge: Cambridge University Press, 1995), 101. This view also underpins Marek Hermann's survey of *DNR* in 'Zwischen heidnischer und christlicher Kosmologie. Isidor von Sevilla und seine Weltanschauung', *Analecta Cracoviensia* 34 (2002): 311–28.

or punishments – a message that Isidore's pagan models likewise conveyed, in their own religious idiom. Isidore seems to be concerned to present the phenomena of the natural world as *symbols*, not as *omens*. Apocalyptic forecasting, then, is akin to judicial astrology, something that Isidore has no time for.[28] Still, for the most part Isidore keeps the realm of allegory distinct and subordinate to his description of the natural order of things.[29]

APPEAL TO REASON

Isidore seems to have taken the position that the kind of 'superstition' he was combating – be it popular or clerical anxiety about natural phenomena, or the biblical literalism of Cosmas Indicopleustes' supporters – could be defeated with a multi-pronged approach. First, he identifies God's plan for his creation with regularity and order that can be described rationally. He quotes with approval (ch. 12.5) Plato's statement that 'the Artificer of the world is rational', and argues that God placed the sun in the middle of the planets 'by divine reason' because the sun is the most 'noble' of the planets (ch. 23.2). Chapter 13, 'The Seven Planets and their Revolutions', opens with an explicit quotation from Ambrose's *Hexaemeron*: the Bible is in accord with secular philosophy that there are numerous 'heavens' within the single zone of 'heaven'; for astronomers, these are the erratic orbits of the planets. Isidore then transitions to a silent quotation from Hilary's commentary *On the Psalms*: God made the heaven 'in accordance with a rational plan, in a particular order' to encompass two zones: a spherical upper heaven that is the home of the angels and the lower heaven of the planets with their complex movements. These are separated by the 'waters above the firmament', which is the subject of the next chapter. But Isidore has used this opportunity to make the point that however unpredictable the motion of the planets might be – and hence eclipses – they are natural events, comprehensible by human reason.

---

28 In this respect, his method mirrors the ecclesiological and allegorical exegesis of the book of Revelation promoted by Apringius, Tyconius and Augustine as a response to exegetes who saw it as a coded timetable and *dramatis personae* of the actual end-times; see the introduction to Faith Wallis's translation of *Bede: Commentary on Revelation*, TTH 58 (Liverpool: Liverpool University Press, 2013), esp. pp. 14–22 and scholarship cited therein.

29 Fontaine, *Isidore et la culture*, 568; Barbara Obrist, 'Le Diagramme isidorien de l'année et des saisons: Son contenu physique et les représentations figuratives', *Mélanges de l'École française de Rome: Moyen âge* 108/1 (1996): 111–12.

At times this appeal to reason can be quite subtle. In chapter 15, Isidore enlists Ambrose to dismiss the view of 'the philosophers' that the sun is only accidentally, and not innately hot. It is noteworthy, however, that Ambrose's argument is not from biblical authority; instead, he points out that reason and experience show that the sun attracts and consumes moisture, so it must be essentially fiery. Even where science seems to conflict with revelation over the 'waters above the heavens' (ch. 14), Ambrose shows how thoroughly the defence of the Bible's authority was actually based on a secular and classical model of the cosmos. For the Fathers, themselves the products of the ancient system of education, found virtually no information in the Bible about the structure of the universe or the physics of the natural world, and so (in the main) took it for granted that unless otherwise stated the truths of Greek *physica* were what the Scriptural account intended.[30] In Isidore's day, however, there were challenges to that view.

Secondly, and perhaps more controversially, Isidore tends to place the pagan poets and philosophers on the same plane of authority as the Fathers, and even the Bible. For example, in chapter 32, Job and Vergil stand shoulder to shoulder to explain how compressed air turns into clouds; in chapter 43, Clement's and Lucan's views on the size of the ocean are yoked without comment. Sometimes the Fathers serve as sources of purely factual information, e.g., Clement on why the sea does not increase in size, though it receives all the waters of the world (ch. 41), and Ambrose on why the sea is salt (ch. 42). The two possible scientific explanations of the moon's phases in chapter 18 are presented in an explicit quotation from Augustine (one which includes an interesting reference to an experimental demonstration using a ball). On the other hand, in the following chapter, Hyginus is the source of information on the course of the moon; Ambrose is the source of additional material on lunar calendars and the effect of the moon on moisture, but he is cited silently. More striking is chapter 36, where a definition of wind drawn without acknowledgement from Augustine gives the impression that it represents the views of Lucretius.

It could be argued that Isidore is obeying the dictates of the genre of didactic writing about natural phenomena, which preferred to chain multiple

---

30 For a penetrating analysis of the adoption of classical science into Patristic thinking, see Hervé Inglebert, *Interpretatio Christiana: Les mutations des savoirs (cosmographie, géographie, ethnographie, histoire) dans l'Antiquité chrétienne, 30–630 après J.-C.* (Paris: Institut d'études augustinienne, 2001), ch. 1. Isidore's views on the waters above the firmament seem more rationalist than those of Augustine; see Inglebert, 217–18 and n. 105. This issue is discussed at greater length in our commentary on ch. 14.

explanations rather than reduce options to a single definitive truth.[31] But if *On the Nature of Things* aimed to promote a rational view of nature to counter 'rustic' superstition, why did Isidore hedge about eclipses; and if a secondary target was the planetary speculations of Priscillianist heretics,[32] why did he hesitate concerning whether the stars are animate? It is worth looking at these two passages in detail. In chapter 20, Isidore explains the eclipse of the sun through a silent quotation from Hyginus. He then adds another view which he says is held by 'the natural philosophers and wise men of the world', namely that a solar eclipse happens 'if the hole of the air, through which the sun pours its rays, is contracted or shut off by some exhalation (*spiritu*)'. The source of this information has not been traced, and Isidore may mean nothing more than that a heavy cloud cover can blot out the light of the sun almost to the same degree that an eclipse can, as anyone who has experienced the build-up to a severe thunderstorm or a tornado can attest. But the reference to a *spiritus* which blocks the sun invokes something more arbitrary (cf. ch. 39 on pestilence), or even intentional. This can even be said to be reinforced by the following paragraph, where Isidore switches to what 'our doctors' say about the allegorical meaning of the solar eclipse. Here he invokes the eclipse that occurred during Christ's crucifixion. Isidore assumes that this was not a natural eclipse (as indeed it could not have been, since Christ was supposedly crucified at the full moon of Passover), but he does not spell this out. Instead, he interprets the eclipse as divine anger at the Jews for their sacrilege. If Isidore were seeking to allay superstition about eclipses by emphasizing their entire naturalness, he seems to have been working at cross purposes here. However, this may not be the case. The eclipse, even the anomalous eclipse of the Crucifixion, is a *symbol* of what was happening at the time of the event, not an *omen* of the future. And that seems to be Isidore's point in his allegorical interpretations. Chapter 27, 'Whether the Stars have a Soul', would seem to be an opportunity to refute one of the principal claims of the followers of Priscillian, bishop of Avila (†385), namely rigorous astral determinism, and a belief that the heavenly bodies were agents in the cosmic war of good/ spiritual and evil/material; the Zodiac signs in particular were identified with evil.[33] Yet Isidore is surprisingly neutral on the topic. He quotes

---

31 Liba Taub, *Aetna and the Moon: Explaining Nature in Ancient Greece and Rome* (Corvallis: Oregon State University Press, 2008), 49; Taub, *Ancient Meteorology* (London and New York: Routledge, 2003), 117.

32 Fontaine, *Isidore: Genèse*, 308.

33 On Priscillianist astrology, see Tim Hegedus, *Early Christianity and Ancient*

Augustine who lays out the problem, but provides no answer; Ecclesiastes and Vergil, however, are in agreement that the sun is a spirit. But if the stars have souls, what becomes of them at the general resurrection? Isidore ends his chapter there, without resolution.[34] This apparent agnosticism is echoed in chapter 39 on pestilence. Here Isidore intimates that pestilence is a divine 'chastisement and reproof' for human sin, though its mechanism is the corruption of the air, either by local flood or drought, or by 'plague-bearing seeds' carried on the wind. In short, Isidore does not seem to be rigorously focused on rationalizing the creation, or divesting it of vitality, agency, and the contingency. He has left too many doors open to Priscillianists and even pagans.

WIDER ENDS: A CHRISTIANIZED ERUDITION?

There is, however, another way to approach *On the Nature of Things* which situates it within the broader trajectory of Isidore's and the Spanish Church's concerns, while securing its audience in both the clergy and the educated elite. This comes into focus if we compare *On the Nature of Things* to Isidore's crowning work, the *Etymologies*. Isidore worked on the *Etymologies* for many years, but its first version was completed by 620–621, since it is also dedicated to King Sisebut. In other words, it was probably in train at or shortly after the completion of *On the Nature of Things*. Isidore did not, at least at first, call his book *Etymologies*. In his first dedicatory letter to Sisebut, he refers to it as 'the work concerning

---

Astrology (New York: Peter Lang, 2007), 339–51. On Isidore's concern with Priscillianism and its astrological doctrines, see Henry Chadwick, *Priscillian of Avila: The Occult and the Charismatic in the Early Church* (Oxford: Clarendon Press, 1976), 82–83, 231–32; on his involvement in combating *clerical* astrology (NB not 'popular superstition'), see Fontaine, 'Isidore de Séville et l'astrologie', *Revue des études latines* 31 (1953), esp. pp. 278–82. What Isidore understood by 'astrology' has been matter for debate, notably because (like many Latin authors) his use of the terms *astrologia* and *astronomia* is fluid. Fontaine (*loc. cit.*) argues that Isidore intended to separate the science of the stars from astrological prediction, and that his ideas evolved over time, culminating in the distinction in *Etym.* 3.27 between 'natural astrology' (modelling future positions of celestial bodies) and 'superstitious astrology' (nativities and elections: cf. *Etym.* 3.71.39). He rejects the argument of M.L.W. Laistner ('The Western Church and Astrology during the Early Middle Ages', *Harvard Theological Review* 34 (1941), esp. pp. 264–68), that Isidore had little awareness of judicial astrology.

34 Fontaine, 'Isidore de Séville et l'astrologie', 284–85. See also Commentary on ch. 27.

the origin of certain things (*opus de origine quarundam rerum*)[35] – a title which resonates closely with *De natura rerum*. Isidore often closely elides the meanings of etymology/*origo* and *natura*: in the opening section of *Etym*. 12, he observes that Adam gave the animals their true and original names 'according to their natural situation (*iuxta condicionem naturae cui serviret*)'.[36]

Jacques Fontaine argues that Isidore's aim was to renovate and to give a Christian vocation to the ancient traditions of encyclopaedic erudition. In Hellenistic scholarly tradition, the teacher provides an *enarratio* or explanation of the allusions in the classical texts; these explanations would be assembled by *compilatio* – a process of combining material from other sources and supplying clarification and expansion – and catalogued in his *commentarius* or notebook. These compendia, many of which (like Suetonius' *Prata*) had names evoking flowery fields in which bees gather nectar for honey, became themselves resources to be exploited and rearranged under new headings, for new purposes. Isidore's encyclopaedism is thus firmly rooted in this classical model. Augustine's *On Christian Learning* (*De doctrina christiana*) offered Isidore a mandate of sorts, in appealing for a suite of reference books for the biblical exegete and Christian preacher: a lexicon of foreign names in the Bible, a manual of history, and a collection of 'explanations of whatever unfamiliar geographical locations, animals, herbs and trees, stone, and metals' are mentioned in Scripture, along with an exposition of numbers (*De doctrina christiana* 2.39.59). Isidore, however, perhaps because he felt less nervous about the prestige and magnetic allure of classical culture than Augustine and his contemporaries, interpreted this mandate in a more generous sense: he would reinvent polymathesis in a Christian mode, using the materials and techniques of the ancient *grammaticus*, and adding in Christian erudition. Over the door to his library, Isidore inscribed these verses: 'You see these meadows (*prata*) filled with thorns, and an abundance of flowers? If you do not wish to gather thorns, gather roses' (*Versus* 2.3–4).[37] The

---

35 *Etym.* Ep. VI, ed. Lindsay, lines 20–21; Fontaine *Isidore et la culture*, 11 n. 1.

36 *Etym.* 12.1, ed. André, 37–39. Isidore's source here is Tertullian; see p. 38 n. 1. *Origo* is an important word in Isidore's lexicon: e.g., grammar is the *origo* of the liberal arts. God is the *origo* of all things (*Diff.* 2.18). In ancient scholarly tradition, the arts have 'originators' or inventors (Varro's *De origine linguae latinae*). Isidore's title for *De officiis* is *De origine officiorum*, and for *Historia Gothorum* is *De origine Gothorum*.

37 Isidore, *Versus* (*Isidori Hispalensis versus*), ed. José María Sánchez Martín, CCSL 113A (Turnhout: Brepols, 2000), 213.

technique of *compilatio* permitted Isidore to gather and recompose *both* ancient scholia on Vergil *and* Patristic writings; hence they often appear on the same stage, and at times the Christian character of his source may even be suppressed.[38] We have already alluded to Isidore's tendency to 'reverse engineer' secular knowledge by lifting nuggets of scientific information from Christian sources, while leaving the Christianity behind.

The *Etymologies* sheds light on some of the deeper motivations driving *On the Nature of Things*. Its first dedication to King Sisebut announces the book's vocation to improve the culture of the lay and ecclesiastical elites of the kingdom. It seeks global knowledge through the encyclopaedic principles of *grammatica* and, in particular, etymology. Etymology also plays a crucial role in *On the Nature of Things*, beginning with chapter 2 on 'Night' which derives its name *nox* from *nocendo* ('harming') because it harms (*noceat*) the eyes. Each division of the night has a distinctive name that intertwines with its key characteristics. Etymology explains why months have their distinctive names, and roots them in the nature and history (ch. 4); it roots the word for year (*annus*) in its cyclical nature (*anus*, 'ring': ch. 6); it literally inscribes (*caelatum*) the constellations onto the meaning of 'heaven' (*caelum*). Like the *Etymologies*, *On the Nature of Things* seeks to reinforce correct Latin by restoring meaning to words and promoting good usage. Finally, both works aim to guide curiosity concerning profane knowledge into the channels of *doctrina christiana*.

---

38 Fontaine, *Isidore: Genèse*, ch. 16. On Isidore's intellectual technique of 'reverse engineering', see, e.g., Fontaine's analysis of his use of Christian sources in the treatment of rhetoric in the *Etymologies*: Fontaine, *Isidore et la culture*, 330–31.

A WORK OF COMPOSITE CONSTRUCTION

Discussion of the structure and aims of *On the Nature of Things* is closely bound up with uncertainty and debate about various states and alterations of the text as it has come down to us in the manuscript tradition. Even the title is contested. Since the edition of Margerin de la Bigne in 1580, Isidore's work has been almost universally known as *De natura rerum*. The title echoes Lucretius' great epic poem *De rerum natura* and points forward to Isidore's medieval successors Bede, Hrabanus Maurus, Alexander Neckam, and Thomas of Cantimpré, who borrowed the title for their own purposes. But is this what Isidore called his treatise? The title *De natura rerum* does not appear in any of the twelve manuscripts that survive from the seventh and eighth centuries, which are either untitled, or go under the rubric *De mundo* or *Liber rotarum*. When the title *De natura rerum* surfaces in the ninth century and later, it is typically in manuscripts of one type of the long recension (for which, see below), and is no more common that other titles of the period like *De astronomia* or *De astra* [sic] *caeli*.[1] Its unquestioned acceptance by modern editors is doubtless owing to its presence in the ninth-century folios attached to the seventh-century Visigothic manuscript of *On the Nature of Things* in Escorial R.II.18. The best argument for taking it as an authentic Isidorian title comes from the description of Isidore's works drawn up by Braulio of Zaragoza after Isidore's death in 636.[2] Among them, Braulio says, was: 'One book *De natura rerum*, addressed to King Sisebut, in which he resolved certain obscure matters concerning the elements, relying on his study of both the Doctors of the Church and the philosophers'. The

---

1 A feminine declension biform of *astrum* is unrecorded elsewhere, but the title *De astra caeli*, with its apparent feminine ablative singular form, *astra*, is found in a number of MSS. It may have begun as a blunder; its persistence could be explained as a frozen echo of the recurrent phrase, '*sicut astra caeli*' (where *astra* is neuter accusative plural), in Deut. 10:22; 28:62.

2 Braulio, *Renotatio librorum domini Isidori*, ed. Martín, CCSL 113B: 201–02. Trans. Barney, et al., *Etymologies*, 7–9.

fact that this particular title is primarily associated with manuscripts of the long recension might suggest that Isidore made a definitive decision about what to call his work only in the course of making final revisions (if indeed Isidore was responsible for the long recension).

TEXT AND IMAGE

The labile character of the text tradition also affects the book's famous illustrations. Isidore inserted a number of diagrams, most of them circular, into his text. As a result, as noted above, *On the Nature of Things* was widely known in the Middle Ages under the title *Liber rotarum*, 'the Book of Wheels'. These diagrams, or '*rotae*', are of interest both in themselves and in the way they were interpreted, redesigned, and added to over the centuries.[3] The diagrams are generally not thought to have originated with Isidore, although their ancient prototypes have not directly survived.[4]

The first problem is to identify the diagrams that go back to Isidore and reflect his intentions. It would be well to bear in mind that diagrams in medieval manuscripts could be, and were, handled separately from the texts with which they were associated. Scribes routinely left spaces for their insertion and filled or arranged for artists to fill the spaces later (if they ever did), sometimes copying from a different manuscript, sometimes altering the diagrams to suit their understanding of what they meant, sometimes bungling through ineptitude at drawing, and sometimes adding new figures.[5]

In his edition, Fontaine prints seven diagrams, which he reproduced in the form of line drawings from an eighth-century manuscript, Munich 14300.[6] These are (with Fontaine's labels):

1. *Rota mensium*, the wheel of the months, at the end of chapter 4.

2. *Rota anni*, the wheel of the year, between chapter 7.4 and chapter 7.5.[7]

---

3 Michael Gorman, 'The Diagrams in the Oldest Manuscripts of Isidore's "De natura rerum" with A Note on the Manuscript Traditions of Isidore's Works', *Studi Medievali* 42 (2001): 529–34.

4 Fontaine, *Traité*, 15–18.

5 See Bruce S. Eastwood, 'The Diagram of the Four Elements in the Oldest Manuscripts of Isidore's "De natura rerum"', *Studi Medievali* 42 (2001): 548.

6 Fontaine, *Traité*, xiii.

7 See Obrist, 'Le Diagramme isidorien'.

3. *Rota circulorum mundi*, the wheel of the circles of the world, between chapter 10.2 and chapter 10.3.

4. *Cybus elementorum*, the cube of the elements, between chapter 11.1 and chapter 11.2.

5. *Rota mundi, anni, hominis*, the wheel of the microcosm, at the end of chapter 11.

6. *Rota planetarum*, the wheel of the planets, at the end of chapter 23.

7. *Rota uentorum*, the wheel of the winds, between chapter 37.4 and chapter 37.5.[8]

Although scribes occasionally omitted some or all of the diagrams from *On the Nature of Things*, these seven are found in most manuscripts. Diagrams 1–6 are explicitly introduced in the text of their chapters by phrases like, 'in accordance with what the accompanying diagram shows' (Diagram 1), which confirms Isidore's intention to insert them. Isidore introduces Diagram 4 with the words, 'I have represented [these elements] by the picture below'. Fontaine takes this as evidence that the 'cube of the elements' was Isidore's own design.[9]

Fontaine omitted from his edition a diagram, or rather, a set of figures, which we refer to as Diagram 5A. It is a feature of *On the Nature of Things* 18.7 and consists of representations of the seven phases of the moon. Each is introduced by a textual formula, so there is no question but that Isidore intended them to be drawn in. Although they are of less intrinsic interest than the other diagrams, most scribes elected to preserve them.[10] Fontaine also omitted the T-O map which most manuscripts of the medium and

8 See Barbara Obrist, 'Wind Diagrams and Medieval Cosmology', *Speculum* 72 (1997): 33–84.

9 Fontaine, *Traité*, 17. Gorman, 'Diagrams', 532 concurs with Fontaine's suggestion. See our commentary on ch. 11 for further discussion.

10 Including the scribe of Munich 14300 from which Fontaine took his illustrations. A few MSS include one or more additional diagrams, the most significant of which we identify as follows, in keeping with Fontaine's numeration:

1A: the seasons, in ch. 7.3 (in one short-recension MS).

6A: the four colours of the rainbow, in ch. 31 (in several medium- and long-recension MSS).

7A: the mountains of Sicily, after ch. 47 (in several long-recension MSS).

There is no reason to attribute any of these to Isidore. They probably all arose outside of Spain in the Carolingian period.

long recensions add at the end of the new final chapter.[11] A T-O map is a schematic diagram which represents the world as a circle (= O) with East at the top. The circle is divided horizontally. The upper half represents Asia, and the lower half, bisected vertically (= T), represents Europe on the left and Africa on the right. It is open to question whether the inclusion of the T-O map can be attributed to Isidore. Michael Gorman concludes that it 'was arbitrarily introduced into a few manuscripts ... and was not part of Isidore's original work'.[12] However, this all depends on what we take Isidore's 'original' work to be, which relates to the question we take up below of Isidore's work habits and the environment in which *On the Nature of Things* was written.

Diagram 7, the wheel of the winds, illustrates some of the complexities of this problem, and will bring us to the question of how the text of *On the Nature of Things* was expanded over time. The nature of the winds and particularly their names exerted a peculiar fascination on the minds of scholars of late Antiquity and the Middle Ages. Isidore devoted two chapters to the subject: chapter 36, 'The Nature of the Winds', and chapter 37, 'The Names of the Winds'. And yet, surprisingly, there are good reasons, the details of which we take up in the Commentary, to believe that Isidore did not compose chapter 37 with a wind diagram in mind. This in turn bears on the question of the recensions of *On the Nature of Things*.

FONTAINE'S THEORY OF THREE RECENSIONS

Complete manuscripts of *On the Nature of Things* can be sorted reliably into one or another of three recensions: 'short', 'medium', or 'long'. The labels are those of the text's editor, Jacques Fontaine.[13] The short recension consists of 46 chapters (chs. 1–43 and 45–47 of the long recension). The medium recension adds a new final chapter, *De partibus terrae*, 'The Parts of the Earth', generating a text of 47 chapters (chs. 1–43 and 45–48 of the long recension). The long recension adds yet another chapter, *De nominibus maris et fluminum*, 'The Names of the Sea and the Rivers', which it inserts between chapters 43 and 45, making in all 48 chapters of text (chs. 1–48).

---

11 His sole reference to it is in his description of Escorial R.II.18, where he observes: 'On remarque à la fin du ch. 48 (fo. 24v) une représentation sommaire des trois parties du monde' (Fontaine, *Traité*, 20).

12 Gorman, 'Diagrams', 530 n. 5.

13 Fontaine, *Traité*, 38–39 ('recension courte', 'recension moyenne', 'recension longue').

An obvious consequence of the way the two chapters were added to form the medium and the long recensions is that the first 43 chapters are numbered alike in the three recensions, but chapters 44–46 of the short and medium recensions become chapters 45–47 in the long recension, and chapter 47 of the medium recension becomes chapter 48 in the long. In the discussion of the recensions that follows we refer to chapters 44(–), 45(44), 46(45), 47(46), and 48(47), putting the chapter number of the short and/ or medium recension in parentheses after the chapter number of the long recension.

Fontaine identified two other defining features of the three recensions.[14] The first is Sisebut's *Epistle*, which Sisebut addressed to Isidore on receiving a copy of *On the Nature of Things*. Fontaine regards the inclusion of the *Epistle*, which is normally placed after chapter 47(46), as diagnostic of the short and medium recensions. The second is a long paragraph (36 lines in Fontaine's edition) inserted into chapter 1.3,[15] which Fontaine takes to be characteristic of the long recension. Fontaine refers to this paragraph as the 'mystical' addition, because the text begins with the word *mystice* (we retain the label hereafter without quotation marks).

We have selected four additional minor features, which are either never or only sporadically present in manuscripts of the short recension, to track. These are: (1) a sentence, *Hiemisphaeria ... sub terra*, 'There are two hemispheres, one of which is above the earth and the other below', in chapter 12.3 (Fontaine, *Traité*, 219); (2) a phrase, *hic et Aparctias* (sometimes + *dicitur*), 'It is also called Aparctias', in chapter 37.1 (*Traité*, 295); (3) a quotation specifically attributed to Solinus: '"The philosophers", as Solinus puts it, "say that the world was formed from four elements and is moved by a certain spirit like some animal, and just as exchanges of breath take place in our bodies, so"', in chapter 40.1 (*Traité*, 305); and (4) a sentence, *Asiam ... Maeotis appellatur*, 'The river Don divides Asia from Europe; it thrusts itself in two branches into the marsh which is called Maeotis', in chapter 48(47).2.

Each of these features (ch. 44(–), ch. 48(47), the mystical addition, Sisebut's *Epistle*, *Hiemisphaeria*, *Aparctias*, 'Solinus', and *Maeotis*) may be regarded as an 'addition' to Isidore's first draft of *On the Nature of Things*, if we accept Fontaine's narrative of the way the text was expanded over time, either by Isidore himself or by other interpolators. This applies

---

14 Fontaine, *Traité*, 38.
15 Fontaine, *Traité*, 175–77.

even to Sisebut's *Epistle*, which was supposedly added by the king to the draft that Isidore sent him. Fontaine's edition incorporates all these features, despite his doubts about the authenticity of several of them, and thus, in effect, presents a synthetic version of the long recension, a fact which has exposed him to criticism for including features not from the pen of Isidore.[16]

SINGLE OR MULTIPLE AUTHORSHIP?

The question that immediately arises concerns the responsibility for, and authorship of, these alterations to *On the Nature of Things* – specifically, the eight features enumerated above, as well as the placement and introduction (or lack thereof) of Diagram 7. Should such changes be attributed to Isidore himself, to persons working under his supervision, or to later revisers?[17] Put in the simplest terms, Fontaine's guarded answer is that the medium recension of 47 chapters can be regarded as Isidore's revision of his original work, but that other changes that result in the long recension, in particular the addition of chapter 44(–) and the insertion of the mystical addition, are probably the work of later interpolators. Fontaine speculated that ch. 44(–) drew upon material that represented 'une érudition plus proprement irlandaise', and that the author of the chapter could well have been Archbishop Ecgbert of York (†766).[18] Somewhat more cautiously, he concluded that in all likelihood Northumbrian scholars were responsible for the mystical addition and chapter 44, as well as for the elimination of Sisebut's *Epistle*.[19]

Fontaine's speculative conclusions have tended to harden into accepted facts.[20] While his arguments cannot be dismissed, our investigation into

16 Fontaine prints the *Epistle* (*Traité*, 328–35), but separately from the text of *DNR*. José Carlos Martín, 'Réflexions sur la tradition manuscrite de trois œuvres d'Isidore de Séville', *Filologia mediolatina* 11 (2004): 227, criticizes Fontaine's decision to include non-Isidorian features and calls for a new edition omitting ch. 44(–) and the mystical addition.

17 'The works of Isidore are essentially *compilationes*, and as such they lent themselves to being modified each time they were copied. ... In fact, it could be said that Isidore's works were designed to accommodate new material'. Gorman, 'Diagrams', 538.

18 Fontaine, *Traité*, 80.

19 Fontaine, *Traité*, 83.

20 See, e.g., J.N. Hillgarth, 'Visigothic Spain and Early Christian Ireland', *Proceedings of the Royal Irish Academy. Section C: Archaeology, Celtic Studies, History, Linguistics, Literature* 62 (1962): 188; Michael W. Herren, 'Classical and Secular Learning among the

the manuscript tradition of *On the Nature of Things* inclines us to look to seventh-century Spain rather than Ireland or Anglo-Saxon England as the likely breeding ground for the changes that have become defining features of the three recensions. Whether Isidore himself was responsible for, and/ or approved of, each of these changes is a separate question. As we have observed, intellectual life enjoyed a vigorous renewal in Spain in the late sixth and seventh centuries. Spain was in contact with Constantinople and literary texts were imported from there, as well as from Rome and North Africa. The picture of Isidore's scriptorium in Seville that begins to emerge from the evidence is of a workshop staffed by active, intellectually engaged scholars, who had access to a library far exceeding the resources available anywhere in the British Isles in the seventh or eighth centuries.[21] *On the Nature of Things*, like Isidore's immensely more complex and extensive *Etymologies*, is a compilation of bits and pieces drawn from many sources. As M.L.W. Laistner, remarking on the substantial overlap between items in *On the Nature of Things* and Books 3, 5, and 13 of the *Etymologies*, justly observed, 'it may be assumed that the author had formed a large collection of excerpts grouped under their appropriate headings, even as a modern compiler might devote his earlier labours to making a card-index'.[22] Yet, while the products of extraction (collections of *sententiae*, monastic florilegia, etc.) are plentiful, the methods Isidore actually used to store and retrieve notes, or to index books, are barely visible and deserve study. Pliny's amassing of huge quantities of extracts, described in the preface of the *Natural History*, depended on what Rex Winsbury aptly calls an 'enabling infrastructure' of literate slave-secretaries.[23] While Isidore

Irish before the Carolingian Renaissance', *Florilegium* 3 (1981): 131; Martín, 'Réflexions sur la tradition manuscrite', esp. p. 224 and n. 45.

21 On the richness of late-Antique culture and resources in seventh-century Spain and the abundance of Isidore MSS, see above, pp. 6–7, and Bernhard Bischoff, 'Die europäische Verbreitung der Werke Isidors von Sevilla', in Bernhard Bischoff, *Mittelalterliche Studien: Ausgewählte Aufsätze zur Schriftkunde und Literaturgeschichte*, vol. 1 (Stuttgart: Hiersemann, 1966), 172–73.

22 M.L.W. Laistner, *Thought and Letters in Western Europe AD 500 to 900*, rev. edn (Ithaca, NY: Cornell University Press, 1957), 123–24 at p. 124.

23 Rex Winsbury, *The Roman Book: Books, Publishing and Performance in Classical Rome* (London: Duckworth, 2009), 162. For a rare example from the Patristic period of a way of indexing a text using stichometric count (a unit of text measuring sixteen syllables), see Richard H. Rouse and Mary A. Rouse, 'Donatist Aids to Bible Study: North African Literary Production in the Fifth Century', in Richard H. Rouse and Mary A. Rouse (eds.), *Bound Fast with Letters: Medieval Writers, Readers, and Texts* (Notre Dame, IN: University of Notre Dame Press, 2013), 24–49.

commanded the resources of an episcopal scriptorium, we know little about its size, composition, and activities, and can only speculate as to how, and to what degree, these men contributed to his compositions. Isidore was certainly the instigator, the shaper, the architect of these works, but, given the Bishop of Seville's many other concerns and obligations, he must have had to rely heavily on his staff of 'research assistants' to dig out or at least note and 'file' material for his use. How much supervision Isidore exercised over the final product is impossible to say. Once the first draft had been launched, he may have left some of the work of revision in the hands of his staff. And the work of revision may have continued in Seville, or even in Toledo, after his death. Analysis of the three recensions suggests that *On the Nature of Things*, like the *Etymologies*, was in Isidore's lifetime an open-ended work in progress, subject to being updated and expanded. The extant manuscripts of *On the Nature of Things* shed light (flickering to be sure) on the stages of the development from the short to the long recension (see Table 1, pp. 36–41).[24]

THE SHORT RECENSION: TWO TYPES

The hypothetical first draft of the short recension can be described as consisting of 46 chapters, not including chapter 44(–) and chapter 48(47).[25] It presumably lacked the mystical addition in chapter 1, the *Hiemisphaeria* sentence in chapter 12, the *Aparctias* phrase in chapter 37, the 'Solinus' quotation in chapter 40, *and* the *Epistle* of Sisebut. Diagram 7, the diagram of the winds, was apparently selected for inclusion in chapter 37 after the completion of the first draft, possibly without instructions for its placement.[26]

This first draft gave rise to two major types of the short recension, which are characterized on the one hand by the placement and treatment of Diagram 7, and on the other by the presence or absence of the *Epistle*.[27]

24 For a complete list, see Inventory of Manuscripts, below, pp. 66–69.

25 We follow Fontaine in assuming that the short recension preceded the medium and the long recensions, and is not an abbreviation of either of these. No serious argument has been advanced against this assumption.

26 For the reasoning behind this statement, see Commentary 37.

27 Table 1 lists the MSS on which our claims, here and below, are based. As more MSS become available online, the interested reader will be able to update our findings. Especially to be regretted is the absence from the Chart of the Bern MSS of the eighth and ninth

In Type I, Diagram 7 is placed at the end of chapter 37 (after ch. 37.5) and it includes the inscription KOCMOC (i.e., 'cosmos') in the centre of the diagram; in addition, the diagram is announced by an introductory formula: 'if the skilled reader investigates the likeness of the circle, he may learn clearly and openly about the quality [*de qualitate*] of the winds under the axis of heaven'.[28] Sisebut's *Epistle* is not added after chapter 47(46). Given that the *Epistle* is never found in that position in any extant manuscript with this formula (which we will refer to by the shorthand phrase, '*de qualitate*'), it would be rash to conclude that it was present in the missing folios of Paris 6400G, despite Fontaine's assertion to the contrary.[29] The same is probably true of another Fleury fragment, Bern A.92/20. A tenth-century Type I manuscript, Cologne 83(II), adds the *Epistle*, but as a preface to *On the Nature of Things*, rather than after chapter 47(46) where it is found in all Type II manuscripts that include it. Milan H 150 inf., which is not included in our Table, is another Type I manuscript that lacks the *Epistle*.

In Type II, Diagram 7 is placed between chapters 37.4 and 37.5, it does not include the KOCMOC inscription, nor is there an introductory formula.[30] Sisebut's *Epistle* is frequently inserted after chapter 47(46). But the *Epistle*, it would seem, was not considered an essential component of the text, and was sometimes omitted. This is probably the explanation for its absence from the ninth-century Cologne 99 and Escorial E.IV.14 and M.II.23 from the thirteenth and fourteenth centuries, which in all other respects are Type II manuscripts. The fundamental distinction between Types I and II can be traced right through to the long recension.

Where and when the changes to the lost archetype were made that led to Types I and II of the short recension, respectively, cannot be determined with certainty, but the very early dates of the oldest manuscripts of both recensions strongly supports the hypothesis that they occurred at the earliest stages of the formation of the text, and probably in the scriptorium in Seville.

---

centuries (Bern A.92/20 and 224 are included on the basis of incomplete information), which it is to be hoped will in the near future become available through the Swiss online project *e-codices*. Although Fontaine considered the *Epistle* of Sisebut to be a diagnostic feature of both the short and medium recensions, this is not tenable. Manuscripts of the two recensions that omit the *Epistle* actually outnumber those that display it.

28 See Appendix 2a for the Latin text.

29 Fontaine, *Traité*, 25 and n. 2.

30 The thirteenth-century medium-recension MS, Florence 22 dex. 12, which inserts some lines from the *Epistle* between chs. 47(46) and 48(47), places Diagram 7 after ch. 37.5.

TABLE 1 CHART OF THE THREE RECENSIONS
OF ISIDORE'S *DE NATURA RERUM*

Manuscript	Date	Origin	Title	Introduction to Rota 7	Rota 7
Short recension T1					
Paris 6400G (f)	vii or viii	Fleury?	*De mundo*	De qual.	after 37.5
Laon 423	viii[med]	Laon	*Liber rotarum*	De qual.	yes
Cologne 83(II)	798/805	Cologne	*De libro rotarum*	De qual.	after 37.5
Bern A.92/20 (f)	viii/ix	Fleury?	defective	De qual.	after 37.5
Bern 224	ix[1/3]	Strasbourg?	*De mundo*	De qual.	after 37.5
Laon 422	ix[1/3]	North-east France	lacks title	De qual.	in 37.4
Avranches 109	xi	France	lacks title	De qual.	after 37.5
New York Plimpton 251	xii[4]	Spain?	lacks title	De qual.	after 37.5
Medium recension T1					
Cambrai 937 (f)	viii	Northern France	lacks title	De qual.	after 37.5
Long recension T1					
St Gall 240	ix[in]	Chelles	lacks title	De qual.	after 37.5
Paris NA 448	xi	Besançon	lacks title	De qual.	after 37.5
Short recension T2					
Karlsruhe 339/1 and Paris 6413(f)	viii[med]	Chelles?	defective	no	yes
Bamberg 61	viii[2]	Monte Cassino	lacks title	no	yes
Munich 14300	viii[ex]	Salzburg	lacks title	no	yes
Escorial R.II.18 (f)	ix[1]	Cordoba?	*De natura rerum*	n/a	n/a

KOCMOC	Epistle	*Aparctias*	*Hiemisphaeria*	Solinus quotation	Ch. 48	*Maeotis*	T-O	Ch. 44	Mystical addition
yes	n/a	no	no	no	n/a	n/a	n/a	no	no
yes	no	no	yes	no	no	n/a	no	no	no
yes	precedes	no	no	no	no	n/a	no	no	no
yes	n/a	n/a	n/a	n/a	n/a	n/a	n/a	n/a	n/a
yes	no	info?	info?	info?	no	n/a	no	no	no
yes	no	no	no	no	no	n/a	no	no	no
yes	no	no	no	no	no	n/a	no	no	no
yes	no	no	yes	no	no	n/a	no	no	no
yes	n/a	no	no	no	n/a	n/a	ch. 17	no	no
yes	no	no	no	no	yes	no	yes	yes	no
yes	no	no	no	no	yes	no	yes	yes	no
no	yes	no	no	no	no	n/a	no	no	no
no	yes	yes	yes	yes	no	n/a	no	no	no
no	yes	no	no	no	no	n/a	no	no	no
n/a	n/a	n/a	no	n/a	n/a	n/a	n/a	n/a	no

Manuscript	Date	Origin	Title	Introduction to Rota 7	Rota 7
Cologne 99	ix[2/4]	Western Germany	*Liber testimonium*	no	yes
Paris 6649	ix[med]	Reims?	*Liber primus*	no	yes
Paris 7533	ix		*De astronomia*	no	yes
Florence 27 sin.9	xi		*De astronomia*	no	after 37.5
Vic 44	1064	Spain	*Liber astrologius*	no	yes
Escorial E.IV.14	xiii	Spain	*De naturis rerum*	no	yes
Escorial M.II.23	xiv	Spain	*De astronomia*	no	yes
Escorial K.I.12	xiv[ex]	Spain	*De astronomia*	no	yes

Medium recension T2

Escorial R.II.18 (f)	vii[ex]	Cordoba?	defective	no	yes
Lisbon 446	xiii[in]	Alcobaça	*De naturis rerum*	Rotatim	after 37.5
Florence 22 dex.12	xiii		*De natura rerum*	Rotatim	after 37.5
Paris 15171	xiii?	Northern France?	*De natura rerum*	Rotatim	yes

Long recension T2

Basel F.III.15f	viii[i]	England	lacks title	no	space
St Gall 238	750/770	St Gall	*Liber rotarum*	no	no
Paris 10616	796/799	Verona	*Liber sancti hisidori*	no	yes
Basel F.III.15a (f) and Copenhagen 19.VII (f)	viii[ex]	Fulda	*Liber sancti isidorii*	no	yes
Weimar 414a (f)	viii[ex]	Fulda?	n/a	n/a	n/a
Besançon 184	ix[in]	Eastern France	*De astra celi*	Quorum	yes
Vatican City 1448	c.810	Trier	*Liber rotarum*	no	space

KOCMOC	Epistle	*Aparctias*	*Hiemisphaeria*	Solinus quotation	Ch. 48	*Maeotis*	T-O	Ch. 44	Mystical addition
no	no	no	yes	no	no	n/a	yes	no	no
no	yes	no	no	no	no	n/a	no	no	no
no	yes	no	no	no	no	n/a	no	no	no
no	no	yes	yes	no	no	n/a	no	no	no
no	yes	no	no	no	no	n/a	no	no	no
no	no	n/a	no	no	no	n/a	no	no	no
no	no	no	no	no	no	n/a	no	no	no
no	yes	no	no	no	no	n/a	no	no	no
no	yes	yes	n/a	no	yes – added	yes	yes	no	n/a
no	no	no	no	no	yes	yes	no	no	no
no	yes	no	no	no	yes	in marg.	yes	no	no
no	no	no	no	no	yes	yes	yes	no	no
n/a	no	yes	yes	yes	yes	no	space	yes	yes
n/a	no	yes	yes	yes	yes	no	no	yes	yes
no	no	yes	yes	yes	yes/ Vegetius	no	yes	yes	yes
no	no	yes	n/a	yes	yes	no	yes	yes	n/a
n/a	n/a	n/a	n/a	n/a	n/a	n/a	n/a	n/a	yes
no	no	yes	yes	yes	yes	no	yes	yes	yes
n/a	no	yes	yes	yes	yes	no	space	yes	yes

Manuscript	Date	Origin	Title	Introduction to Rota 7	Rota 7
Bamberg Msc.N.1	ix$^{1/3}$	Eastern France	*De astra caeli*	Quorum	yes
Brussels 9311-9319	ix$^{1/3}$	St-Amand	*De natura rerum*	no	yes (N)
Karlsruhe 229	821	Abruzzo	*lacks title*	no	yes
Vatican City 834	ix	Lorsch	*De astra celi*	Quorum	yes
Verdun 26	ix$^{2/3}$	St-Vanne	*De astra celi*	Quorum	yes
Munich 396	ix^{ex}	Brittany?	*De natura rerum*	no	yes
Exeter 3507	960/986	Southern England	*De natura rerum*	no	yes
London Cotton Domitian I	970	Canterbury?	*De natura rerum*	no	yes
London Harley 3099	1130/1140	Munster-bilsen	lacks title	no	no
Florence S. Marco 582	xii		lacks title	no	yes
Florence 29.39	xiii		*De astris celi*	no	yes

Zofingen Metamorphosis

Zofingen Pa 32	ix^1	St Gall	*De rerum natura*	no	after 37.5
Einsiedeln 167	ix^2	St Gall?	*De rerum natura*	no	after 37.5
London Harley 2660	1136	West Rhineland	*De rerum natura*	no	after 37.5
Einsiedeln 360	xii	Engleberg	*De rerum natura*	n/a	n/a
London Harley 3035	1496	Western Germany	*De rerum naturibus*	no	after 37.5

KOCMOC	Epistle	*Aparctias*	*Hiemisphaeria*	Solinus quotation	Ch. 48	*Maeotis*	T-O	Ch. 44	Mystical addition
no	no	yes	yes	yes	yes	no	with legends	yes	yes
no	no	yes	yes	yes	yes	no	yes	yes	yes
no	no	no	no	yes	yes	no	yes	yes	no
no	no	yes	yes	yes	yes	no	with legends	yes	yes
no	no	yes	yes	yes	yes	no	with legends	yes	yes
no	no	yes	yes	yes	yes	no	yes	yes	yes
no	no, but?	yes	yes	yes	yes	no	yes	yes	yes
no	no, but?	yes	yes	yes	yes	no	yes	yes	yes
no	no	yes	yes	yes	yes	no	no	yes	yes
no	no	yes	yes	yes	yes/ Vegetius	no	yes	yes	yes
no	no	yes	yes	yes	yes	no	with legends	yes	yes
no	no	yes	yes	yes	yes/ Vegetius	no	no	yes	yes
no	no	yes	yes	yes	yes/ Vegetius	no	no	yes	yes
no	no	yes	yes	yes	yes/ Vegetius	no	no	no	yes
n/a	no	n/a	yes	yes	yes/ Vegetius	no	no	yes	yes
no	no	yes	yes	yes	yes/ Vegetius	no	no	no	yes

It is possible, for example, that someone was bothered by the absence of an introduction for Diagram 7 and composed the *de qualitate* formula. A copy of this Type I short recension manuscript reached Merovingian Gaul by the end of the seventh or the beginning of the eighth century, where it became the copy-text for Paris 6400G, the earliest extant Type I manuscript.[31]

The scribe in Seville who prepared the presentation copy sent to the king in Toledo apparently included Diagram 7, which he inserted between chapters 37.4 and 37.5, but he did not supply an introduction for the diagram, and the diagram did not contain the KOCMOC inscription. King Sisebut replied with his *Epistle*, which someone (Sisebut? Isidore?) attached to the end of *On the Nature of Things* – after chapter 47(46), *De monte Aetna*. The resulting Type II short recension (46 chapters + *Epistle* + Diagram 7 without an introduction or KOCMOC) survives in Spain in the seventh-century folios of Escorial R.II.18, though, oddly, the *Epistle* is not attributed to Sisebut, something which is true of the *Epistle* in all the other Type II manuscripts as well.

THE MEDIUM RECENSION

The addition of a new final chapter, *De partibus terrae*, to *On the Nature of Things* defines the medium recension. The new chapter is typically accompanied by a diagram of the known world, known as a T-O map. Fontaine argues that Isidore himself composed the chapter.[32] It draws on the same sources he used throughout *On the Nature of Things* (Ambrose, Augustine, Hyginus). We concur with Fontaine's conclusion that there is no serious doubt about its authenticity, although the fact that it consists of verbatim quotations rather than the slightly modified adaptations and paraphrases that characterize Isidore's usual method might suggest that Isidore gave the job of working up the chapter to one of his assistants.

Fontaine based his hypothesis of a medium recension of 47 chapters on the fragmentary evidence of Escorial R.II.18 and Cambrai 937. Three

---

31 A case might be made that Type I originated in France, but this seems unlikely, because it would necessitate either the assumption that an unmodified copy of the archetypal first draft somehow reached France or a complicated explanation of how a Type II manuscript became the basis for the creation of a quite different Type I manuscript in France.

32 Fontaine, *Traité*, 41–42.

complete manuscripts of which he was unaware – Lisbon 446, Florence 22 dex. 12, and Paris 15171 – confirm his hypothesis.[33]

Escorial R.II.18 illustrates the transition from the short to the medium recension. It is a composite codex. Fols. 9r–24v are from a Spanish manuscript of the latter part of the second half of the seventh century – the oldest extant manuscript of *On the Nature of Things*. Its provenance is Oviedo, but it may have originated in Cordoba,[34] or possibly even in Toledo or Seville itself.[35] The first 15 chapters and the beginning of chapter 16 are lost (the loss is made up on fols. 1r–8v of the composite codex by a ninth-century manuscript of chapters 1–16). The surviving seventh-century folios begin with chapters 16–46 of the short recension (fols. 9r–23r). Up to the end of chapter 46, the scribe used Visigothic uncial; he then switched to a stately majuscule script for the *Epistle* of Sisebut, perhaps as befitting the work of a king (fols. 23v–24r). Yet, as we have pointed out above, Sisebut is not identified as the author of the poem. The *Epistle* is followed in turn by the new chapter, *De partibus terrae*, here numbered as chapter 47. This chapter, again in Visigothic uncial, is written in a slightly later second hand (fol. 24v).[36] At the foot of fol. 24v are inscribed side-by-side two very simple T-O maps. The one on the left appears to be the original and the one on the right a later copy (but why the original should have been so placed as to allow the two maps to be symmetrically positioned is another puzzle).[37] Diagram 7, which omits the KOCMOC inscription, is placed after chapter 37.4 and is not preceded by an introductory formula.

Escorial R.II.18, therefore, began as a Type II short-recension manuscript. Then, probably before the end of the seventh century and certainly before the Arab invasion of 711, a second scribe was inspired to add the new final chapter, *De partibus terrae*, converting the short-recension manuscript into a medium-recension one.

Cambrai 937, Fontaine's other piece of evidence for the medium recension, is a nearly complete eighth-century manuscript of *On the Nature*

---

33 Another medium-recension MS is the eleventh-century Vatican, Reg. lat. 1573, about which we lack detailed information.

34 Fontaine, *Traité*, 70.

35 Bischoff, 'Die europäische Verbreitung', 173 n. 8.

36 Fontaine, *Traité*, 20–21; see reproduction in Veronika von Büren, 'Le "De Natura Rerum" de Winithar', in Carmen Codoñer and Paulo Farmhouse Alberto (eds.), *Wisigothica After M.C. Díaz y Díaz* (Florence: SISMEL/Edizioni Galluzzo, 2014), 399.

37 See G. Menéndez-Pidal, 'Mozárabes y Asturianos en la cultura de la Alta Edad Media en relación especial con la historia de los conocimientos geográficos', *Boletín de la Real Academia de la Historia* 134 (1952), 168 and plate 2; von Büren, 'Le *DNR* de Winithar', 400.

of Things from northern France, which breaks off at the end of fol. 62v in the first section of chapter 47(46), *De monte Aetna*. However, chapter 48(47) is listed in the *capitula* as chapter 47, sufficient proof that it once followed the missing sections of chapter 47(46). However, unlike Escorial R.II.18, Diagram 7 is placed at the end of chapter 37 and introduced by the *de qualitate* formula. It includes the KOCMOC inscription. That is, Cambrai 937 apparently derives from a Type I manuscript, with the addition of the new final chapter of the medium recension. Fontaine points out that two quaternions are lost between fol. 62v and fol. 63r, on which begins Isidore's *De origine officiorum* in the middle of book I, chapter 16. According to his calculations, the lacuna is just sufficient to have contained the remainder of chapter 47(46), chapter 48(47), the *Epistle* of Sisebut, and the missing chapters of *De origine officiorum*.[38] However, since no existing *de qualitate* manuscript includes the *Epistle* (with the anomalous exception of Cologne 83(II), discussed above), as with Paris 6400G,[39] it would be rash to assume its presence in Cambrai 937.

Lisbon 446, an early-thirteenth-century Portuguese codex from Santa Maria of Alcobaça, is a complete manuscript of the medium recension. Both the *capitula* and the chapters are numbered 1 to 47. It bears a distant relationship to the medium-recension type represented by Cambrai 937. Diagram 7 is placed at the end of chapter 37, but it is introduced by an introductory formula (*rotatim*) that differs from the *de qualitate* one used in Cambrai 937 (see Appendix 2c), and the diagram lacks the KOCMOC inscription. Whether the *rotatim* formula was a development from the *de qualitate* formula or an independent response to the perceived need for an introduction for the diagram must remain an open question. An extended additional passage on the winds follows Diagram 7.[40] Sisebut's *Epistle* is not included.

Florence 22 dex. 12 from thirteenth-century Italy intriguingly exhibits all the features of Lisbon 446, including the *rotatim* formula, except for the fact that it inserts lines 39–61 of Sisebut's *Epistle*, without attribution, between chapter 47(46), and chapter 48(47), precisely where the *Epistle* is found in the Escorial manuscript. The odd arrangement – chapter 47(46) + *Epistle* + chapter 48(47) – begins to look intentional. Apparently, the

---

38 Fontaine, *Traité*, 26.

39 For which, see above (p. 35).

40 Oxford, Ashmole 393, a fragment consisting of the first 37 chapters of *DNR*, concludes with an abbreviated version of the *rotatim* formula. It is probably another MS of the medium recension.

Type II medium recension circulated outside of Spain and at some point influenced the exemplar of Florence 22 dex. 12.

A third complete medium-recension manuscript, Paris 15171, may be from northern France and is uncertainly dated to the thirteenth century. Like Lisbon 446 and Florence 22 dex. 12, it employs the *rotatim* formula, although Diagram 7 is placed after chapter 37.4 and the arrangement of the formula and an accompanying paragraph is somewhat different (see Appendix 2c). There is no trace of Sisebut's *Epistle*.

The new final chapter of the medium recension contains a sentence that is never, to the best of our knowledge, found in that chapter in manuscripts of the long recension. The sentence, 'The river Don divides Asia from Europe; it thrusts itself in two branches into the marsh which is called Maeotis',[41] faithfully reproduces the text of Hyginus that Isidore has been quoting. In itself, its omission can easily be explained as a classic case of haplography: the preceding sentence from Hyginus ends with the same word, and the scribe's eye skipped from the first to the second *appellatur*. But since the long recension is necessarily based on the medium recension, how to account for the omission of the *Maeotis* sentence in manuscripts of the long recension is a puzzle. The omission would seem to imply that all copies of the long recension, however much they vary from each other in other respects and display features associated with different branches of the manuscript tradition, derive from a single manuscript of the medium recension in which the haplography occurred. Perhaps it would be better to assume that the original draft of chapter 48(47) omitted the *Maeotis* sentence, which was added to the exemplar of chapter 48(47) in Escorial R.II.18 and (at several removes) in the three *rotatim* manuscripts. It is frustrating not to know whether the lost chapter of the Type I medium recension Cambrai 937 included *Maeotis*. The puzzle must remain unresolved.

The medium recension certainly came into being in Visigothic Spain before being exported to regions as far apart as the Atlantic coast of the Iberian Peninsula, the Loire valley, and Tuscany, as evidenced by Lisbon 446, Paris 15171, and Florence 22 dex. 12. The wide geographical separation of these late manuscripts imply the loss of a number of earlier copies.

---

41 Ch. 48(47).2: *Asiam ab Europa Tanais diuidit, bifariam se coniciens in paludem quae Maeotis appellatur.* This sentence is found in all MSS of the medium recension that we have been able to check: Escorial R.II.18; Florence 22 dex. 12 (in the margin); Lisbon 446; Paris 15171. Ch. 48(47) is missing from Cambrai 937. We lack information concerning Vatican City Reg. 1573.

THREE SPANISH INTERPOLATIONS?

The *Hiemisphaeria* sentence in chapter 12, the *Aparctias* phrase in chapter 37, and the 'Solinus' quotation in chapter 40 turn up infrequently in manuscripts of the short and medium recensions, but are very nearly universal in the long recension. The three features track so consistently together that it seems likely that they entered the manuscript tradition at roughly the same time. The prevailing assumption is that this occurred in connection with the formation of the long recension, perhaps in the British Isles, and that their very occasional appearances in manuscripts of the short and medium recensions must be due to 'contamination'. But a closer look at the evidence suggests an alternative explanation.

The *Aparctias* phrase in Escorial R.II.18 proves that a version of the short recension incorporating the *Aparctias* phrase was circulating in Spain in the second half of the seventh century. There is no mystery about the source of the phrase. Any alert reader of *On the Nature of Things* could spot it in the inscriptions of Diagram 7, the diagram of the winds. José Carlos Martín regards it as a scribal gloss, which entered into the manuscript tradition sometime after Isidore's last revision.[42] However that may be, the important point is that the insertion must have taken place in Spain.

Wherever *Aparctias* appears in the manuscript tradition the *Hiemisphaeria* sentence appears with it.[43] This strongly suggests (though it does not prove) that *Hiemisphaeria* would have been found in the missing folio of Escorial R.II.18 that contained chapter 12. The source for *Hiemisphaeria* is Cassiodorus, *Institutions* 2.7.2. Isidore makes limited but significant use of Cassiodorus.[44] He quotes Cassiodorus on *Hiemisphaeria* more extensively in *Etymologies* 3.43, and he draws from this same section in *On the Nature of Things* 16. So *Hiemisphaeria* too could have originated in seventh-century Spain.

The close association of the *Aparctias* phrase and the *Hiemisphaeria* sentence with the 'Solinus' quotation raises the possibility of a Spanish origin for the latter as well. Isidore borrowed extensively from Solinus'

---

42 Martín, 'Isidorus Hisp.', 358–59.

43 The only time *Aparctias* appears without *Hiemisphaeria* is in a fourteenth-century collection of extracts from *DNR*, Barcelona 569.

44 Isidore knew Book II of the *Institutions*, but not Book I; see Laistner, *Thought and Letters*, 102 and Fontaine, *Isidore: Genèse*, 334.

Collectanea rerum memorabilium for his *Etymologies*,[45] but the only certain borrowing in *On the Nature of Things* occurs here in chapter 40.1, which may suggest that a text of Solinus came into his possession shortly before he completed the first draft. Initially, Isidore borrowed only the second half of the sentence in which Solinus offered a possible explanation for the tides. But subsequently, he or someone else inserted the first half of the sentence and added, *ut ait Solinus*, 'as Solinus says'.[46] This circumstance prompted Martín to conclude that the first half of the sentence might have been an Anglo-Saxon insertion of the eighth century.[47] Although Solinus' *Collectanea* had only a limited circulation in Anglo-Saxon England, it was known to Aldhelm and Bede,[48] so Martín's conjecture cannot be ruled out.[49] But a more plausible scenario in our view is that Isidore or one of his assistants decided his original partial quotation from Solinus might misleadingly suggest that the earth is a living soul; completing the quotation reminds the reader that this is just an analogy. In brief, Isidore may have had second thoughts, which led him to make this change in the course, possibly, of preparing a revision of the long recension.[50]

THE LONG RECENSION: CHAPTER 44(–) AND THE MYSTICAL ADDITION

The essence of the long recension is a new chapter, *De nominibus maris et fluminum*, 'The Names of the Sea and the Rivers', which is always inserted between chapter 43 and chapter 45(44), making in all a work of 48 chapters. Fontaine observes that the new chapter differs in style and content from analogous chapters in *On the Nature of Things*, i.e., chapter 12, 'Heaven and Its Name'; chapter 26, 'The Names of the Stars'; and chapter 37 'The Names of the Winds'. It lacks the religious allegory which these earlier

45 See the *Index locorum* in Theodor Mommsen's edition of Solinus, *Collectanea rerum memorabilium* (Berlin, 1895), 245–48.

46 Isidore never cites Solinus by name in the *Etymologies*. See Stahl, *Roman Science*, 215.

47 Martín, 'Isidorus Hisp.', 359.

48 Michael Lapidge, *The Anglo-Saxon Library* (Oxford: Oxford University Press, 2006), 187 and 225.

49 Herren, 'Classical and Secular Learning', 139, includes Solinus in his provisional list of sources known to the Irish before *c*.800.

50 See Commentary on ch. 40.

chapters incorporate and consists for the most part of simple definitions that look like glosses. On the other hand, the sources of chapter 44(–) are typical of Isidore. Some of the definitions are paralleled in books 13 and 14 of the *Etymologies*; others are unique to *On the Nature of Things*. Suetonius Tranquillus' *Meadows* (*Prata*) had already been cited in chapters 37 and 38 and would be again in the *Etymologies*. Naevius, Atta, and Pacuvius are likewise quoted in the *Etymologies*. The definition of 'blind flood' taken from Atta is backed up with a phrase borrowed from a letter of the emperor Augustus. Augustus is not cited elsewhere either in *On the Nature of Things* or the *Etymologies*, but neither is there any trace of his letters in the British Isles: if Augustus' work could be found anywhere, it would probably be the library of Seville.

In support of the claim that the long recension 'could have been of Irish provenance', William D. McCready appeals to Herren's edition of the *Hisperica Famina*, a collection of Hiberno-Latin poems which employs a strange, learned vocabulary. In it, McCready says, 'Herren comments on the word *tollus*, of Old Irish derivation. With the exception of Isidore *DNR* 44.5, it occurs only in Hiberno-Latin texts'.[51] Herren (who more guardedly speaks of 'one possible exception') and McCready refer to the sentence in chapter 44(–).5: 'Cascades [*tulli*] are chutes of water, like those that are in the river Anio where its pitch is the steepest'.[52] The etymological relationship, if any, between *tollus*, which occurs several times in the *Hisperica Famina*,[53] and is said to be of Old Irish derivation, and the classical Latin *tullius*, meaning much the same thing, which can be traced back to Ennius in the 2nd century BC and is found in Pliny, can be left to the specialists. The source of Isidore's definition of *tulli(i)* in chapter 44(–) is the *De significatu uerborum* of Sextus Pompeius Festus (late 2nd century AD), as Fontaine points out in his apparatus.[54] Isidore or an interpolator may have got this

---

51 William D. McCready, 'Bede, Isidore, and the *Epistola Cuthberti*'. *Traditio* 50 (1995): 89 n. 50; see Michael Herren (ed.), *The Hisperica Famina: I. The A-Text. A New Critical Edition with English Translation and Philological Commentary* (Toronto: Pontifical Institute of Mediaeval Studies, 1974), 134 and also Herren, 'On the Earliest Irish Acquaintance with Isidore of Seville', in Edward James (ed.), *Visigothic Spain: New Approaches* (Oxford: Clarendon Press, 1980), 245–46 and 250.

52 *Tulli aquarum proiectus, quales sunt in Aniensi flumine quam maxime praecipiti.*

53 Herren's note is on line 60 of the A-text.

54 *OLD*, s.v. *tullius*, cites Ennius and Pliny, as well as Festus. Festus, *Sexti Pompei Festi de verborum significatu quae supersunt cum Pauli epitome*, ed. Wallace M. Lindsay (Leipzig: Teubner, 1913): '*Tullios alii dixerunt esse silanos, alii riuos, alii uehementes proiectiones sanguinis arcuatim fluentis, quales sunt Tiburi in Aniene*'.

definition at second hand, but it does not prove an Irish origin of chapter 44(–). The chapter, which resembles an undigested transcription of reading notes, can readily be explained as an Isidorian afterthought. Isidore was deeply engaged in the *Etymologies* project. Perhaps he composed chapter 44(–); more probably he left its implementation to one of his assistants.[55]

In addition to chapter 44(–), most long-recension manuscripts include the mystical addition in chapter 1. However, there are at least four manuscripts of the long recension in which it is lacking. Three: Bern 610, Karlsruhe 229, and St Gall 240 date from the ninth century; the fourth, Paris NA 448, which is related to St Gall 240, is from the eleventh century.

We do not have sufficient data on Bern 610 to draw any firm conclusions. Karlsruhe 229 can be removed from consideration. It is demonstrably a mixed-recension manuscript, copied partly from a short-recension text and partly from a long.[56]

However, St Gall 240, a manuscript created at Chelles at the beginning of the ninth century, tells another story. The fact that its *capitula* list 48 numbered chapters confirms that it was intended from the start as a long-recension text. And yet it lacks not only the mystical addition but also the *Hiemisphaeria*, *Aparctias*, and 'Solinus' features that are typically found in the long recension.

It was a Type I version of the short recension that gave rise to the medium-recension exemplar that lies behind Cambrai 937. Point by point the features found in Cambrai 937 correspond to those in the long-recension St Gall 240, as Table 1 shows. It is impossible to attribute such a coincidence to chance. St Gall 240 descends from the same medium-recension type that is represented by Cambrai 937. The question that remains is where and how the transformation occurred. The nuns of Chelles, or their predecessors somewhere in the line of transmission, may have deliberately conflated a Cambrai-type manuscript with a long-recension manuscript, taking the first 43 chapters of *On the Nature of Things* from the former and adding

---

55 Arévalo, the eighteenth-century editor of Isidore, argues in favour of Isidore's authorship of ch. 44(–). Faustino Arévalo, *S. Isidori Hispalensis Episcopi ... Opera Omnia*, vols. 1–2, *Isidoriana* (Rome, 1797) 1:663.

56 It begins with the *capitula* of the short recension and omits not only the mystical addition but also the *Hiemisphaeria* sentence of ch. 12 and the *Aparctias* phrase of ch. 37. But, then, somewhere between ch. 37 and ch. 40, the scribe evidently shifted to a copy-text of the long recension, which included the 'Solinus' quotation in ch. 40 and, of course, chs. 44(–) and 48(47). The fact that chs. 33 and 34 appear out of order and unnumbered after ch. 37 suggests that the scribe's short-recension copy-text was defective, which may be the reason he shifted to another.

chapters 44(–) to 48(47) from the latter. This is the explanation favoured by Fontaine, in which case St Gall 240 and Paris NA 448 are, like Karlsruhe 229, essentially mixed-recension manuscripts.[57] Or, the new chapter 44(–) may have been inserted into a Cambrai-type manuscript of the medium recension in Spain before the mystical addition was conceived of, creating an 'unexpanded' version of the long recension that led ultimately to St Gall 240 and Paris NA 448.

The mystical addition was inserted into a version of the long recension that included the *Hemisphaeria*, *Aparctias*, and 'Solinus' features without the *de qualitate* formula or the placement of Diagram 7 at the end of chapter 37. It was the last substantial alteration to the textual tradition of *On the Nature of Things* and there is no reason to doubt that it was added in Spain.[58] The paragraph is entirely consistent with Isidore's style, his method of close paraphrase, his practice of making inventories. It forms a logical complement to the uncontestably authentic final section of 1.3, which likewise opens with a similar adverb, *'Prophetice'* ('in a prophetic sense). Isidore's principal source is Jerome's *Commentaria in Zachariam*, which he relies on elsewhere in *On the Nature of Things*. It is a matter of opinion as to whether the additional allegory brought in by the paragraph overburdens the chapter. One might counter that the brief amount of allegorization in the original draft of the chapter hardly does justice to the importance of 'day' in the Christian interpretation of the first six days of Creation. Isidore may have been inspired to expand the chapter by his investigation into liturgical feasts for *Etym.* 6.18.

---

57 Jacques Fontaine, 'La Diffusion de l'œuvre d'Isidore de Séville dans les scriptoria helvétiques du haut moyen âge', *Revue suisse d'histoire* 12 (1962): 317 and n. 15.

58 *Pace* Hillgarth, 'Visigothic Spain and Early Christian Ireland', 188, who postulates an Irish or Irish-influenced English origin for the long recension and thus, by implication, for the mystical addition.

OUT OF SPAIN AND INTO THE FUTURE

The changes that we have chronicled took place rapidly and were essentially complete by the time the earliest manuscripts that have survived were being copied. Seventh-century Spain is the most plausible locus for these alterations and transformations. Their nature is such that we can easily imagine them being authorized by Isidore or members of his *familia*, who were acquainted with his aims and had access to his sources. One addition, however, was composed by a different author, namely the *Poem on the Moon (Carmen de luna)*, or *Epistle*, addressed by King Sisebut to Isidore. The *Epistle* is the most substantial and at the same time the most puzzling of these various transformations. Fontaine makes the point that it can be regarded as a literary pendant that corresponds to Isidore's dedicatory prose epistle.[1] Since this text also constitutes the first evidence of the reception of *On the Nature of Things*, it calls for some closer examination here. A translation of the poem can be found in Appendix 1.

Sisebut begins with an amusing prologue, somewhat reminiscent of Vergil's First Eclogue, teasingly contrasting Isidore's idle leisure with his own anxious cares at home and on the battlefield (1–8). He imagines that Isidore may be composing a poem on the nature of eclipses in the stock imagery of pastoral poetry ('laurels', 'ivy', 'Phoebus' locks'), a project about as likely to succeed, he remarks, as the tortoise is to outrun the hound (9–14).

Nevertheless, he, Sisebut, will attempt to give the true explanation of the eclipse of the moon, avoiding popular folktales. This he does at considerable length and in great detail, in poetic language that is obscure almost to the point of impenetrability (15–57). Then he offers to do the same for the eclipse of the sun; his explanation takes all of four lines (58–61). The poem ends abruptly here, leaving the impression that it might be unfinished.[2]

---

1 Fontaine, *Traité*, 151.

2 Fontaine, *Traité*, 154, prefers to see this short description of the solar eclipse as 'une sorte de conclusion plutôt que l'amorce d'un développement inachevé'.

This is a curious response to *On the Nature of Things*. Isidore's treatise is, of course, written in prose (unlike the *De rerum natura* of his great predecessor Lucretius). Sisebut is the one who risks poetry, and there is certainly a long classical tradition of poetic treatment of natural history.[3] But Isidore has already explained the nature of both the solar eclipse (ch. 20) and the lunar eclipse (ch. 21) in straightforward terms, without resorting to poetic metaphor (or folktales). Sisebut's motive for composing the poem has been variously interpreted as an expression of longing for the contemplative life, a contribution to a national Spanish culture, an allegory of human rulership reflected in cosmology, and, most recently, as evidence of a certain rivalry between the king and Isidore himself for supremacy in cultural authority.[4] Certainly the relationship between the two men was never straightforward. If (as Yitzak Hen suggests) Sisebut was trying to outdo Isidore in his *Epistle*, Isidore responded with thinly veiled criticism of Sisebut's policy of forced conversion of Jews in his *Sententiae* and in *On the Catholic Faith against the Jews* (*De fide catholica contra Iudeos*, c.614–615). After the king's death, Isidore expressed his disapproval more frankly in his *History of the Goths* – the context of the somewhat qualified praise of Sisebut's learning cited above.[5]

Sisebut's *Epistle* was an integral appendage to *On the Nature of Things* in Type II of the short recension.[6] We have had occasion to remark on two

3 Taub, *Aetna and the Moon*, ch. 2.

4 Contemplative life: Vincenzo Recchia, *Sisebut di Toledo, il 'Carmen de luna'*, Quaderni di 'Vetera christianorum' 3 (Bari, 1971) and Recchia, 'Sul Carmen de luna di Sisebuto di Toledo', *Invigilata Lucernis* 20 (1998): 201–19. Spanish national culture: Laurens Johan van der Lof, 'Der Mäzen König Sisebutus und sein "De eclipsi lunae"', *Revue des études augustiniennes* 18 (1972): 145–51. Cosmic and political order: Stephen C. McCluskey, *Astronomies and Cultures in Early Medieval Europe* (Cambridge: Cambridge University Press, 1998), 124. Hen on the other hand sees it as an episode in a 'literary duel' with Isidore (see following note).

5 Yitzak Hen, 'A Visigothic King in Search of an Identity: "Sisebutus Gothorum gloriosissimus princeps"', in Richard Corradini, Matthew Gillis, Rosamond McKitterick, and Irene van Renswoude (eds.), *Ego Trouble: Authors and their Identities in the Early Middle Ages* (Vienna: Verlag der Österreichischen Akademie der Wissenschaften, 2010), 94–98.

6 Our translation of Sisebut's *Epistle* in Appendix 1 is based on Fontaine's edition (*Traité*, 328–35). For other editions and the manuscript tradition, see *Traité*, 159–61. A recent study of the diffusion and use of the text is provided by Paulo Farmhouse Alberto, 'King Sisebut's *Carmen de luna* in the Carolingian School', in Paulo Farmhouse Alberto and David Paniagua (eds.), *Ways of Approaching Knowledge in Late Antiquity and the Early Middle Ages: Schools and Scholarship* (Nordhausen: Traugott Bautz, 2012), 177–205.

oddities in connection with the *Epistle*. First, when it does appear in either the short or medium recension it is 'nearly' always placed at the end of chapter 47(46). Second, Sisebut is 'almost' never identified as its author. Were it not for Cologne 83(II), 'nearly' and 'almost' could be deleted from these two observations. Unlike other manuscripts that include the *Epistle*, Cologne 83/(II), which is a Type I manuscript, displays the *Epistle* as a prologue to *On the Nature of Things* instead of as an appendage to chapter 47(46), and unlike them it names Sisebut as the poet. The Cologne manuscript raises the possibility that a Type I version of *On the Nature of Things* circulated on the Continent with the *Epistle* affixed in this initial position, naming Sisebut as author, and indeed there is some corroborating evidence for this. In the ninth century, a scribe working in north-eastern France tacked lines 31–44 of the *Epistle* onto the end of Bede's *On the Nature of Things*, prefacing the lines with the remark, 'the sun is eighteen times greater than the earth, as Sisebut, the king of the Goths, confirms, writing in heroic verses to his teacher Saint Isidore, bishop of the city of Seville, to explain why the sun beyond the earth does not touch [illuminate] the moon when it is eclipsed'.[7] A tenth-century glossary now in the Vatican quotes one line from the poem with the comment that it was found 'in the prologue of Sisebut to Isidore'.[8] Several long-recension manuscripts of 'the English type' from the tenth century and later (see Appendix 6b) include the comment that a poem beginning, 'Perhaps in the woods you are leisurely composing idle songs' (i.e., Sisebut's *Epistle*), was sometimes inserted as a prologue to *On the Nature of Things*. But the comment also makes it clear that the true identity of the author of the poem was unknown at least in that time and place in England, just as it had been in Aldhelm's time (see below, p. 56).

So we are left with a mystery – why was the *Epistle* routinely presented, so far as we now know, without a hint as to its author, even,

There is an English translation in John R.C. Martyn, *King Sisebut and the Culture of Visigothic Spain, with Translations of the Lives of Saint Desiderius of Vienne and Saint Masona of Mérida* (Lewiston, NY: Edwin Mellen Press, 2008), 111–16.

7 Amiens 222 (cf. Kendall/Wallis, 44 and 65; online: *Gallica*), fol. 27v: '*Decies octies maior est sol quam terra sicut Sisebutus Gottorum Rex confirmat, scribens uersibus heroicis suo magistro Sancto Isidoro Spalense ciuitatis Episcopo, quare sol ultra terram non tangit lunam quando deficit*'.

8 Max Manitius, *Geschichte der lateinischen Literatur des Mittelalters*, vol. 1, *Von Justinian bis zur Mitte des 10. Jahrhunderts* (Munich: C.H. Beck, 1911), 188: '*Illud in prologo sesebuti ad ysidorum*'. Manitius also notes that the Carolingian grammarian Clemens Scottus cited two lines from the poem under Sisebut's name.

apparently, in Spain? Given what is known of Sisebut's pride in his literary accomplishments, it is unlikely that the king himself desired anonymity. More likely is the hypothesis that his authorship was deliberately effaced. If the latter were the case, it might suggest political antagonisms that would only be plausible in the context of the Visigothic monarchy. Sisebut was succeeded by his son, Reccared II, whose fate is obscure and who reigned for just a few months, and then in 621 by Suinthila, who came from a different noble family. Suinthila was overthrown in 631 by Sisenand. Isidore heaped praises on Sisebut and Suinthila in their turn (at least, while they were alive), but was silent concerning Sisenand.[9] Furthermore, around 620, Isidore dedicated an early draft of the *Etymologies* to the king under the same rubric he used for *On the Nature of Things*, 'Isidore to his Lord and Son Sisebut'.[10] But some manuscripts substitute Braulio's name for the king's, which was 'perhaps a change made by Isidore after Sisebut's death in 621'.[11] Isidore's previous association with Sisebut may have become a liability; indeed, Isidore's criticism of the king in the *Chronicle* might suggest that he was distancing himself from Sisebut after his death.[12]

Judging from the statistics compiled by Michael Gorman, *On the Nature of Things* was one of the most widely copied texts in Western Europe in the period between 650 and 800. Twelve manuscripts, eight of them complete, exist from this period, more than for the *Etymologies* or for any of the major works of Augustine.[13] For the Middle Ages as a whole, at least 90 manuscripts of the complete work (some fragmentary) are extant, as well as another 35 containing excerpts. Its popularity indicates that it filled a real need, which may relate to Isidore's radical decision to separate analysis of natural phenomena from the Creation narrative of Genesis.[14]

---

9 For details, see Collins, *Visigothic Spain*, 75–81.

10 For the text, see *Etymologiae*, ed. Lindsay, Epistola VI.

11 Barney, et al., *Etymologies*, 413 n. 10.

12 Hen, 'Visigothic King', 98.

13 See Gorman, 'Diagrams', 535. Isidore's appeal was not limited to *DNR*: J.N. Hillgarth, 'Ireland and Spain in the Seventh Century', *Peritia* 3 (1984): 5 points out that 'more manuscripts of Isidore are preserved from before 800 than of any other Latin author, apart from Augustine, and this despite the fact that the distribution of Isidore's works only began about 615'. Ian Wood, in a forthcoming paper, 'The Problem of Late Merovingian Culture', in Gerald Schwedler and Raphael Schwitter (eds.), *Exzerpieren–Kompilieren– Tradieren. Transformationen des Wissens zwischen Spätantike und Frühmittelalter*, lists 17 Merovingian MSS of various of Isidore's works.

14 See above, p. 10.

An Italian scribe in the eleventh century went one step further than Isidore by systematically stripping out the allegorical passages from the text, leaving only secular natural history. It would be interesting to learn more about the environment in which this manuscript (Florence Plut. 27 sin. 9) was copied.

The story of the diffusion of *On the Nature of Things* on the Continent and in the British Isles has been ably narrated by Jacques Fontaine, Bernhard Bischoff, Michael Gorman, and others.[15] Our purpose here is to trace the fortunes and further development of the various types we have described.

IRELAND AND ANGLO-SAXON ENGLAND

A number of Isidore's works, including *On the Nature of Things*, reached Ireland in the seventh century, and were exploited *inter alia* by the *Hisperica Famina* and in computistical texts.[16] We know that *On the Nature of Things* was imported into Great Britain late in the seventh century. Circumstantial evidence suggests that *On the Nature of Things* was available in Canterbury. At some point in the years between 670 and 690, scholars at the school established there by Archbishop Theodore and Abbot Hadrian compiled a glossary of terms from various learned works. The glossary comes to us at one remove in the form of the late-eighth-century Leiden Glossary, written at St Gall.[17] It contains two sets of *lemmata* in whole or in part from *On the Nature of Things*.[18] Neither

---

15 Fontaine, *Traité*, 69–84; Fontaine, 'La Diffusion de l'œuvre d'Isidore'; Fontaine, 'La Diffusion carolingienne du *De natura rerum* d'Isidore de Séville d'après les manuscrits conservés en Italie', *Studi medievali* 7 (1966):108–27; Bischoff, 'Die europäische Verbreitung'; Gorman, 'Diagrams'.

16 Herren, *Hisperica Famina*, 20; Herren, 'On the Earliest Irish Acquaintance'; Hillgarth, 'Visigothic Spain and Early Christian Ireland', 186–87; Hillgarth, 'Ireland and Spain in the Seventh Century', 8–9; Dáibhí Ó Cróinín, 'A Seventh-Century Irish Computus from the Circle of Cummian', in *Early Irish History and Chronology* (Dublin: Four Courts Press, 2003), 121.

17 *A Late Eighth-Century Latin–Anglo-Saxon Glossary Preserved in the Library of Leiden University (MS. Voss. Q°Lat. N°. 69)*, ed. John Henry Hessels (Cambridge: Cambridge University Press, 1906). See Michael Lapidge, 'The School of Theodore and Hadrian', *ASE* 15 (1986): 54–55 and Lapidge, *Anglo-Saxon Library*, 175–76; David W. Porter, 'Isidore's *Etymologiae* at the School of Canterbury', *ASE* 43 (2014): 7–44.

18 Leiden Glossary 27.1–33 and 44.1–29: *A Latin–Anglo-Saxon Glossary*, 22 and 45–46. Porter, 'Isidore's *Etymologiae*', 39 and 43 shows that the definition of three of the headwords

includes any glosses of terms from chapter 44(–) or chapter 48(47), from the mystical addition in chapter 1, or from the *Epistle* of Sisebut, which prevents us from drawing a firm conclusion about the recension on which they were based. However, the first set is introduced under the rubric, *In libro rotarum*,[19] and the second set, largely concerned with chapter 12, *De caelo*, quotes the *hiemisphaeria* sentence from chapter 12.3.[20] The Canterbury manuscript of *On the Nature of Things* therefore had an affinity with manuscripts that display some form of the title *liber rotarum* and contain the *hiemisphaeria* sentence. The mid-eighth-century long-recension St Gall 238 and Vatican City Pal. 1448, written at Trier at the beginning of the ninth century, as well as the mid-eighth-century Type I short-recension Laon 423, share these features. Laon 423 may have acquired them through contamination with a long-recension manuscript.[21] But the St Gall Leiden Glossary makes it seem likely that the Canterbury glosses were based ultimately on the same (Spanish?) exemplar that lies behind St Gall 238 and Vatican City Pal. 1448.

Aldhelm's copy of *On the Nature of Things* at Malmesbury included the *Epistle* of Sisebut, and therefore we can assume that it was either a Type II short-recension text or a text of the medium recension based on it. Aldhelm received his early education from the Irish monk Máeldub at Malmesbury, which was an Irish foundation, but he also studied as a young man with Abbot Hadrian at Canterbury, where he might first have encountered *On the Nature of Things*. As was almost always the case in Type II manuscripts, Aldhelm's copy lacked a rubric identifying Sisebut as the author of the *Epistle*. He assumed the poem was by Isidore because he quoted the second line of the *Epistle* to show that Isidore understood the metrical principle of elision.[22]

Bede, at the twin Northumbrian monasteries of Wearmouth and Jarrow, established by Benedict Biscop, knew and used Isidore's *On the Nature of Things* as the foundation for his own didactic texts, *On the Nature of*

---

in the first set (27.8, 27.13, 27.16) and one from the second (44.16) are from the *Etymologies*. However, the definitions of the first 9 headwords in the second set (44.1–9) are taken verbatim from *DNR* 12.3–4. Porter is mistaken in deriving 44.4 and 44.8 from the *Etymologies*.

19   Leiden Glossary 27: *A Latin–Anglo-Saxon Glossary*, 22.

20   Leiden Glossary 44.8: '*Hiemisperia duo sunt quorum alterum est super terram alterum sub terra*'. *A Latin–Anglo-Saxon Glossary*, 45.

21   See Gorman, 'Diagrams', *stemma codicum*, 541.

22   Aldhelm, *De metris et enigmatibus ac pedum regulis 9, in Opera*, ed. Rudolf Ehwald, MGH: AA 15 (1919):81, '*Isidorus uero uocales elisit ita: Argutusque inter latices et musica flabra*'.

Things and *On Times*.[23] His copy of Isidore's work certainly included chapter 48(47), which demonstrates that it must have been either a medium- or a long-recension manuscript.[24] It is true that Bede could have borrowed the final sentence of his *On the Nature of Things*, chapter 51, which appears in Isidore's chapter 48(47).2, either from Isidore or from Isidore's source, Augustine. But it is simply not credible to suppose that Bede independently created a final chapter for his work that covers the same ground as Isidore's new final chapter, and quotes the same sentence from Augustine. It seems likely that Bede's copy of Isidore's work displayed the title *De natura rerum*, since Bede appropriated it for one of his own treatises. This may be regarded as slender confirming evidence that it was in fact the long recension that had come into his hands.[25]

TRAFFIC BETWEEN SPAIN AND ITALY

How and from where did a copy of *On the Nature of Things* reach the school at Canterbury? Conceivably it came from Ireland or directly from Spain, though neither of these paths seems as likely as one from Italy. Canterbury did not have close connections with Ireland.[26] There is no evidence of communication between Spain and Canterbury in the latter part of the seventh century. Fragments of a folio from book 11 of Isidore's *Etymologies*, which are now in the monastic library at St Gall, were apparently written at Bobbio in the seventh century.[27] Bobbio was an Irish foundation, which might suggest transmission from Ireland, but

23 See Kendall/Wallis, 7–13.

24 Despite claims for the short recension expressed by Wesley M. Stevens, 'Scientific Instruction in Early Insular Schools', in Michael W. Herren (ed.), *Insular Latin Studies: Papers on Latin Texts and Manuscripts of the British Isles, 550–1066* (Toronto: Pontifical Institute of Mediaeval Studies, 1981), 100 and n. 58; Stevens, 'Sidereal Time in Anglo-Saxon England', in Calvin B. Kendall and Peter S. Wells (eds.), *Voyage to the Other World: The Legacy of Sutton Hoo* (Minneapolis: University of Minnesota Press, 1992), 136–37; McReady, 'Bede, Isidore, and the Epistola Cuthberti', 84–85 (see von Büren, 'Le *DNR* de Winithar', 393 n. 21). Our argument in favour of the medium ('second') recension (Kendall/ Wallis, 10, influenced by Fontaine, *Traité*, 74) now seems to us overconfident.

25 See above, pp. 52–53.

26 See Herren, 'Scholarly Contacts between the Irish and the Southern English in the Seventh Century', esp. pp. 30–35, 45–46, and 53, for what evidence there is of Irish influence at Canterbury.

27 Gorman, 'Diagrams', 536.

the fragments are best taken as evidence that some of Isidore's works had begun to circulate in Italy in the Lombard era in or shortly after Isidore's lifetime, perhaps from Rome. Archbishop Theodore was accompanied and guided on his trip from Rome in 669 by that indefatigable collector of books, Benedict Biscop, who was in fact put in charge of the abbey of Saints Peter and Paul in Canterbury where the school was established, until Hadrian, who had set out with them but was detained by Ebroin, the mayor of the palace in Neustria, and did not arrive until a year later, could take up his duties as abbot.[28] Perhaps Theodore and Hadrian found their scholarly resources still inadequate to their needs; Bede tells us that Benedict returned to Rome in 671 and 'brought back a good many books of divine teaching'.[29] If Isidore's *On the Nature of Things* had been available in Rome at the time, it would surely have seemed a useful addition to the library at Canterbury, for Theodore and Hadrian gave 'their hearers instruction not only in the books of holy Scripture but also in the art of metre, *astronomy*, and ecclesiastical computation'.[30]

Bamberg 61, a Type II short-recension manuscript from the second half of the eighth century, which was prepared at Monte Cassino, is the only short- or medium-recension manuscript to contain the 'Solinus' quotation (it also includes *Aparctias* and *Hiemisphaeria*). Assuming that the features *hiemisphaeria*, *Aparctias*, and 'Solinus' did arise in seventh-century Spain, how can their presence in Bamberg 61 be explained? Fontaine's explanation is visually depicted in his *stemma codicum* (*Traité*, 70 *bis*) and 'Carte de la diffusion du *Traité* en Europe' (*Traité*, 84–84 *bis*). In his reconstruction, Bamberg 61 I descends from a Frankish archetype of the short recension, which was influenced at some stage by a long-recension manuscript stemming from a Germanic archetype, which in turn went back to an English archetype. Gorman accepts Fontaine's reconstruction.[31] This *stemma* is convoluted, but it fits the facts and may be correct; indeed, if the 'Solinus' quotation in chapter 40.1 is an Irish or Anglo-Saxon insertion it almost has to be correct. But that an insular scribe would have recognized a sentence that begins 'some say' (*quidam aiunt*) as being a quotation

---

28 Bede, *Lives of the Abbots of Wearmouth and Jarrow*, ch. 3 (ed. Grocock and Wood, 27–28); *Ecclesiastical History*, 4.1; Bernhard Bischoff and Michael Lapidge, *Biblical Commentaries from the Canterbury School of Theodore and Hadrian*, Cambridge Studies in Anglo-Saxon England 10 (Cambridge: Cambridge University Press, 1994), 123–34.

29 Bede, *Lives of the Abbots*, ch. 4 (ed. Grocock and Wood, 30, trans. p. 31).

30 *Bede's Ecclesiastical History* 4.2 (Colgrave and Mynors, 333–35). Our emphasis.

31 Gorman, 'Diagrams', *stemma codicum*, 541.

from Solinus and then coupled it with a second borrowing from the same passage seems less likely than the scenario sketched above, namely that the quotation was completed and its author's name attached to it in Isidore's own library.

A more economical explanation in our view is that the presence of these features in Bamberg 61 was the result of direct communication between Visigothic Spain and Lombard Italy in the seventh century. Veronika von Büren argues that a short-recension manuscript of *On the Nature of Things* closely related to Escorial R.II.18 was carried by Visigothic emigrants to northern Italy in the seventh century.[32] Churchmen, most often bishops, circulating between Italy and France in the second half of the eighth century, introduced this version into France. At the same period, *On the Nature of Things* was brought into the collection of computistical texts at Monte Cassino 'under the auspices of Paul the Deacon'.[33] In short, the Monte Cassino manuscript, Bamberg 61, may owe its existence to a 'corrected copy' (that is, one that inserted the new *hiemispheria, Aparctias,* and 'Solinus' interpolations) of a Type II short-recension manuscript that reached Italy from Spain in the seventh century.

GAUL

Type I of the short recension of *On the Nature of Things* reached the valley of the Loire by the end of the seventh or beginning of the eighth century.[34] It was copied there at that time, possibly at Fleury, in Paris 6400G, a fifth- and sixth-century palimpsest of African origin.[35] The text of *On the Nature of Things* in Paris 6400G (upper palimpsest) conceivably owes its existence to a Spanish exemplar from before the end of the seventh century. At any rate, the Loire valley is the earliest known point of entry of *On the Nature of Things* into Merovingian Gaul, which, as Ian Wood has convincingly shown, had a wider and deeper manuscript culture than has previously been suspected. At least 11 of Isidore's works are known to have been

---

32 It would be the functional equivalent of Fontaine's α.

33 von Büren, 'Le *DNR* de Winithar', 402–03.

34 If the twelfth-century MS, NY Plimpton 251, is indeed Spanish, it might be possible to see it as late evidence of the origin of Type I in Spain. However, it is just as likely to be the result of the reimportation of *DNR* into Spain from France in the wake of the *Reconquista*.

35 Fontaine, *Traité*, 25. Paris 6400G is known to have been at Fleury from the ninth century; it was 'very possibly' copied there in the abbey's earliest years (Fontaine).

copied at various centres there before the end of the Merovingian period.[36] Fleury's interest in the work can be traced in the ninth century and beyond, whether in copying and disseminating Type I of the short recension,[37] or in acquiring the long recension,[38] or in making extracts.[39] It may have been a transit point for other Type I manuscripts as well,[40] which had moved north and east by early in the ninth century (Laon 422 and Cologne 83(II)). Type II of the short recension makes its first recorded appearance outside of Spain in the mid-eighth century at the double monastery of Chelles. The spread of *On the Nature of Things* into Gaul seems likely to be attributable as much to the flight of Spanish clerics and monks with what they could carry from their libraries northward through Septimania after the Arab invasion of 711 as to demand by Gallic scholars prior to that time for Isidore's work. Such an assumption, if true, might go part way towards explaining the geographical concentration of much of the short recension in northern France.

GERMANY AND SWITZERLAND: THE ZOFINGEN METAMORPHOSIS

The oldest extant manuscript of the expanded long recension appears to be Basel F.III.15f, which was apparently copied in England during the first half of the eighth century. Copies of this recension may have survived in

36 Wood, 'The Problem of Late Merovingian Culture'. In addition to *DNR*, the works of Isidore include *De Ecclesiasticis Officiis, De Legibus, Differentiae, Etymologies, Excerpta Patrum, Liber Proemiorum, Questions on the New Testament, Sententiae, Soliloquies, and Synonyma*: Wood (n. 29).

37 Cf., possibly, Avranches 109 (s. xi); Bern A.92/20; Bern 224; Bern 249; Milan H 150 inf.; Vatican City Reg. 25; possibly, Vatican City Reg. 1260. Any or all of these may descend independently from the same exemplar that lies behind Paris 6400G.

38 Cf. Bern 610.

39 Cf. Berlin Phill. 1833 (s. x/xi); possibly, Bern 417; London Harl. 3017; possibly, Paris 5543.

40 The *stemma codicum* constructed by Fontaine (Traité, 70 *bis*) and extended by Gorman ('Diagrams', 541) shows all the *de qualitate* MSS descending from a common exemplar δ, which implicitly ('Familia Gallica') and visually (Fontaine, Traité, 84 *bis*) they locate in France. As should be entirely evident by now, we believe that Type I of the short recension, like the other recensions and major branches of *DNR*, originated in seventh-century Spain. This does not invalidate the Fontaine/Gorman *stemma*, but it does suggest that their postulated seventh-century exemplars, α, β, γ, δ, ν, and probably π could all be interpreted as Spanish (see 'Carte de la diffusion du Traité en Europe' in Fontaine, *Traité*, 84 bis).

southern England (at Canterbury?), because it resurfaces in the shape of the 'English type' in the second half of the tenth century.[41] Basel F.III.15f was carried from England to Fulda in Germany, which was founded by a disciple of Boniface in 744. Both Basel F.III.15a + Copenhagen 19.VII and the Weimar fragment (Weimar 414a) of the end of the eighth century in Anglo-Saxon minuscule may have been produced in the scriptorium of Fulda. This was at a time (*c*.800) when a library catalogue prepared at Fulda shows that the monastery possessed a copy of Isidore's *Liber rotarum*.[42] The title probably indicates a long-recension text.

The long-recension St Gall 238 was prepared at St Gall *c*.760–780 by the scribe Winithar, who would later become abbot.[43] The abbey of St Gall is important not only for its library and scriptorium, but also because it was 'a meeting-place of Italian, Frankish and Insular (English as well as Irish) culture'.[44] It was probably there that the most interesting post-Isidorian transformation of *On the Nature of Things* took place, giving rise to what we will refer to as the 'Zofingen metamorphosis' – the result of a deliberate decision to incorporate *On the Nature of Things* into the text of Isidore's *Etymologies*. The decision was probably taken in the first half of the ninth century at St Gall, the time and place of the production of Zofingen Pa 32, and may have been related to an ambitious project to create an encyclopaedia on an even grander scale than the *Etymologies*.[45]

The Zofingen redactor's principal model for the text of *On the Nature of Things* was, it seems, an expanded long-recension text, Paris 10616, prepared a few years earlier between 796 and 799 in Verona.[46] Paris 10616 exhibits the normal chapter order of the long recension, but it enlarges the final chapter, chapter 48(47), with an extract from Vegetius and a section *De Trinitate* from the *Etymologies* (*Etym.* 7.4) (see Appendix 5a). The editor may have felt that a concrete example of the way the division of the

---

41 See Appendix 6b.

42 Lapidge, *Anglo-Saxon Library*, 151–53 ('*Sancti Esidori r[..]arum*').

43 For Winithar's career, see von Büren, 'Le *DNR* de Winithar'.

44 W.M. Lindsay, 'The Affatim Glossary and Others', *Classical Quarterly* 11 (1917):195.

45 Veronika von Büren, 'Isidore, Végèce et Titanus au VIIIᵉ siècle', in Pol Defosse (ed.), *Hommages à Carl Deroux*, vol. 5, *Christianisme et moyen âge: Néo-latin et survivance de la latinité* (Brussels: Latomus, 2003), 43 believes Winithar may have brought the model for the Zofingen metamorphosis to St Gall from Italy. She situates the impetus for the encyclopaedic project, the *Liber glossarum*, in the north of Italy, probably Verona, 'dans ce milieu composé de *peregrini*, c'est-à-dire des exilés wisigoths et insulaires et des autochthones'.

46 The first part of this MS, containing a commentary on Genesis, is now Paris, Bibliothèque nationale lat. 10457. See von Büren, 'Isidore, Végèce et Titanus', 40 and n. 10.

world into three parts affects climate and men (Vegetius), together with an implicit reminder of the allegorical significance of the threefold division of one world (*De Trinitate*), would be an appropriate way of bringing *On the Nature of Things* to a conclusion. He signalled the allegory by inscribing the words *patrem / filium / spiritum / sanctum* in the four corners of a T-O map, which he inserted between the Vegetius extract and the *De Trinitate* chapter. The Zofingen redactor kept the Verona manuscript's enlarged version of chapter 48(47) (Vegetius + *De Trinitate*), though he dropped the T-O map.

The first task of the Zofingen redactor was to determine where *On the Nature of Things* would best fit into the 20-book structure of the *Etymologies*. His obvious choice might be thought to have been books 13 and 14, which cover much the same ground as *On the Nature of Things*. Perhaps he considered repetition an obstacle. At any event, book 3 ('On mathematics') and book 5 ('On laws and on times') provided the openings that he was looking for. He inserted *On the Nature of Things* after the section *De astronomia*, 'On astronomy', in book 3 (*Etym.* 3.24–71 (Gasparotti and Guillaumin, *Etym.* 3.23–70)) and transferred a section from book 5 of the *Etymologies*, which he labelled *De temporibus*, 'On times' (= *Etym.* 5.28–39), to the end of *On the Nature of Things*. 'On times' concludes with Isidore's shorter chronicle of the six ages of the world. The presence at St Gall of manuscripts of Bede's *The Reckoning of Time* (*De temporum ratione*) (with its Greater Chronicle), *On the Nature of Things* (*De natura rerum*), and *On Times* (*De temporibus*) (with its Lesser Chronicle) may have suggested the idea of associating 'On times' with *On the Nature of Things*. Interestingly, the redactor prefaced the section 'On astronomy' with the words, *Incipit excarpsum de libro rotarum sancti ysidori episcopi* ('Here begins an extract from the Book of Wheels of Saint Isidore the bishop'). In effect, he created a new book 4 of the *Etymologies* with the title *Liber rotarum* (no doubt borrowed from Winithar's St Gall 238).[47] The new *Liber rotarum* in turn is made up of three parts – *De astronomia*, (the old) *De rerum natura* (note the inversion in the title), and *De temporibus*.

The *capitula* (fol. 56r–v) for *De rerum natura* precede the preface and list 59 chapters. The final chapter heading ('Recapitulation of the above') may have been intended to introduce 'On times'. The Zofingen redactor

---

47 The three-part *Liber rotarum* is labelled book 4 (fol. 50r), but so is the book 'On medicine', which follows the section 'on times' (fol. 81v). The redactor evidently was reluctant to disturb the 20-book structure of the *Etymologies*.

subdivided and rearranged the 48 chapters of the long recension into 58 chapters. The new chapters with their equivalents in the regular order of *On the Nature of Things* are given in Appendix 6a. Apart from the subdivision of chapter 26 ('The Names of the Stars') into nine chapters, the chief difference in the Zofingen order is the displacement of chapters 15–21 to the end of the text. It may be that the redactor felt that these chapters on the sun and the moon, the heavenly bodies by which time is measured (Gen. 1:14), would constitute an appropriate bridge to the new chapter 59, 'On Times'. The only chapters that are individually moved out of place are the two chapters, 44(–) and 48(47), that convert the short recension into the long – an almost certain indication that the Zofingen redactor was acquainted with the short recension.

FROM THE CAROLINGIAN PERIOD TO THE AGE OF PRINT

The success of Bede's three scientific treatises, *On the Nature of Things* and *On Times*, for which Isidore's *On the Nature of Things* had provided the framework, and *The Reckoning of Time*, his greatly enlarged and more detailed treatment of the *computus,* appears to have diminished somewhat the demand for Isidore's *On the Nature of Things*.[48] Bede's treatises were copied and recopied at Fleury.[49] There, and elsewhere, they were often accompanied by excerpts from Isidore's *On the Nature of Things* rather than the full text; the reverse is not often the case.[50] Unlike Bede, Isidore did not generate an extensive tradition of glosses and commentaries in the Carolingian period. This may be because the Carolingians were interested primarily in the *computus – On the Nature of Things* was heavily excerpted

---

48 Eastwood, 'Early Medieval Cosmology', 307. For the complicated history of the reception of Bede's *DTR* on the Continent, see James T. Palmer, 'The Ends and Futures of Bede's *De temporum ratione*', in Peter Darby and Faith Wallis (eds.), *Bede and the Future* (Farnham: Ashgate, 2014), 150–59.

49 Judging from the lists in Charles W. Jones (eds.), *De temporum ratione*, CCSL 123B (Turnhout: Brepols, 1977), 242–56 and Kendall/Wallis, 44–66, there are still extant 11 MSS of one or more of Bede's three treatises that were certainly or possibly copied at Fleury. The actual number is probably higher.

50 Bede *DTR/DNR/DT* complete, Isidore excerpt: Oxford St John's 17; Paris 5239; Paris 5543; Strasbourg 326; Bede *DTR/DNR* complete, Isidore excerpt: Baltimore 73; Trier 2500; Bede *DTR* complete, Isidore excerpt: Cambridge CCC 291; Bede *DNR/DT* complete, Isidore excerpt: Vatican City Reg. 123; Bede *DNR* complete, Isidore excerpt: Vatican City Reg. 309.

into Carolingian computistical treatises and compilations[51] – or because, having rediscovered Macrobius and Martianus Capella and Calcidius, they preferred to engage with more sophisticated texts on cosmology. Somewhat later, much the same process of reduction can be found in manuscripts of Bede's works. The fortunes of Isidore's *On the Nature of Things* also parallel those of Bede's works in that from a low-water mark in the tenth century (11 MSS) there is a rise in the number of copies that peaks in the twelfth century (22 codices) and then drops off steeply in the thirteenth (7), fourteenth (6), and fifteenth centuries (4).

The various types of the three recensions that developed in Visigothic Spain and endured with further modifications over the centuries are reflected in the first printed editions. Unsurprisingly, the version of the work that would be reproduced depended on the manuscript(s) available locally to its editor. Günther Zainer, the editor of the *editio princeps* of *On the Nature of Things*, published in Augsburg in 1472, took for his copy text a 58-chapter, long-recension manuscript of the Zofingen metamorphosis.

A century later, in Paris, Margerin de la Bigne brought out a version in 47 chapters (1580), part of the first complete edition of the works of Isidore. It was based on a manuscript of the medium recension, possibly Paris 15171. Both the manuscript and the printed edition include the poem *De Ventis* immediately following chapter 37. This late-antique poem circulated apart from *On the Nature of Things* and is first found in Bern 611 (late eighth/ early ninth century). A translation and commentary is included in Appendix 4. Another complete edition of Isidore's works appeared in Madrid in 1599 under the direction of several editors. Juan Grial, who edited *On the Nature of Things*, utilized four manuscripts, one of which was Escorial R.II.18.[52] Evidently, Grial overlooked or chose to ignore the additional chapter added to Escorial R.II.18, because his is a short-recension text of 46 chapters. The Paris edition of 1601 of the complete works, edited by Jacques du Breul, utilized the 47-chapter medium-recension text of Margerin de la Bigne for *On the Nature of Things* (including the poem *De Ventis* after chapter 37).

A Spanish Jesuit, Faustino Arévalo, prepared yet another complete edition of Isidore's works. Volume 7, containing *On the Nature of Things*,

---

51 See the entry under 'Isidor von Sevilla *De natura rerum*' in the source index of Arno Borst's comprehensive edition of Carolingian computistical literature, *Schriften zur Komputistik im Frankenreich von 721 bis 818*, MGH: Quellen zur Geschichte des Mittelalters 21 (Hanover: Hahnsche Buchhandlung, 2006), 3:1489–90.

52 At a guess, the other three were the short-recension MSS, Escorial E.IV.14, M.11.23, and K.I.12.

appeared in Rome in 1803. Arévalo had access to manuscripts both in Spain and in Italy. It is not always clear which they were, but one of them was Escorial R.II.18. His text combines features drawn from one or another of the three Recensions. It lacks the mystical addition in chapter 1 and *Hiemisphaeria* in chapter 12 (probably Arévalo was guided here by the short-recension ninth-century folios of Escorial R.II.18); it inserts *Aparctias* in chapter 37.1, omits 'Solinus' in chapter 40.1, and includes chapter 48(47) with the feature *Maeotis* (here Arévalo was following the short-/medium-recension, seventh-century folios of Escorial R.II.18, silently omitting Sisebut's *Epistle*).[53] Arévalo included chapter 44(–) under the mistaken impression that it had never before been printed. He claimed to have taken this chapter from the 'best' (*optima*) of many (long-recension) manuscripts available to him (presumably in Rome). The result is an eclectic version of *On the Nature of Things* in 48 chapters that does not correspond to any version of the work in the manuscript tradition.

With Gustav Becker, we enter into the modern era of textual criticism. For his edition, published in Berlin in 1857, Becker consulted 13 short- and long-recension manuscripts, principally from collections in Germany and Switzerland. He collated two of these in their entirety – the long-recension Bamberg Msc. Nat. 1 (Becker's *A*) and the short-recension Bamberg Patr. 61 (Becker's *B*). For chapters 44(–) and 48(47) and the mystical addition, which are necessarily missing from Bamberg Patr. 61, and for several selected passages, he added collations from various other manuscripts. In the end, Becker's printed text is an excellent representation of the expanded version of the long recension.

Jacques Fontaine based his edition, which is the basis of our translation and commentary, on a complete collation of 16 manuscripts from the ninth century and earlier, and from libraries from all parts of Europe. It is an inclusive text, eclectic, like Arévalo's, but none the worse for that. It remains the definitive edition of *On the Nature of Things*.

---

53 Arévalo includes the three omitted features in footnotes. He prints the *Epistle* separately in an appendix, *Opera Omnia*, 7:183–85, under the impression that it was a poem by Isidore.

INVENTORY OF MANUSCRIPTS AND EDITIONS OF ISIDORE'S *ON THE NATURE OF THINGS*

Here we offer a comprehensive list of all the manuscripts and editions of *On the Nature of Things* that have been traced to date. The terms 'short', 'middle', and 'long' recensions, as well as our shorthand designations of the various features that differentiate the different states of the text (e.g., 'mystical addition', 'Solinus', etc.) are explained above. The absence of chapter 44(–) and the mystical addition from manuscripts of the short and medium recensions is assumed without further notice, as is the absence of chapter 48(47) from manuscripts of the short recension. Likewise, the absence of the *Epistle* of Sisebut from manuscripts of the long recension is not recorded. For references to Diagrams 1A, 6A, and 7A, see Introduction, p. 29 n. 10.

The abbreviation *DNR* (*De natura rerum*), when used alone, refers to Isidore's *On the Nature of Things* (Bede's work of the same title is always preceded by his name).

Manuscripts personally examined by Calvin B. Kendall are indicated by the initials 'cbk'. Digitized facsimiles accessible on the Internet are indicated by the rubric 'online', and identified by an abbreviation keyed to the list of 'Digitized Manuscript Repositories and Catalogues' at the end of the Bibliography.

MANUSCRIPTS OF ISIDORE'S *DE NATURA RERUM*

1. AVRANCHES, Bibliothèque municipale 109, fols. 97r–112r, s. xi, France. Lacks title. Short recension. Omits *Hiemisphaeria* in ch. 12.3. Omits *Aparctias* in ch. 37.1. Adds formula of introduction for Diagram 7 after ch. 37.5 (see Appendix 2a). Omits 'Solinus' in ch. 40.1. Diagrams 1–7 + 5A. KOCMOC inscription in Diagram 7, which is displaced to end of chapter. No T-O map. No *Epistle* of Sisebut.
Catalogue général 10, 51; Fontaine, *Traité*, 37. Online: *BVMM*.

2. BALTIMORE, Walters Art Gallery 73, fols. 2v and 7v, s. xii[2], England. Excerpts, similar to Oxford, St John's College 17. Fol. 2v: Diagram 6 + second variant wheel of planets + excerpted legends (not Isidore's) displayed as plain text on the side. Fol. 7v: ch. 11 + Diagrams 4–5.

The codex includes excerpts from Bede's *DTR* and *DNR*.

Bober, 'An Illustrated Medieval School-Book'; Wallis, 'Calendar & Cloister'. Online: *The Digital Walters*.

3. BAMBERG, Staatsbibliothek, Msc. Nat. 1 (HJ.IV.17), fols. 1v–43r, s. ix[1/3], eastern France. *Incipit liber de astra caeli sancti hisidori spalensis episcopi*. Long recension. Includes mystical addition in ch. 1. Inserts *Hiemisphaeria* in ch. 12.3. Inserts *Aparctias* in ch. 37.1. Adds formula of introduction for Diagram 7 in ch. 37 (see Appendix 2b). Inserts 'Solinus' in ch. 40.1. Diagrams 1–7 + 5A. T-O map (fol. 43v), with legends as in Appendix 5b.

Becker, *DNR*, xxiii–xxiv; Leitschuh, *Katalog* 1.2:408; Beeson, *Isidor-Studien*, 70; Fontaine, *Traité*, 37; Bischoff, *Katalog*, no. 220; Eastwood, 'The Diagram of the Four Elements', 555; Suckale-Redlefsen, *Die Handschriften*, 24–26. Online: MDZ. Becker's *A*.

4. BAMBERG, Staatsbibliothek, Msc. Hist. 3 (E.III.14), fol. 191r–v, s. xi, Italy. Excerpt: somewhat expanded text of ch. 37.1–4. *Ventorum quattuor cardinales sunt. Primus cardinalis septentrio. ... Dictus est autem chorus. Eo quod omnem spiritum uentorum ipse concludat.*

Leitschuh, *Katalog* 1.2:123; Colgrave and Mynors, *Bede's Ecclesiastical History*, lxv.

5. BAMBERG, Staatsbibliothek, Patr. 61 (HJ.IV.15), fols. 82v–103r, s. viii[2], Monte Cassino. Lacks title. Short recension. Inserts *Hiemisphaeria* in ch. 12.3. Inserts *Aparctias* in ch. 37.1. Inserts 'Solinus' in ch. 40.1. Diagrams 1–4; 6–7 + 5A. The *Epistle* of Sisebut appears on fols. 102v–103r.

The codex includes Cassiodorus' *Institutions*.

Becker, *DNR*, xxiv; Leitschuh, *Katalog* 1.1:426; Beeson, *Isidor-Studien*, 69–70; Mynors, *Cassiodori Institutiones*, x–xi; Lowe, *CLA* 8 (no. 1029); Fontaine, *Traité*, 28–29; Bischoff, *Katalog*, no. 234a; Gorman, 'Diagrams', 540; Martín, 'Sisebut', 403; Martín, 'Isidorus Hisp.', 353. Fontaine's *C*.

6. BARCELONA, Biblioteca de Catalunya 569, fols. 24r–33v, s. xiv, Spain? Excerpts from short or medium recension, used to form an anonymous

work. *Incipit expositio computo de diusionibus temporum.* The first untitled chapter begins: *Diuisionibus temporum quot sunt ... xiiii* [cf. *PL* 90, 653]. It is followed (fol. 25r) by a chapter entitled *De momento.* The excerpts from *DNR* begin on fol. 25v: chs. 1–3; 6; 8–12; 27–43; 46(45). Chs. 3 and 6 are only partly drawn from Isidore. There is no preface and no *capitula.* Omits mystical addition in ch. 1.3. Omits *Hiemisphaeria* in ch. 12.3. Inserts *Aparctias* in ch. 37.1. Omits 'Solinus' in ch. 40.1. No diagrams; no T-O map. No *Epistle* of Sisebut.

Mateu Ibars, *Colectánea paleográfica* 1:476; cbk.

7. BASEL, Universitätsbibliothek Lat. F.III.15a + COPENHAGEN, Universitetsbibliothek, Frag. 19.VII, s. viii^cx, probably Fulda. Detached fragments of same MS. Anglo-Saxon minuscule. Long recension.

Copenhagen fragment: two strips cut from a single folio. Strip 1 recto: title (*Incipit liber sancti isidorii*) to preface 1 (... *et quaedam ex rerum na<tura>*); strip 2 recto: preface 1 (*et rationem dierum* ...) to preface 2 (... *naturam noscere super<stitiosae>*). Strip 1 verso (in two columns): *capitula* of chs. 1–5 and chs. 25–32; strip 2 verso: *capitula* of chs. 9–15 and chs. 37–45(44). The latter includes the rubric for ch. 44(–) (*De nominibus maris et fluminum*).

Basel fragment, fols. 1v–16v: ch. 15.3 (*scriptum est uobis autem* ...) to end. A skip on fol. 4r from ch. 19.2 to ch. 22.2. Inserts *Aparctias* in ch. 37.1. Inserts 'Solinus' in ch. 40.1. Diagrams 6–7. T-O map (fol. 16v).

Becker, *DNR*, xxv; Beeson, *Isidor-Studien,* 67–68; Lowe, *CLA* 7 (no. 842); Lowe, *CLA* 10 (no. 842); Fontaine, *Traité,* 31, 35–37; Stevens, 'Compotistica et Astronomica', 44; Bischoff, *Katalog,* no. 271; Gorman, 'Diagrams', 540; Martín, 'Isidorus Hisp.', 353–54; Martín, 'Réflexions sur la tradition manuscrite', 208–09 and n. 14 (pp. 211–12). Fontaine's *F* (Basel). Online: E-manuskripter (Copenhagen). Fontaine's *Z* (Copenhagen).

8. BASEL, Universitätsbibliothek Lat. F.III.15f, fols. 1r–13r, s. viii^l, England; prov. Fulda. Lacks title. Long recension. In the *capitula,* ch. 44(–) is numbered 43: hence chs. 45(44)–48(47) are numbered 44–47; there are no numbers in the text. Includes mystical addition in ch. 1. Inserts *Hiemisphaeria* in ch. 12.3. Inserts *Aparctias* in ch. 37.1. Inserts 'Solinus' in ch. 40.1. Spaces left for Diagrams 1–7 + 5A (space for figures of 5A at end of ch. 18.7; text for 5A omitted). Space for T-O map.

Becker, *DNR*, xxv; Beeson, *Isidor-Studien,* 68; Lowe, *CLA* 7 (no. 848); Fontaine, *Traité,* 32; Stevens, 'Compotistica et Astronomica', 44; Gneuss and Lapidge, *Anglo-Saxon Manuscripts*, no. 786; Gorman, 'Diagrams', 540; Martín, 'Isidorus Hisp.', 353. Online: *e-codices*. Fontaine's *A*.

9. BASEL, Universitätsbibliothek Lat. F.III.15k, fols. 1r–21r, s. ix$^{1/3}$, Benediktbeuern; prov. Fulda. *Rotarum ysidori*. Short recension. Diagrams: 4 (fol. 7r), 7 (fol. 18r).
The codex includes excerpts from Bede's *DTR*.

Becker, *DNR*, xxv; Beeson, *Isidor-Studien*, 68; Fontaine, *Traité*, 37; Fontaine, 'La Diffusion de l'œuvre d'Isidore', 309–10; Bischoff, *Katalog*, no. 282; Gorman, 'Diagrams', *stemma codicum*, 541; Eastwood, 'The Diagram of the Four Elements', 555; Eastwood, *Ordering the Heavens*, 164, n. 135. Becker's *E*.

10. BERLIN, Staatsbibliothek zu Berlin, Preussischer Kulturbesitz, Hamilton 689, fols. 1r–8v, s. xi, perhaps northern Italy. Fragment: long recension, arranged here in 50 chapters (the first three chapters lost through damage to the folios). Vegetius interpolation at end of ch. 48 (see Appendix 5a). Diagrams 1–7 + 7A (fol. 8v, mountains of Sicily: *Ericus, Nebrodes, Neptunus, Ethna*).

Fontaine, *Traité*, 37; Boese, *Die Lateinischen Handschriften*, 332–33.

11. BERLIN, Staatsbibliothek zu Berlin, Preussischer Kulturbesitz, Phillipps 1830 (Rose 129), fol. 3v, s. ix/x, Metz. *Hic quoque potest uideri quo ordine spirant uenti.* Excerpt: abbreviated text of ch. 37.1–4 within the 12 segments of a diagram of the winds (= Diagram 7) with a T-O map in its centre; the text has been rearranged to fit into the appropriate points of the compass. Inserts *Aparctias* in ch. 37.1 (*qui et aparchias*). The poem *De Ventis* (see Appendix 4) appears beneath the diagram.

Bischoff, *Katalog*, no. 436; Obrist, 'Wind Diagrams', 50–52 and fig. 10.

12. BERLIN, Staatsbibliothek zu Berlin, Preussischer Kulturbesitz, Phillipps 1833 (Rose 138), s. x/xi, Fleury. Excerpts.

Jones, *Bedae Pseudepigrapha*, 112; Mostert, *The Library of Fleury*, 47 (BF012); Wallis, 'Calendar & Cloister'.

13. BERLIN, Staatsbibliothek zu Berlin, Preussischer Kulturbesitz, lat. fol. 307, fol. 2r, s. xii^2. Excerpt: ch. 38.
Jordanus.

14. BERN, Burgerbibliothek A.92/20, 2 fols., s. viii/ix, probably region of the Loire, Fleury(?). Fragment: short recension, chs. 33.2–36.2; 37.4–38.2. Adds formula of introduction for Diagram 7 after ch. 37.5 (see Appendix 2a). KOCMOC inscription in Diagram 7 (fol. 2r), which is displaced to end of chapter.

Hagen, *Catalogus*, 128; Beeson, *Isidor-Studien*, 68; Lowe, *CLA* 7 (no. 857); Fontaine, *Traité*, 37; Fontaine, 'La Diffusion de l'œuvre d'Isidore', 310–11; Homburger, *Die Illustrierten Handschriften*, 41–42 and ill. 12; Mostert, *The Library of Fleury*, 56 (BF065); Bischoff, *Katalog*, no. 523; Obrist, 'Le Diagramme isidorien des saisons', fig. 13; Gorman, 'Diagrams', *stemma codicum*, 541.

15. BERN, Burgerbibliothek F.219/2, fols. 1r–4r, s. ix$^{1/3}$, Loire region. Fragment: short recension, from ch. 29.2 (*culpam suam admonitus ...*) to ch. 47(46).3. The *Epistle* of Sisebut follows *DNR* on fol. 4r.

Hagen, *Catalogus*, 271; Becker, *DNR*, xxv; *Anth. Lat.* 1.2:3; Beeson, *Isidor-Studien*, 68; Fontaine, *Traité*, 37; Fontaine, 'La Diffusion de l'œuvre d'Isidore', 309; Bischoff, *Katalog*, no. 556; Gorman, 'Diagrams', *stemma codicum*, 541; Alberto, 'Sisebut's *Carmen*', 184. Becker's *K*.

16. BERN, Burgerbibliothek 212(I), fol. 109r, s. ix$^{1/3}$, Mainz(?). Excerpt: abbreviated text of ch. 37.1–4 within the 12 segments of a diagram of the winds (= Diagram 7) with a T-O map in its centre; the text has been rearranged to fit into the appropriate points of the compass. Inserts *Aparctias* in ch. 37.1 (*qui et aparchias*). The poem *De Ventis* (see Appendix 4) precedes the diagram. Identical to Karlsruhe 106.

Mynors, *Cassiodori Institutiones*, xxxi; Homburger, *Die Illustrierten Handschriften*, 85–86 and ill. 74; Obrist, 'Wind Diagrams', 49–51 and fig. 9.

17. BERN, Burgerbibliothek 224, fols. 164v–174r, s. ix$^{1/3}$, Strasbourg(?). *Explicit liber de mundo* (fol. 174r). Short recension. Adds formula of introduction for Diagram 7 after ch. 37.5 (see Appendix 2a). Diagrams 1–7 + 5A. KOCMOC inscription in Diagram 7, which is displaced to end of chapter. The *Epistle* of Sisebut is omitted.

Hagen, *Catalogus*, 274; Becker, *DNR*, xxv; Fontaine, *Traité*, 37; Fontaine, 'La Diffusion de l'œuvre d'Isidore', 312; Homburger, *Die Illustrierten Handschriften*, 65–69; Mostert, *The Library of Fleury*, 66 (BF123); Bischoff, *Katalog*, no. 557; Eastwood, *Ordering the Heavens*, fig. 3.13. Becker's *F*.

18. BERN, Burgerbibliothek 249, fols. 47v–64v, s. ix^{in}, Fleury. *Explicit liber de mundo* (fol. 64v). Short recension. Diagrams 1–5, 7 (Diagram 6 is lost). The *Epistle* of Sisebut is omitted.

Hagen, *Catalogus*, 285; Becker, *DNR*, xxv; Beeson, *Isidor-Studien*, 29, 68; Fontaine, *Traité*, 37; Fontaine, 'La Diffusion de l'œuvre d'Isidore', 311; Díaz y Díaz, *Index Scriptorum*, 32; Homburger, *Die Illustrierten Handschriften*, 47–50; Mostert, *The Library of Fleury*, 67–68 (BF133); cf. Bischoff, *Katalog*, no. 3719 (+ Orléans, Bibliothèque municipale 185 [162]). Becker's **G**.

19. BERN, Burgerbibliothek 417, fols. 61v–98v, AD 826(?), vicinity of Tours, Fleury(?). *Incipit liber de astra celi*. Excerpts from the long recension: chs. 1–27, 29, 31–32, 34, 38, 43, 45(44)–48(47); chs. 20 and 21 are combined into one. Includes the mystical addition in ch. 1 (fol. 63r). Diagrams 1–6. T-O map, with legends as in Appendix 5b.

The codex includes excerpts from Bede's *DTR*.

Hagen, *Catalogus*, 372; Becker, *DNR*, xxv; Beeson, *Isidor-Studien*, 69; Jones, *Bedae Pseudepigrapha*, 113; Fontaine, *Traité*, 37; Fontaine, 'La Diffusion de l'œuvre d'Isidore', 313; Homburger, *Die Illustrierten Handschriften*, 69–71; Mostert, *The Library of Fleury*, 76 (BF185); Bischoff, *Katalog*, no. 592. Becker's **H**.

20. BERN, Burgerbibliothek 610, fols. 11v–47r, s. ix^2, vicinity of Tours, Fleury(?); prov. Fleury. *Incipit liber primus Bedae* [sic] *de compoto*. Long recension(?). *Capitula: I De diebus II De nocte ... XL* [sic] *De pestilentia XLI* [sic] *De oceano*. In the text, after the chapter *De oceano*, erroneously numbered ch. 41 in the *capitula*, come (unnumbered) *QVVR MARE CONCRESCAT* [= ch. 41], QVVR MARE AMARAS HABEAT AQVAS [= ch. 42], DE NILO [= ch. 43], DE POSSITIONE TERRE (fol. 44v) [= ch. 44(–)(?)], DE POSITIONE TERRE (fol. 45r) [= ch. 45(44)], DE TERRAE MOTV [= ch. 46(45)], DE MONTE ETHNA XLVII [= ch. 47(46)], and XLVIII DE PARTIBVS TERRAE [= ch. 48(47)]. No mystical addition in ch. 1. Diagrams 1–7. No T-O map. The *Epistle* of Sisebut is omitted.

The codex includes Bede's *DTR*, *DNR*, and *DT*.

Hagen, *Catalogus*, 478; Beeson, *Isidor-Studien*, 69; Jones, *Bedae Pseudepigrapha*, 113; Fontaine, *Traité*, 37; Fontaine, 'La Diffusion de l'œuvre d'Isidore', 313–14; Homburger, *Die Illustrierten Handschriften*, 128–30; Mostert, *The Library of Fleury*, 81 (BF221); Bischoff, *Katalog*, no. 609.

21. BERN, Burgerbibliothek 676, fols. 137r–170v, s. xv.
De causis naturalium (= *De natura rerum*).
Hagen, *Catalogus codicum*, 499.

22. BESANÇON, Bibliothèque municipale 184, fols. 1v–55v, s. ix^{in}, eastern France; prov. Murbach. *Incipit liber de astra celi sancti hisidori spalensis episcopi.* Long recension. Includes mystical addition in ch. 1. Inserts *Hiemisphaeria* in ch. 12.3. Inserts *Aparctias* in ch. 37.1. Adds formula of introduction for Diagram 7 in ch. 37 (see Appendix 2b). Inserts 'Solinus' in ch. 40.1. Diagrams 1–7. T-O map.
Beeson, *Isidor-Studien*, 65; Fontaine, *Traité*, 33–34; Bischoff, *Katalog*, no. 623; Gorman, 'Diagrams', 540; Martín, 'Isidorus Hisp.', 353. Fontaine's *B*.

23. BORDEAUX, Bibliothèque municipale 609, fols. 1r–23r, s. xii$^{1/2}$, Italy. *Incipit sancti Ysidori ... de astronomia liber.* Short recension. No diagrams. The *Epistle* of Sisebut follows *DNR* on fol. 23r–v.
Delpit, *Catalogue des manuscrits*; *Catalogue général* 23:292; Fontaine, *Traité*, 38; Alberto, 'Sisebut's *Carmen*', 186.

24. BOULOGNE, Bibliothèque municipale 188 (360), fol. 30r, s. x, St-Bertin. Excerpt: abbreviated text of ch. 37.1–4 within the 12 segments of a diagram of the winds (= Diagram 7) with a T-O map in its centre; the text has been rearranged to fit into the appropriate points of the compass. The text is 'virtually identical' to that of Berlin 1830.
Eastwood, *Ordering the Heavens*, 411–17 and fig. 6.19.

25. BRUSSELS, Bibliothèque Royale Albert Ier 5447–5458, s. xii, Belgium. Medium recension(?). Lacks ch. 44(–).
Becker, *DNR*, xxvi; Díaz y Díaz, *Index Scriptorum*, 32; Fontaine, *Traité*, 38.

26. BRUSSELS, Bibliothèque Royale Albert Ier 9311–9319 (1322), fols. 67r–89v, s. ix$^{1/3}$, vicinity of St-Amand. *Incipit liber esidori de natura rerum.* Long recension. In the *capitula*, eight paragraphs of ch. 26 are separately numbered chs. 27–34, and the remaining chapter numbers are altered to 35–56. The chapters in the text are numbered normally 1–48. Includes mystical addition in ch. 1. Inserts *Hiemisphaeria* in ch. 12.3.

Inserts *Aparctias* in ch. 37.1. Inserts 'Solinus' in ch. 40.1. Diagrams 1–7 (7 oriented N) + 5A. T-O map.

Becker, *DNR*, xxv–xxvi; Beeson, *Isidor-Studien*, 26, 71; Díaz y Díaz, *Index Scriptorum*, 32; Fontaine, *Traité*, 37; Bischoff, *Katalog*, no. 728; Eastwood, 'The Diagram of the Four Elements', 555. Online: *Belgica*. Becker's *L*.

27. CAMBRAI, Bibliothèque municipale 937 (836), fol. 1v (?), s. ix/x, northern France; prov. Fleury(?), cathedral of Cambrai. Excerpt: ch. 48 of *DNR*; presumably this was to replace what was lost in later part of codex (see next). Vegetius interpolation at end of ch. 48 (see Appendix 5a).

Fontaine, *Traité*, 26 ('un lecteur de *nat*. A d'ailleurs rajouté en minuscule caroline [au IXe ou Xe siècle] le texte de ce chapitre manquant sur le V° du f° 1 du ms'); the chapter does not appear in the image of fol. 1v in online: *BVMM*. Fontaine's *E²*.

28. CAMBRAI, Bibliothèque municipale 937 (836), fols. 38v–62v, s. viii, northern France; prov. Fleury(?), cathedral of Cambrai. Lacks title. Fragment: medium recension. Preface to ch. 47(46).1 (*... stratu sulphore*). Chs. 45(44)–47(46) are numbered 44–46. Ch. 48(47) listed in the *capitula* as ch. 47; text of ch. 48(47) was added in the ninth or tenth century on fol. 1v (see previous). Omits *Hiemisphaeria* in ch. 12.3. Omits *Aparctias* in ch. 37.1. Adds formula of introduction for Diagram 7 after ch. 37.5 (see Appendix 2a). Omits 'Solinus' in ch. 40.1. Diagrams 1–7 + 5A and 6A (colours of rainbow). KOCMOC inscription in Diagram 7, which is displaced to end of chapter. T-O map, fol. 48v, opposite ch. 17 (Gorman, 'Diagrams', p. 530, n. 5). *Epistle* of Sisebut probably never included (Fontaine, *Traité*, 26, argues that it was).

Beeson, *Isidor-Studien*, 27, 65; Lowe, *CLA* 6 (no. 744); Fontaine, *Traité*, 26–27; Mostert, *The Library of Fleury*, 84 (BF243); Bischoff, *Katalog*, no. 809a; Gorman, 'Diagrams', 540; Eastwood, 'The Diagram of the Four Elements', 555–56 and fig. 7; Martín, 'Isidorus Hisp.', 353. Fontaine's *E*.

29. CAMBRIDGE, Corpus Christi College 291, fol. 132r–132v, s. xi/xii, Canterbury, St Augustine's. Excerpt: ch. 23 only. Omits Diagram 6.

The codex includes Bede's *DTR*.

Gameson, *Manuscripts*, no. 74; Gneuss and Lapidge, *Anglo-Saxon Manuscripts*, no. 85. Online: *Parker Library on the Web*.

30. CAMBRIDGE, Magdalene College F.4.26, fols. 1r–17v, s. xiii (James); s. xii (Díaz y Díaz), prov. Coventry Cathedral. Long recension, gap between chs. 7 and 10. Diagrams, plus diagram of the world on fol. 18.

James, *Catalogue of Manuscripts of Magdalene College*, 53–54; Díaz y Díaz, *Index Scriptorum*, 32.

31. CAVA DE' TIRRENI, Archivio e Biblioteca della Badia della SS. Trinità 3, fol. 150v, s. xi, southern Italy. Excerpt, with some material from *DNR*. Incipit, *dies est solis presentia quo usque ad occasum perueniat* (ch. 1.1). Explicit, *suprema dicitur ad horam nonam usque ad occasum solis* (variant version of the pseudo-Bedan *De diuisionibus temporum* [cf. *PL* 90, 656]).

The codex includes excerpts from Bede's *DTR* and *DNR*.

HMML.

32. COLOGNE, Erzbischöfliche Diözesan- und Dombibliothek 83(II), fols. 126r–143r, Cologne, written for Hildebald, bishop of Cologne, AD 798–805. *Incipit epistola sisebuto* [sic] *regis gothorum missa ad isidorum de libro rotarum*. Lacks separate title for *DNR*. Short recension. Omits *Hiemisphaeria* in ch. 12.3. Omits *Aparctias* in ch. 37.1. Adds formula of introduction for Diagram 7 after ch. 37.5 (see Appendix 2a). Omits 'Solinus' in ch. 40.1. Diagrams 1–7 + 5A. KOCMOC inscription in Diagram 7, which is displaced to end of chapter. No T-O map. The *Epistle* of Sisebut precedes *DNR* (fol. 126r–v).

Codex contains Bede's *DTR*.

Beeson, *Isidor-Studien*, 70; Jones, *Bedae Pseudepigrapha*, 115; Lowe, *CLA* 8 (no. 1154); Fontaine, *Traité*, 27–28, 160–61; Bischoff, *Katalog*, no. 1907; Gorman, 'Diagrams', 540; Martín, 'Sisebut', 403; Martín, 'Isidorus Hisp.', 353. Online: *CEEC*. Fontaine's *K*.

33. COLOGNE, Erzbischöfliche Diözesan- und Dombibliothek 99, fols. 53r–82r, s. ix$^{2/4}$, western Germany(?). *Incipit liber testimonium isidori; explicit liber primus*. Short recension. The *capitula* list 46 numbered chapters. Spaces below chapter headings numbered *xliii* and *xlui* filled with unnumbered titles of, respectively, ch. 44(–) (*de nominibus maris et fluminum*) and ch. 48 (*de partibus terrae*). Inserts *Hiemisphaeria* in ch. 12.3. Omits *Aparctias* in ch. 37.1. Omits 'Solinus' in ch. 40.1. Diagrams 1–7 + 5A. T-O map added at foot of fol. 82r after ch. 47(46). The *Epistle* of Sisebut is omitted.

Fontaine, *Traité*, 37; Bischoff, *Katalog*, no. 1913. Online: *CEEC*.

[COPENHAGEN, Universitetsbibliothek, Frag. 19.VII, see BASEL, Universitätsbibliothek Lat. F.III.15a]

34. DIJON, Bibliothèque municipale 446, fols. 2r–19r, s. xii²ᐟ³, Burgundy; prov. Cîteaux. *Incipit salutatio Ysidori Spalensis ad Sisebutum.* Short recension(?). Diagrams 1–3, 5–6, space for 4. The *Epistle* of Sisebut follows on fol. 19r–v.

Catalogue général 5:105–06; Fontaine, *Traité*, 38; Zaluska, *Manuscrits enluminés de Dijon*, 142.

35. DIJON, Bibliothèque municipale 448, fols. 37v–47v, s. xi¹ᐟ², perhaps St-Pierre and St-Paul de Bèze (Côte-d'Or); prov. St-Bénigne. Long recension. A diagram of the winds appears on fol. 75r, with *De Ventis* (see Appendix 4) inscribed within it. The opening three lines of the poem are placed in the central disk of the diagram.

The codex includes Bede's *DTR*, *DNR*, and *DT*.

Catalogue général 5:106–09; Fontaine, *Traité*, 38; Zaluska, *Manuscrits enluminés de Dijon*, 30–33; Obrist, 'Wind Diagrams', 52 and fig. 11.

36. EDINBURGH, University Library 123, fols. 144v–154v, s. xii, Metz(?). *Incipit prefatio Sancti Ysidori hispalensis episcopi de responsione mundi et astrorum ordinatione ad sesibutum philosophum.* Excerpt: preface to ch. 21. Ends: ... *in toto orbe diffundit. Amen. Amen. Explicit.*

Borland, *A Descriptive Catalogue*, 195–97; Díaz y Díaz, *Index Scriptorum*, 32.

37. EINSIEDELN, Stiftsbibliothek 167, s. ix², St Gall(?); prov. Einsiedeln. Long recension, divided into 58 chapters (Zofingen type, see Appendix 6a). Vegetius interpolation (see Appendix 5a) [+ *De Trinitate* (*Etym.* 7.4) unconfirmed] at end of ch. 48(47). Direct copy of Zofingen Pa 32.

DNR is inserted between books 3 and 4 of the *Etymologiae*, exactly as in Zofingen Pa 32. The codex includes Bede's *DT*.

Beeson, *Isidor-Studien*, 13, 69; Fontaine, *Traité*, 37; Gorman, 'Diagrams', *stemma codicum*, 541; von Büren, 'Isidore, Végèce et Titanus', 40.

38. EINSIEDELN, Stiftsbibliothek 360 (77), fols. 34v–45v, s. xii, prov. Engelberg. *Incipit liber sancti ysidori de rerum natura.* Long recension, divided into 58 chapters (Zofingen type, see Appendix 6a). The *capitula*

(which precede title and preface) and chapters are numbered and ordered exactly as in Zofingen P 32 (*q.v.*). Some folios are trimmed with loss of text; text illegible in places. Chs. 16.3–19.1 (between fols. 44v and 45r) and chs. 34–38.2 (between fols. 42v and 43r) are lost. Includes mystical addition in ch. 1. Inserts *Hiemisphaeria* in ch. 12.3. Inserts 'Solinus' in ch. 40.1. Vegetius interpolation (see Appendix 5a) at end of ch. 48(47). Diagrams 1–5; 7A (mountains of Sicily). Omits diagram and text for 6. No T-O map.

DNR is inserted between books 3 and 4 of the *Etymologiae*, exactly as in Zofingen Pa 32 (but with omission of *De Trinitate*), with comprehensive title *De libro rotarum*.

Online: *e-codices*.

39. ERLANGEN, Universitätsbibliothek 186, fols. 233v–243r, s. xii^2, Germany. *Incipit liber ysidori de mappa mundi*. Short recension. An additional chapter, *De natura solis*, begins on fol. 243v; the poem *De Ventis* (*Primus a parthias*, line 4) (see Appendix 4) appears on fol. 246r.

Fischer, *Handschriften* 1:207–09; Fontaine, *Traité*, 38.

40. ESCORIAL, Real Biblioteca de San Lorenzo E.IV.13, fols. 38v–49r, s. xii, Spain. *Incipit liber de astra caeli sancti hisidori*. Excerpts, from the medium recension? No *capitula*; chapters are unnumbered: chs. 20–21, 29–30, 33–38, 40–43, 45(44)–48(47). Adds formula of introduction for Diagram 7 after ch. 36.3 (see Appendix 2d). Omits *Aparctias* in ch. 37.1. Omits text of ch. 37.5. Inserts 'Solinus' in ch. 40.1. Diagram 7 (before ch. 37). T-O map (embedded in ch. 48; *hispania* inscribed beneath *Europa*). The *Epistle* of Sisebut is omitted.

Antolín, *Catálogo* 2:100–01; Díaz y Díaz, *Index Scriptorum*, 54–55; Fontaine, *Isidore et la culture*, 453 n. 1; Fontaine, *Traité*, 38; cbk.

41. ESCORIAL, Real Biblioteca de San Lorenzo E.IV.14, fols. 100r–113v, s. xiii, Spain. *Incipit libellus domini ysidori de naturis rerum*. Short recension. Omits *capitula*. Omits *Hiemisphaeria* in ch. 12.3. Ch. 36 breaks off after *habeantur in locis* (36.2); ch. 37.1–4 omitted (but space is left for all the missing text). Ch. 37.5 follows Diagram 7. Omits 'Solinus' in ch. 40.1. Diagrams 1–7, 5A, 6A. The *Epistle* of Sisebut is omitted.

Antolín, *Catálogo* 2:102–03; cbk.

42. ESCORIAL, Real Biblioteca de San Lorenzo K.I.12, fols. 1r–13v, s. xiv^{ex}, Spain. *Incipit prefatio sancti ysidori yspan[i]ensis episcopi de*

astronomia. Short recension. Omits *Hiemisphaeria* in ch. 12.3. Omits *Aparctias* in ch. 37.1. Omits 'Solinus' in ch. 40.1. Diagrams 1–7, 5A. The *Epistle* of Sisebut follows on fol. 13v. *Explicit liber primus. Incipiunt capitula secundi libri.*

Antolín, *Catálogo* 2:507–08; Alberto, 'Sisebut's *Carmen*', 186; cbk.

43. ESCORIAL, Real Biblioteca de San Lorenzo M.II.23, fols. 1r–15r, s. xiv, Spain. *Incipit prephatio sancti ysidori yspaniensis episcopi de astronomia.* Short recension. Omits *Hiemisphaeria* in ch. 12.3. Omits *Aparctias* in ch. 37.1. Diagram 7 follows ch. 37.5 (no introduction). Omits 'Solinus' in ch. 40.1. Diagrams 1–7, 5A. The *Epistle* of Sisebut is omitted.

Antolín, *Catálogo* 3:76; cbk.

44. ESCORIAL, Real Biblioteca de San Lorenzo R.II.18, fols. 1r–8v, s. ix[1], region of Cordoba; prov. Oviedo. Part of a composite codex (see two next). Visigothic minuscule and cursive. *Incipit liber de natura rerum domini isidori hispalensis episcopi directus ad sisebutum regem.* Fragment: short recension(?). Title to ch. 16.3 (*... superior sit a luna*). No *capitula.* Omits mystical addition in ch. 1.3. Omits *Hiemisphaeria* in ch. 12.3. Diagrams 1–5.

Antolín, *Catálogo* 3:481; Fontaine, *Traité*, 22–23; Gorman, 'Diagrams', 540; Martín, 'Isidorus Hisp.', 354; cbk. Fontaine's *O* [where appropriate we refer to this as *O₁*; see 46].

45. ESCORIAL, Real Biblioteca de San Lorenzo R.II.18, fols. 9r–24v, s. vii[ex], doubtless Cordoba; prov. Oviedo. The oldest surviving MS of *DNR*. Fragment: medium recension. Chs. 16–47(46) in Visigothic uncial are on fols. 9r–23r; no ch. 44(–). The *Epistle* of Sisebut appears in majuscule on fols. 23v–24r. Ch. 48 (numbered 47) of *DNR* (in a slightly later hand) is on fol. 24v. Inserts *Aparctias* in ch. 37.1. Omits 'Solinus' in ch. 40.1. Inserts *Maeotis* in ch. 48(47).2. Diagrams 5A, 6–7. Two T-O maps, side-by-side.

Antolín, *Catálogo* 3:481; Lowe, *CLA* 11 (no. 1631); Fontaine, *Traité*, 20–22; Stevens, 'Sidereal Time', 149, n. 57; Gorman, 'Diagrams', 540; Martín, 'Sisebutus', 403; Martín, 'Isidorus Hisp.', 354; cbk. Fontaine's *H*.

46. ESCORIAL, Real Biblioteca de San Lorenzo R.II.18, fols. 33v–34r, s. ix[1](?), prov. Oviedo (fols. 25–34 have a different origin from either of the above; see Martín, 'Isidorus Hisp.', p. 354 n. 185, and the references therein). Excerpt: ch. 37. Omits *Aparctias* in ch. 37.1. No Diagram 7.

Excerpts from the *Etymologies* (14.3–5; 9.6–7) precede ch. 37; the poem *De Ventis* (*Versi de supra nominatis uentis*) (see Appendix 4) follows on fol. 34v.

Anth. Lat. 1.2:6; Baehrens, *Poetae Latini minores* 5:383; Antolín, *Catálogo* 3:481–82; Fontaine, *Traité*, 22–23; Martín, 'Isidorus Hisp.', 354; cbk. Fontaine's *O* [where appropriate we refer to this as *O₂*; see previous].

47. EXETER, Cathedral Library 3507, fols. 67r–97v, *c*.960–986, Canterbury, or possibly Exeter; prov. Exeter. Long recension (English type, see Appendix 6b). Includes mystical addition in ch. 1. Inserts *Hiemisphaeria* in ch. 12.3. Inserts *Aparctias* in ch. 37.1. Inserts 'Solinus' in ch. 40.1. Diagrams 1–7 (two versions of 3; 4 rendered in circular form) + 5A and 6A (colours of the rainbow). T-O map (reversed). The *Epistle* of Sisebut is omitted (but cf. Appendix 6b).

Ker, *Medieval Manuscripts*, 2:814; Stevens, 'Scientific Instruction', 99–100 and n. 57; Stevens, 'Sidereal Time', 136; Gneuss and Lapidge, *Anglo-Saxon Manuscripts*, no. 258; *ASMMF* 22:57–66. Microfiche: *ASMMF* 22.

48. FLORENCE, Biblioteca Medicea Laurenziana, S. Marco 582, fols. 145v–154r, s. xii. *Incipiunt capitula libri xxiii.* Long recension. *Capitula* number 50 chapters, giving two numbers to ch. 25 and subdividing ch. 26 into two; text chapters are unnumbered. Includes mystical addition in ch. 1. Inserts *Hiemisphaeria* in ch. 12.3. Inserts *Aparctias* in ch. 37.1. Inserts 'Solinus' in ch. 40.1. Vegetius interpolation at end of ch. 48 (see Appendix 5a). Diagrams 1–3, 5–7 + 5A and 7A (mountains of Sicily). T-O map.

Fontaine, *Traité*, 37. Online: *Biblioteca Firenze*.

49. FLORENCE, Biblioteca Medicea Laurenziana, S. Marco 604, fols. 106v–117r, s. ix. Excerpts: chs. 1–4; 6; 12; 15; 17–19; 38; 41; 29–31. Includes mystical addition in ch. 1. Inserts *Hiemisphaeria* in ch. 12.3. Omits Diagram 1 from ch. 4. Includes Diagram 5A (one moon); leaves space for 6A (colours of the rainbow) in ch. 31. The *Epistle* of Sisebut precedes *DNR* on fol. 106r–v.

Arévalo, *Isidoriana* 1:664–65; Fontaine, *Traité*, 160–61. Online: *Biblioteca Firenze*. Fontaine's *F* (*Epistle* only)

50. FLORENCE, Biblioteca Medicea Laurenziana, Plut. 16.39, fols. 98r–99v, s. ix^in. Excerpt: ch. 37.1–4 (*Ventorum primus est … in india*

serena (fol. 99r). Inserts *Aparctias* in ch. 37.1. The poem *De Ventis* (see Appendix 4) with the rubric *Versus xii uentorum* begins fol. 99r (lines 1–6); Diagram 7 (the wheel of the winds) is placed at the top of fol. 99v, followed by lines 7–8 of *De Ventis*. The remainder of the manuscript is lost.

The excerpt is preceded on fols. 96v–98r by *Etym.* 3.29–39 (Gasparotti and Guillaumin, *Etym.* 3.28–38) with chapters numbered in margin from *uiiii* to *xuiiii*. Ch. 37 is numbered *xx* and *De Ventis* is numbered *xxi*.

Anth. Lat. 1.2:6; Baehrens, *Poetae Latini minores* 5:383. Online: Internet culturale; *Biblioteca Firenze*.

51. FLORENCE, Biblioteca Medicea Laurenziana, Plut. 22 dex.12, fols. 35v–44r, s. xiii. *Incipit liber beati ysidori yspalensis episcopi de natura rerum ad sisebutum regem.* Medium recension (both *capitula* and chapters are numbered). Omits *Hiemisphaeria* in ch. 12.3. Adds formula of introduction for Diagram 7 at head of ch. 37 and after Diagram 7 another passage (see Appendix 2c). Omits *Aparctias* in ch. 37.1. Diagram 7 follows ch. 37.5. Omits 'Solinus' in ch. 40.1. Inserts *Maeotis* in ch. 48(47).2 (in margin). Diagrams 1–7 + 5A (phases of the moon). A diagram possibly illustrating an eclipse is inserted halfway into the text of ch. 48(47) (same diagram in Paris 15171). T-O map. The *Epistle* of Sisebut, lines 39–61, without attribution, is inserted between ch. 46 (*de monte ethna*) and ch. 47 (*de partibus terrae*).

Online: *Biblioteca Firenze*.

52. FLORENCE, Biblioteca Medicea Laurenziana, Plut. 27 sin. 9, fols. 173v–184r, s. xi. *Incipit libellus ysidori iunioris illustris de astronomia siue de natura rerum.* Abbreviated short recension. Passages containing allegorical interpretations are systematically omitted from chapters. Inserts *Hiemisphaeria* in ch. 12.3. Inserts *Aparctias* in ch. 37.1. Omits 'Solinus' in ch. 40.1. Diagrams 1 (circle only), 3 (brief inscription), 5A, 6–7; space for 4. Diagram 7 (which follows ch. 37.5) is a different, composite type, emphasizing 4 cardinal winds with their qualities. No T-O map. The *Epistle* of Sisebut is omitted.

Final chapter (*de ethna*) followed immediately by *Etym.* 3.71 (Gasparotti and Guillaumin, *Etym.* 3.70), *De nominibus stellarum, quibus ex causis nomina acceperunt,* in its entirety.

Arévalo, *Isidoriana* 1:446, 661–62; Fontaine, *Traité*, 37. Online: *Biblioteca Firenze*.

53. FLORENCE, Biblioteca Medicea Laurenziana, Plut. 29.39, fols. 1r–19v, s. xiii. *Incipit liber de astris celi.* Long recension. Includes mystical addition in ch. 1. Inserts *Hiemisphaeria* in ch. 12.3. Inserts *Aparctias* in ch. 37.1. Inserts 'Solinus' in ch. 40.1. Diagrams 1–7 + 5A. T-O map, with legends as in Appendix 5b.

A second version of Diagram 7 (fol. 16r) exhibits a severely abbreviated text of ch. 37.1–4 within the 12 segments of the diagram; the text has been rearranged to fit into the appropriate points of the compass, with *Aparctias* inserted (*qui et apartias*). Inscribed in the centre of the second diagram is the formula: *hec sunt signa xii uentorum iuxta ordinem suum distincte proprie et appellatiue.*

Arévalo, *Isidoriana* 1:377. Online: Internet culturale; *Biblioteca Firenze.*

54. FLORENCE, Biblioteca Riccardiana 379/4, fols. 115r–148v, s. xii$^{1/4}$. *Prologus in libro rotarum Ysidori.* Long recension. Vegetius interpolation at end of ch. 48 (see Appendix 5a). Diagrams.

Manus.

55. GRONINGEN, Bibliotheek der Rijksuniversiteit Cd. 3, s. xi.

Fontaine, *Traité*, 37.

56. KARLSRUHE, Badische Landesbibliothek, Karlsruhe 339/1 + PARIS, Bibliothèque nationale lat. 6413, s. viii^{med}, northeast France, probably Chelles. Detached fragments of same MS. Short recension.

Karlsruhe fragment: preface 3 (*et sensus et uerba* ...) to ch. 1.3 (... *caecitatem secundum*) + ch. 6.2 (*ut populi Romani* ...) to ch. 6.6 (... *principium reuoluuntur*), with *capitula* of short recension.

Paris fragment, fols. 1r–36r: ch. 7.3 (*et paulisper relaxato* ...) to ch. 47(46). Omits *Hiemisphaeria* in ch. 12.3. Omits *Aparctias* in ch. 37.1. Omits 'Solinus' in ch. 40.1. Diagrams 2–7 + 5A. The *Epistle* of Sisebut appears on fols. 34r–36r.

Beeson, *Isidor-Studien*, 44, 66; Lowe, *CLA* 5 (no. 567); Fontaine, *Traité*, 23–24; Gorman, 'Diagrams', 540; Martín, 'Sisebut', 403; Martín, 'Isidorus Hisp.', 353. Online: *Gallica.* Fontaine's *D.*

57. KARLSRUHE, Badische Landesbibliothek, Augiensis 106, s. ix, fol. 52v, scribe from Corbie (Bischoff). Excerpt, abbreviated text of ch. 37.1–4 within the 12 segments of a diagram of the winds (= Diagram 7)

with a T-O map in its centre; the text has been rearranged to fit into the appropriate points of the compass. Inserts *Aparctias* in ch. 37.1 (*qui et aparchias*). The poem *De Ventis* (see Appendix 4) precedes the diagram on fol. 52r. Identical to Bern 212/I.

Mynors, *Cassiodori Institutiones*, xxx–xxxiii; Bischoff, *Katalog*, no. 1635. Online: *Carolingian Libraries*.

58. KARLSRUHE, Badische Landesbibliothek, Augiensis 111, fols. 73r–75r, s. ix$^{1/4}$, south-western Germany. Excerpts: ch. 1.1–3 (*INCIPIT DE DIEBVS QVOD IN LIBRO ROTARVM PRENOTATVR ... fidei scientiaque* [sic] *peruenit*) (hence no mystical addition); ch. 3.1–3 (*DE EBDOmada ... explere cursum suum*); ch. 4.1–7 (*DE MENSIBVS ... quinque dies pronuntiantur*); ch. 11.1 (*DE IIIIOR PARTIBVS MVNDI ... a se separantur*); ch. 10.1–2 (*DE V CIRCVLIS MVNDI ... talis distinguitur* [sic] *figura*); ch. 6.1–4 (*DE ANNIS ... reuertebatur antiqua possessio*); ch. 25; ch. 29; ch. 46(45); ch. 34. Diagram 3 (fol. 75r).

Beeson, *Isidor-Studien*, 30, 119; Fontaine, *Traité*, 37; Bischoff, *Katalog*, no. 1639. Online: *Carolingian Libraries*; *BLB*.

59. KARLSRUHE, Badische Landesbibliothek, Augiensis 229, fols. 139v–183v, AD 821(?), Abruzzo, vicinity of Chieti; prov. Reichenau. Long recension. *Capitula* of short recension. Chs. 33 and 34 appear, out of order and unnumbered, after ch. 37. No mystical addition in ch. 1. Omits *Hiemisphaeria* in ch. 12.3. Omits *Aparctias* in ch. 37.1. Inserts 'Solinus' in ch. 40.1. Diagrams 1–4, 5A, 7. T-O map.

The codex includes excerpts from Bede's *DTR*.

Beeson, *Isidor-Studien*, 30, 70; Fontaine, *Traité*, 37; Bischoff, *Katalog*, no. 1719. Online: *Carolingian Libraries*; *BLB*.

60. LAON, Bibliothèque Suzanne Martinet 422, fols. 1r–22v, s. ix$^{1/3}$, north-eastern France; prov. Notre-Dame de Laon. Short recension. *DNR* chs. 36–37 inserted after *DNR* 6 both in the *capitula* and in the text (hence *DNR* chs. 7–35 are numbered as 9–37). Omits *Hiemisphaeria* in ch. 12.3. Omits *Aparctias* in ch. 37.1. Laon 422 adds a variant of the introductory formula for Diagram 7 (see Appendix 2a), and inserts Diagram 7 in the first sentence of ch. 37.4 (here ch. 8). Omits 'Solinus' in ch. 40.1. Diagrams 1–7 + 1A (seasons) and 5A. KOCMOC inscription in Diagram 7. No T-O map. No *Epistle* of Sisebut.

Catalogue général des départements 1:224–25; Beeson, *Isidor-Studien*,

65; Fontaine, *Traité*, 37; Contreni, *The Cathedral School of Laon*, 179; Bischoff, *Katalog*, no. 2114; Eastwood, 'The Diagram of the Four Elements', 555 and fig. 4; Obrist, 'Le Diagramme isidorien des saisons', figs. 7–8; Obrist, 'Wind Diagrams', figs. 25 and 27. Online: *Gallica*.

61. LAON, Bibliothèque Suzanne Martinet 423, fols. 1r–33v, s. viii[med], written at Laon by Dulcia. *Incipit liber rotarum sancti isidori spalensis episcopi*. Short recension. Inserts *Hiemisphaeria* in ch. 12.3. Omits *Aparctias* in ch. 37.1. Adds formula of introduction for Diagram 7 following the diagram (see Appendix 2a). Omits 'Solinus' in ch. 40.1. Diagrams 1–7 + 5A. KOCMOC inscription in Diagram 7. No T-O map. No *Epistle* of Sisebut.

   Catalogue général des départements 1:225; Beeson, *Isidor-Studien*, 27, 65; Lowe, *CLA* 6 (no. 766); Fontaine, *Traité*, 30–31; Contreni, *The Cathedral School of Laon*, 179; Gorman, 'Diagrams', 540. Online: Ville de Laon. Fontaine's *L*.

62. LISBON, Biblioteca Nacional, Alc. CCXII/375, fols. 164r–167r, s. xiii, prov. Santa Maria (Alcobaça). *Incipit liber de naturis rerum ad sisebutum gotorum regem editus a sancto dei uiro isidoro spalensis sedis episcopo.* Fragment of medium recension; incomplete copy of Lisbon 446: chs. 1–7.7 (*... luminum ac tenebrarum alterius uicibus separantur*).

   Black and Amos, *The Fundo Alcobaça* 3:125–27. Online: BNP.

63. LISBON, Biblioteca Nacional, Alc. CCIX/446, fols. 206r–220v, s. xiii[in], prov. Santa Maria (Alcobaça). *Incipit liber de naturis rerum ad sisebutum gotorum regem. Editus a sancto dei uiro ysidoro spalensis sedis episcopo.* Medium recension (both *capitula* and chapters in text are numbered). Omits *Hiemisphaeria* in ch. 12.3. Adds formula of introduction for Diagram 7 at head of ch. 37 and after Diagram 7 another passage (see Appendix 2c). Omits *Aparctias* in ch. 37.1. Diagram 7 follows ch. 37.5. Omits 'Solinus' in ch. 40.1. Inserts *Maeotis* in ch. 48(47).2. Diagrams 1–7 + 5A. No T-O map. The *Epistle* of Sisebut is omitted.

   Ch. 47 (*De partibus terrae*) ends at bottom of fol. 220va. fol. 220vb begins with the rubric *De uerberatione uentorum* (numbered as ch. 48) and the poem *De Ventis* (lines 1–13) (see Appendix 4). A short passage of unidentified material extends to the first two lines of fol. 221r (in single-column format). Below this, another Diagram 1 and *DNR* ch. 5. The lower portion of fol. 221r is blank, as is fol. 221v.

   Black and Amos, *The Fundo Alcobaça* 3:236–238. Online: BNP.

64. LONDON, British Library, Cotton Domitian I, fols. 3r–37r, AD 970, probably Canterbury, St Augustine's. *Incipit liber isidori spalensis aepiscopi de natura rerum.* Long recension (English type, see Appendix 6b). Includes mystical addition in ch. 1. Inserts *Hiemisphaeria* in ch. 12.3. Inserts *Aparctias* in ch. 37.1. Inserts 'Solinus' in ch. 40.1. Diagrams 1–7 (two versions of 3; 4 rendered in circular form) + 5A and 6A (colours of the rainbow). T-O map. The *Epistle* of Sisebut is omitted (but cf. Appendix 6b).

The codex includes ch. 2 of Bede's *DNR*.

Stevens, 'Scientific Instruction', 99–100 and n. 57; Stevens, 'Sidereal Time', 136; *ASMMF* 5:35–41; Gneuss and Lapidge, *Anglo-Saxon Manuscripts*, no. 326; Stevens, private communication. Microfiche: *ASMMF* 5.

65. LONDON, British Library, Cotton Tiberius C I (+ British Library, Harley 3667 = 'Peterborough Computus'), fols. 6v–7r (ch. 11), 10r–v (ch. 4), 11r (ch. 37), 11v (ch. 10), 12v (ch. 23), 1122–1135, Peterborough. Excerpts (exclusively chapters with diagrams). Diagrams 1, 3–4, 6–7.

Similar to Oxford, St John's College 17. The codex includes an excerpt from Bede's *DT*.

Gameson, *Manuscripts*, no. 404; Wallis, 'Calendar & Cloister'; *Production and Use* (database).

66. LONDON, British Library, Cotton Vitellius A XII, fols. 48r–64r, s. xi^{ex}, Salisbury. Long recension (English type, see Appendix 6b). Includes mystical addition in ch. 1. Inserts 'Solinus' in ch. 40.1. 5 diagrams. T-O map (reversed).

Fontaine, *Traité*, 37; Stevens, 'Scientific Instruction', 99–100 and n. 57; Stevens, 'Sidereal Time', 136; Gameson, *Manuscripts*, no. 419; Gneuss and Lapidge, *Anglo-Saxon Manuscripts*, no. 398; Stevens, private communication.

67. LONDON, British Library, Harley 2660, fols. 33r–45r, AD 1136, West Rhineland. *Incipit liber v ysidori de rerum natura ad sisopotum regum.* Long recension, redivided into 58 numbered chapters (Zofingen type, see Appendix 6a), with inadvertent (?) omission of text of ch. 44(–). Includes mystical addition in ch. 1. Inserts *Hiemisphaeria* in ch. 12.3. Inserts *Aparctias* in ch. 37.1. Diagram 7 follows ch. 37.5. Inserts 'Solinus' in ch. 40.1. Vegetius interpolation at end of ch. 48 (see Appendix 5a). Diagrams 1–5; 7. Omits diagram and text for 5A and 6. No T-O map.

The only substantive differences from Zofingen Pa 32 are: (1) Harley 2660 omits ch. 51 (= ch. 44(–), *de nominibus maris uel fluminum*), from the text (ch. 51 is listed in the *capitula*, and chs. 50 and 52 are numbered in the text in accordance with Zofingen Pa 32); (2) space was left for the diagram of the winds, Diagram 7, but it was not filled in until the fifteenth century from a different source; (3) Diagram 7A (mountains of Sicily) is omitted; (4) *De Trinitate* is omitted from ch. 48(47).

DNR is inserted between books 3 and 4 of the *Etymologiae*, as in Zofingen Pa 32, but without comprehensive title *De libro rotarum*.

Online: BL-Digitised Manuscripts.

68. LONDON, British Library, Harley 3017, fols. 88v–94r, s. ix, Fleury (or eastern France). Excerpts (exclusively chapters with diagrams): chs. 7.1–4, 10, 11, and 23. Diagrams 2–6.

Somewhat similar to Oxford, St John's College 17. The codex includes excerpts from Bede's *DTR*, *DNR*, and *DT.*

Mostert, *The Library of Fleury*, 107 (BF382); Bischoff, *Katalog*, no. 2466; Wallis, 'Calendar & Cloister'. Online: BL-Digitised Manuscripts.

69. LONDON, British Library, Harley 3035, fols. 47r–63v, AD 1496, West Germany, probably Eberhardsklausen. *Incipiunt capitula de rerum naturibus.* Long recension (Zofingen type, see Appendix 6a), redivided into 55 numbered chapters, with omission of text of ch. 44(–), *de nominibus maris uel fluminum.* Includes mystical addition in ch. 1. Inserts *Hiemisphaeria* in ch. 12.3. Inserts *Aparctias* in ch. 37.1. Diagram 7 follows ch. 37.5. Inserts 'Solinus' in ch. 40.1. Vegetius interpolation at end of ch. 48 (see Appendix 5a). Diagrams 1–5, 7, and 7A (mountains of Sicily). Omits diagram and text for 5A and 6. No T-O map.

Derivative of Harley 2660(?), but retains Diagram 7A of Zofingen type. Harley 3035 renumbers the 58 chapters as listed in Harley 2660, omitting the Preface from the numeration, failing to assign a number to the chapter, *Vtrum sidera animam habeant*, and omitting the chapter *De nominibus maris et fluminum*, not only from the text (as Harley 2660 does) but also from the *capitula*. *De Trinitate* is omitted from ch. 48(47).

DNR is inserted between books 3 and 4 of the *Etymologiae*, as in Zofingen Pa 32, but without comprehensive title *De libro rotarum*.

Online: BL-Digitised Manuscripts.

70. LONDON, British Library, Harley 3099, fols. 154r–164v, 1130/1140, Munsterbilsen (Belgium); prov. Arnstein. Lacks title. Long recension. Includes mystical addition in ch. 1. Inserts *Hiemisphaeria* in ch. 12.3. Inserts *Aparctias* in ch. 37.1. Inserts 'Solinus' in ch. 40.1. Diagrams 1–5, space for 5A and 6, no 7. No T-O map.

Díaz y Díaz, *Index Scriptorum*, 32. Online: BL-Digitised Manuscripts.

71. MILAN, Biblioteca Ambrosiana C 10 sup., fols. 100r–118r, AD 1341/1342, scribe: Michele Sibenik.

Biblioteca Pinacoteca Accademia Ambrosiana.

72. MILAN, Biblioteca Ambrosiana H 150 inf., fols. 138r–161v, *c*.AD 810, eastern France(?); prov. Bobbio. Lacks title. Short recension. Adds formula of introduction for Diagram 7 after ch. 37.5 (see Appendix 2a). Lacks diagrams (including Diagram 7?). The *Epistle* of Sisebut is omitted (Fontaine).

DNR is part of a computistical miscellany known as the *Liber de computo* or the Bobbio Computus (Warntjes, *The Munich Computus*, xxiv). The hymn, *De ratione temporum*, which has been attributed to Sisebut as well as to Bede, follows *DNR* (Jones, *Bedae Pseudepigrapha*, 92–93).

Beeson, *Isidor-Studien*, 66; Fontaine, *Traité*, 37; Fontaine, 'La Diffusion carolingienne', 119–20; Bischoff, *Katalog*, no. 2621; Gorman, 'Diagrams', *stemma codicum*, 541.

73. MODENA, Biblioteca Estense 988, fols. 80v–95v, s. ix/x. *Libri rotarum sancti hisidori spanensis episcopi.* Fragment: long recension. Preface to ch. 26.6 (... *sed amplius quam*). The *capitula* list 48 chapters. Diagrams 1–6 + 5A.

Fontaine, 'La Diffusion carolingienne', 122–23; *Manus*; Gorman, 'Diagrams', *stemma codicum*, 541.

74. MONZA, Biblioteca Capitolare C 9 (69), fols. 1r–19r, s. x^{1/2}, north Italy. *De astra celi.*

Bischoff, 'Verbreitung', 189; *Manus*.

75. MUNICH, Bayerische Staatsbibliothek CLM 396, fols. 1v–34r, s. ix^{ex}, Brittany or Wales(?). *Incipit interpretatio isidori episcopi de natura rerum.* Long recension. The *capitula* omit ch. 19 and list two chs. 26 (= chs. 28 and 29); hence in the *capitula* ch. 48 is numbered 46. Includes mystical addition

in ch. 1. Inserts *Hiemisphaeria* in ch. 12.3. Inserts *Aparctias* in ch. 37.1. Inserts 'Solinus' in ch. 40.1. Diagrams 1–7 + 5A, 6A, 7A (and others). T-O map. A cosmic diagram, labelled *peritia esidori*, with accompanying text, follows *DNR* on fol. 34v.

Catalogus Codicum Latinorum 1.1:105; Fontaine, *Traité*, 37; Bischoff, *Katalog*, no. 2927. Online: *MDZ*.

76. MUNICH, Bayerische Staatsbibliothek CLM 14300, fols. 1r–22v, s. viii[ex], Salzburg; prov. Regensburg. No title. Short recension. Omits *Hiemisphaeria* in ch. 12.3. Omits *Aparctias* in ch. 37.1. Omits 'Solinus' in ch. 40.1. Diagrams 1–7 + 5A. No T-O map. The *Epistle* of Sisebut appears on fols. 21v–22v.

Becker, *DNR*, xxvi; Beeson, *Isidor-Studien*, 47–48, 71; Lowe, *CLA* 9 (no. 1294); Fontaine, *Traité*, 24–25; Bischoff, *Katalog*, no. 3150; Gorman, 'Diagrams', 540; Martín, 'Sisebut', 403; Martín, 'Isidorus Hisp.', 353. Online: *MDZ*. Fontaine's *M*.

77. MUNICH, Bayerische Staatsbibliothek CLM 14456, fol. 70r, s. ix[2/4], St Emmeram, Regensburg. Excerpt: Diagram 4, with *DNR* 4.7 (lines 56–59) inscribed in centre.

Bischoff, *Katalog*, no. 3202. Online: *MDZ*.

78. MUNICH, Bayerische Staatsbibliothek CLM 16128, fols. 1r–41v, s. viii/ix, Salzburg; prov. St. Nicholas at Passau. *De astronomia.* Short recension. Diagrams: 4 (15r), 5 (16r), 7 (fol. 35v). The *Epistle* of Sisebut appears on fol. 41r–v.

Copied from Munich 14300.

Becker, *DNR*, xxvi; Beeson, *Isidor-Studien*, 48, 71; Lowe, *CLA* 9 (no. 1313); Fontaine, *Traité*, 37, 160; Bischoff, *Katalog*, no. 3295; Martín, 'Sisebut', 403; Obrist, 'Wind Diagrams', fig. 4. Fontaine's *N* (*Epistle* only)

79. NEW YORK, Columbia University Library, Plimpton 251 (Phillipps 2651), fols. 97r–112v, s. xii[4], Spain(?). Lacks title. Short recension. Inserts *Hiemisphaeria* in ch. 12.3. Omits *Aparctias* in ch. 37.1. Adds formula of introduction for Diagram 7 after ch. 37.5 (see Appendix 2a). Omits 'Solinus' in ch. 40.1. Space for Diagrams 1–5 + 5A (first crescent drawn in); inserts 6–7. KOCMOC inscription in Diagram 7, which is displaced to end of chapter. No T-O map. No *Epistle* of Sisebut.

Online: Columbia University Archive (s.v. Plimpton 251).

80. OXFORD, Bodleian Library, Ashmole 393, pt. V, fols. 1r–10v, s. xiv or xv, England. *Incipit liber beatissimi Ysidori Yspaliensis Episcopi. De natura rerum*. Fragment: medium recension(?). Chs. 1–37. Ends: *Si a septembrione uersos rotatim legas in girum inuenies ordines eorundem* (formula of introduction for Diagram 7 at head of ch. 37 [see Appendix 2c]). Two diagrams of the winds (Diagram 7) follow on each page of last leaf.

Black, *Catalogue of the Ashmolean Manuscripts*, 307.

81. OXFORD, Bodleian Library, Auct. F.2.20 (Western 2186), fols. 1r–16r, s. xi², England; prov. Exeter(?). Long recension (English type, see Appendix 6b). Diagrams 1–5. Diagram 4 (fol. 5v), which Isidore depicted as a *figura solida*, is here redesigned as a wheel (cf., Exeter 3507). T-O map.

Díaz y Díaz, *Index Scriptorum*, 32; Stevens, 'Scientific Instruction', 99–100 and n. 57; Murdoch, *Album*, 281 and 383; Gameson, *Manuscripts*, no. 624; Gneuss and Lapidge, *Anglo-Saxon Manuscripts*, no. 536.

82. OXFORD, Bodleian Library, Auct. F.3.14 (Western 2372), fols. 1r–19v, s. xii^i (pre-1125), Malmesbury. Fragment: long recension (English type, see Appendix 6b). Includes mystical addition in ch. 1. Inserts 'Solinus' in ch. 40.1. Diagrams.

The codex includes Bede's *DTR*, *DNR*, and *DT*.

Fontaine, *Traité*, 38; Stevens, 'Scientific Instruction', 99–100 and n. 57; Gameson, *Manuscripts*, no. 626; Stevens, private communication.

83. OXFORD, St John's College 17, fols. 37v (ch. 23), 39r–v (ch. 11), 40r (ch. 10.1–2), 40v (ch. 37), AD 1102–1113, Thorney Abbey. Excerpts (exclusively chapters with diagrams). Omits *Aparctias* in ch. 37.1. Diagram 7 is placed at the end of ch. 37.5. Diagrams 3–7.

The codex includes Bede's *DTR*, *DNR*, and *DT*.

Gameson, *Manuscripts*, no. 794; Wallis, 'Calendar & Cloister'. Online: *Oxford Digital Library*.

84. OXFORD, St John's College 178, fols. 9r–37v, s. xiv, England. Long recension (English type, see Appendix 6b).

Stevens, 'Scientific Instruction', 99–100 and n. 57.

85. PARIS, Bibliothèque nationale lat. 2796, fols. 53v–54r, s. ix^{in}, northern France. Excerpt, ch. 3 only.

Beeson, *Isidor-Studien*, 100, 119; *Bibliothèque nationale: Catalogue général* 3:90–95; Fontaine, *Traité*, 37; Bischoff, *Katalog*, no. 4232.

86. PARIS, Bibliothèque nationale lat. 2990A, fols. 149r–156v, s. ix^1, Saint-Amand? Fragment: ch. 24 (*De lumine stellarum*) to ch. 36 (*... uenti autem interdum*).

Bibliothèque nationale: Catalogue général 3:374; Díaz y Díaz, *Index Scriptorum*, 32; Bischoff, *Katalog* no. 4265.

87. PARIS, Bibliothèque nationale lat. 4860, fols. 98r–107r, s. ix^2, Reichenau; prov. Mainz. Long recension. Ch. 48(47) is followed by two chapters (numbered 49–50, not listed in the *capitula*) entitled *De zonis caeli*, the text of which consists of Vergil's *Georgics* 1.231–45 (see St Gall 238), and *Quomodo terminatur terra oceano* (= the *Epistle*). Diagrams: 1 (fol. 99v), 2 (fol. 100r), 5 (fol. 103v). The *Epistle* of Sisebut (lines 1–33) appears on fol. 107r.

The codex includes Bede's *DTR*, *DNR*, and *DT*.

Fontaine, *Traité*, 37; Bischoff, *Katalog* no. 4860; Kline, *Maps of Medieval Thought*, 14 and nn. 9–11; Alberto, 'Sisebut's *Carmen*', 185.

88. PARIS, Bibliothèque nationale lat. 5239, fols. 138v–139r, s. x, St Martial, Limoges. Excerpt: ch. 37 with the poem *De Ventis* (see Appendix 4). Diagram 7 (fol. 139r). The diagram is oriented with north at the top.

Copied from Paris 5543. See Strasbourg 326. The codex includes Bede's *DTR*, *DNR*, and *DT*.

Díaz y Díaz, *Index Scriptorum*, 32; Obrist, 'Wind Diagrams', 45 and n. 68; Palmer, 'The Ends and Futures of Bede's *DTR*', 150 n. 49.

89. PARIS, Bibliothèque nationale lat. 5543, AD 847, France, Fleury(?). Excerpts. The Poem *De Ventis* (see Appendix 4) appears on fol. 140r.

The codex includes Bede's *DTR*, *DNR*, and *DT*.

Baehrens, *Poetae Latini minores* 5:383; Díaz y Díaz, *Index Scriptorum*, 32; Mostert, *The Library of Fleury*, 207–08 (BF1058); Bischoff, *Katalog* no. 4367; Wallis, 'Calendar & Cloister'.

90. PARIS, Bibliothèque nationale lat. 6400G, fols. 112v–145v, s. vii or viii, Fleury (upper palimpsest)(?); provenance Fleury. *In christi nomine incipit liber sancti esidori de mundo.* Fragment: short recension. *Capitula* of the short recension. The loss of a folio results in a lacuna from ch. 7.6 (*arriditate siccatur ...*) to ch. 9.3 (*... ita declinatur*). A second loss after fol. 145v truncates the text of *DNR* at the beginning of ch. 46(45).3 (*e motum autem illic*). Omits *Hiemisphaeria* in ch. 12.3. Omits *Aparctias* in ch. 37.1. Adds formula of introduction for Diagram 7 after ch. 37.5 (see Appendix 2a). Omits 'Solinus' in ch. 40.1. Diagrams 1–7 + 5A. KOCMOC inscription in Diagram 7, which is displaced to end of chapter. Despite Fontaine's speculation to the contrary (*Traité*, 25 and n. 2), it is doubtful that the *Epistle* of Sisebut followed the lost final chapter (*de monte Aetna*). See Introduction, p. 35.

Beeson, *Isidor-Studien*, 65–66; Lowe, *CLA* 5 (no. 564a); Fontaine, *Traité*, 25–26; Mostert, *The Library of Fleury*, 211–12 (BF1080(2)/1081(2)); Gorman, 'Diagrams', 540; Martín, 'Isidorus Hisp.', 354; Obrist, 'Wind Diagrams', fig. 6. Online: *Gallica*. Fontaine's *P*.

[PARIS, Bibliothèque nationale lat. 6413, see Karlsruhe 339/1]

91. PARIS, Bibliothèque nationale lat. 6414, s. xii.
Fontaine, *Traité*, 38.

92. PARIS, Bibliothèque nationale lat. 6649, fols. 2r–25r, s. ix$^{3/4}$, northeastern France. Lacks title. Short recension. Omits *Hiemisphaeria* in ch. 12.3. Omits *Aparctias* in ch. 37.1. Omits 'Solinus' in ch. 40.1. Diagrams 1–7 + 5A (with additional man in the moon figure). No T-O map. The *Epistle* of Sisebut appears on fols. 24r–25r.

Fontaine, *Traité*, 37; Bischoff, *Katalog* no. 4412; Alberto, 'Sisebut's *Carmen*', 185. Online: *Gallica*.

93. PARIS, Bibliothèque nationale lat. 7533, fols. 87r–101r, s. ix. *Incipit liber sancti Isidori de astronomia.* Omits preface and *capitula*. Short recension. Skip from ch. 3.1 (*... a fando dicta*) to ch. 4.7 (*Dehinc reuerteris ...*) with omission of Diagram 1. Omits *Hiemisphaeria* in ch. 12.3. Omits *Aparctias* in ch. 37.1. Omits 'Solinus' in ch. 40.1. Diagrams 2–7 + 5A. The *Epistle* of Sisebut appears on fols. 100v–101r.

Bischoff, *Katalog* no. 4467; Alberto, 'Sisebut's *Carmen*', 184. Online: *Gallica*.

94. PARIS, Bibliothèque nationale lat. 10616, fols. 1r–93v, Verona, written in the time of bishop Eginon, between 796 and 799. *In nomine domini incipit liber sancti hisidori spaniensis.* Long recension. Includes mystical addition in ch. 1. Inserts *Hiemisphaeria* in ch. 12.3. Inserts *Aparctias* in ch. 37.1. Inserts 'Solinus' in ch. 40.1. Vegetius interpolation (see Appendix 5a) + *De Trinitate* (*Etym.* 7.4) at end of ch. 48(47). Diagrams 1–7 + 5A, 7A (fol. 88v, mountains of Sicily). T-O map, fol. 91r, with *patrem/filium/ spiritum/sanctum* in the four corners; *De Trinitate* follows.

Lowe, *CLA* 5 (no. 601); Fontaine, *Traité*, 29–30; Gorman, 'Diagrams', 540; Martín, 'Isidorus Hisp.', 354. Online: *Gallica.* Fontaine's *V.*

95. PARIS, Bibliothèque nationale lat. 15171, fols. 200r–219v, northern France? (prov. unknown), s. xiii? *Incipit prologus beatissimi ysidori ispalensis archiepiscopi. De natura rerum.* Medium recension. No *capitula.* Omits *Hiemisphaeria* in ch. 12.3. Omits *Aparctias* in ch. 37.1. Adds *rotatim* formula of introduction for Diagram 7 preceded by *Idem flatus* paragraph after ch. 37.4 (see Appendix 2c). Two versions of Diagram 7 are placed between the introductory formula and ch. 37.5. The poem *De Ventis* (see Appendix 4) precedes ch. 38. Omits 'Solinus' in ch. 40.1. Inserts *Maeotis* in ch. 48(47).2. Diagrams 1–7 (two versions of 7) + 5A. A diagram possibly illustrating an eclipse precedes ch. 48(47) (same diagram in Florence 22 dex. 12). T-O map. The *Epistle* of Sisebut is omitted.

Online: *Gallica.*

96. PARIS, Bibliothèque nationale, Nouv. acq. lat. 448, fols. 45r–77r, s. xi, Besançon. Lacks title. Long recension. Confusion in *capitula* after ch. 41: ch. 42 omitted, but added in margin with mark of insertion; chs. 43–44 numbered as xlii and xliii; ch. 45 numbered as xliii; chs. 46–48 numbered as xlv, xlvi, and xlvii. Chapters in text are correctly numbered, except for ch. 43, which is numbered as xlii (as is ch. 42). No mystical addition in ch. 1. Omits *Hiemisphaeria* in ch. 12.3. Omits *Aparctias* in ch. 37.1. Adds formula of introduction for Diagram 7 after ch. 37.5 (see Appendix 2a). Omits 'Solinus' in ch. 40.1. Diagrams 1–7 + 5A. KOCMOC inscription in Diagram 7, which is displaced to end of chapter. T-O map.

Cf. St Gall 240.

Fontaine, *Traité*, 37. Online: *Gallica.*

97. ROUEN, Bibliothèque municipale 26 (A292), fols. 192v–196v, s. ix, Jumièges. Fragments. Diagram 7, fol. 193v.

The codex includes Bede's *DNR* and *DT.*

Beeson, *Isidor-Studien*, 100, 119; *Catalogue général* 1:9; Fontaine, *Traité*, 37; Bischoff, *Katalog* no. 5367; Gneuss and Lapidge, *Anglo-Saxon Manuscripts*, no. 919.3.

98. ST GALL, Stiftsbibliothek 110, pp. 513–516, Verona, probably under bishop Eginon, between 796 and 799. Excerpt: ch. 7.4 (*uer quippe constat* ...) – 7.6. Includes Diagram 2.

The excerpt and surrounding computus are identical to St Gall 225.

Beeson, *Isidor-Studien*, 95, 119; Lowe, *CLA* 7 (no. 907); Fontaine, *Traité*, 34. Online: *e-codices*.

99. ST GALL, Stiftsbibliothek 225, pp. 126–127, AD 760–797 (possibly 773), St Gall. Excerpt: ch. 7.4 (*uer quippe constat* ...) – ch. 7.6. Includes Diagram 2.

The excerpt is embedded without attribution in an incomplete computus. The sentences from ch. 7.4 are inserted at the end of an untitled ch. 15 (*Si uis inuenire quotus annus sit a passione domini* ...). The diagram and Isidore's ch. 7.5 and 7.6 are numbered as ch. 16. Ch. 17 of the computus is entitled, *De sollemnitatibus*. See St Gall 110.

Beeson, *Isidor-Studien*, 29, 119; Lowe, *CLA* 7 (no. 928); Fontaine, *Traité*, 34; Martín, 'Isidorus Hisp.', 354. Online: *e-codices*. Fontaine's *T.*

100. ST GALL, Stiftsbibliothek 238, pp. 312–385, AD 760–780, St Gall, written by Winithar under abbot Othmar. *Incipit liber rotarum sancti hysidori.* Long recension. Various confusions in numbering both in *capitula* and in text. Chs. 19 and 46(45) omitted from *capitula*, but not from text. Comment on God's promise to Noah, signified by the rainbow, added at end of ch. 31; explanation of origin of springs and circulation of water through the earth added at end of ch. 33. Ch. 48 (falsely numbered 49) is followed by a chapter (numbered 50, not listed in the *capitula*) entitled *De ordinacione solis uel reliqua*, the text of which consists of Vergil's *Georgics* 1.231–44. Includes mystical addition in ch. 1. Inserts *Hiemisphaeria* in ch. 12.3. Inserts *Aparctias* in ch. 37.1. Inserts 'Solinus' in ch. 40.1. Diagrams 1–5; omits text for 5A. No T-O map.

Beeson, *Isidor-Studien*, 69; Lowe, *CLA* 7 (no. 934); Fontaine, *Traité*, 32–33; Fontaine, 'La Diffusion de l'œuvre d'Isidore', 315; Gorman, 'Diagrams', 540; Martín, 'Isidorus Hisp.', 354. Online: *e-codices*. Fontaine's *S.*

101. ST GALL, Stiftsbibliothek 240, pp. 116–189, s. ix^{in}, Chelles; prov. St Gall. Lacks title. Long recension. The *capitula* list 48 numbered chapters. No mystical addition in ch. 1 (*propheticae*, the first word after the omission, is glossed, *id est mistice* (p. 119)). Omits *Hiemisphaeria* in ch. 12.3. Omits *Aparctias* in ch. 37.1. Adds formula of introduction for Diagram 7 after ch. 37.5 (see Appendix 2a). Omits 'Solinus' in ch. 40.1. Diagrams 1–7 + 5A. KOCMOC inscription in Diagram 7, which is displaced to end of chapter. T-O map.

Isidore's *Allegoriae sacrae scripturae* precedes *DNR*; it concludes (p. 115): *Explicit domino et filio sisebuto esidorus.*

Beeson, *Isidor-Studien*, 30, 69; Fontaine, *Traité*, 37; Fontaine, 'La Diffusion de l'œuvre d'Isidore', 317–18; Bischoff, *Katalog* no. 5677. Online: *e-codices.*

102. ST GALL, Stiftsbibliothek 855, pp. 415–429, s. ix$^{2/4}$, western Germany. Excerpts: chs. 39–40; ch. 43; chs. 46(45)–47(46). The chapters are unnumbered (but entitled) and run consecutively. Apparently a severely abridged version of the short recension (ch. 39 begins at the top of a verso, following the *Institutions* of Cassiodorus on the recto). Omits 'Solinus' in ch. 40.1. No T-O map. The *Epistle* of Sisebut (pp. 425–429) follows uninterruptedly upon the last word of ch. 47(46) with no indication of a shift. The poem appears to be copied mechanically without comprehension.

The codex includes Cassiodorus' *Institutions* 2.

Beeson, *Isidor-Studien*, 119; Mynors, *Cassiodori Institutiones*, xix–xx; Fontaine, *Traité*, 37, 160; Fontaine, 'La Diffusion de l'œuvre d'Isidore', 318–19; Martín, 'Sisebut', 403; Bischoff, *Katalog* no. 5853. Online: *e-codices.* Fontaine's *G* (*Epistle* only)

103. ST-OMER, Bibliothèque d'agglomération 201, fol. 1r, s. xii, St-Bertin. Excerpts from chs. 36 and 37. A diagram of the winds (*duodecim uenti qui mundi globii circumagunt*) (= Diagram 7) surrounding a diagram of the world, with the figure of the Tower of David in the centre (*Turris Dauid Ierusalem umbilicus orbis*), is above the excerpts.

Online: *Gallica.*

104. SOISSONS, Bibliothèque municipale 128, fols. 133v–157v, s. xv. Diagrams.

Catalogue général des départements 3:108–09.

105. STRASBOURG, Bibliothèque nationale et universitaire 326, fols. 119v–120r, s. x, Angoulême or Limoges. Excerpt: *DNR* 37 and the poem *De Ventis* (see Appendix 4); the diagram of the winds (fol. 120r) follows the poem. Omits *Aparctias* in ch. 37.1. Adds formula of introduction for Diagram 7 after ch. 37.5 (see Appendix 2a). Diagram 7 includes the phrase *qui et apartias*.

Like Paris 5239. The codex includes Bede's *DTR, DNR,* and *DT.*

Obrist, 'Wind Diagrams', 45 and n. 68. Online: *BVMM.*

106. TRIER, Stadtbibliothek 1084/115 4°, fol. 99r, s. xi, St Maximin. Excerpt, abbreviated text of ch. 37.1–4 within the 12 segments of a diagram of the winds (= Diagram 7) with a T-O map in its centre; the text has been rearranged to fit into the appropriate points of the compass. Inserts *Aparctias* in ch. 37.1 (*qui et aparcias*). The T-O map is placed within a square with the inscription: *Mundus IV angulos habet et IV partes diuersas.*

An abbreviated excerpt from Bede, *DNR* 27 (*uenti australes ... altanus in pelago*), is placed beneath the wind-rose.

Franz, *Karolingische Beda-Handschrift,* 64 (no. 33), illus. p. 63; Obrist, 'Wind Diagrams', 58.

107. TRIER, Stadtbibliothek 2500, fol. 20r, about 840, Reims or Laon. *Hic quoque potest uideri quo ordine spirant uenti.* Excerpt, abbreviated text of ch. 37.1–4 within the 12 segments of a diagram of the winds (= Diagram 7) with a T-O map in its centre; the text has been rearranged to fit into the appropriate points of the compass. Inserts *Aparctias* in ch. 37.1 (*qui et aparchias*). The poem *De Ventis* (see Appendix 4) appears beneath the wind-rose.

The codex includes Bede's *DTR, DNR,* and *DT.*

Franz, *Karolingische Beda-Handschrift,* 14–25, illus. 19; Obrist, 'Wind Diagrams', 51–52; Bischoff, *Katalog,* no. 6201.

108. VATICAN CITY, Biblioteca Apostolica Vaticana, Pal. lat. 834, fols. 47v–90v, s. ix[2], eastern France or western Germany; prov. Lorsch. *Incipit liber de astra celi sancti hisidori spalensis episcopi.* Long recension. Includes mystical addition in ch. 1. Inserts *Hiemisphaeria* in ch. 12.3. Inserts *Aparctias* in ch. 37.1. Adds formula of introduction for Diagram 7 in ch. 37 (see Appendix 2b). Inserts 'Solinus' in ch. 40.1. Diagrams 1–7 + 5A. T-O map, with legends as in Appendix 5b. Another diagram, apparently a

later addition, appears in ch. 20 in the right margin of fol. 69r, illustrating the eclipse of the sun. Closely related to Besançon 184.

Arévalo, Isidoriana 1:662; Beeson, *Isidor-Studien*, 66–67; Fontaine, *Traité*, 37; Fontaine, 'La Diffusion carolingienne', 116; Bischoff, *Katalog*, no. 6560. Online: *Bibliotheca Laureshamensis*.

109. VATICAN CITY, BAV, Pal. lat. 1448, fols. 19r–39v, *c.*810, Trier; prov. Mainz. *Incipit liber rotarum sancti Isidori episcopi*. Long recension. Includes mystical addition in ch. 1. Inserts *Hiemisphaeria* in ch. 12.3. Inserts *Aparctias* in ch. 37.1. Inserts 'Solinus' in ch. 40.1. No diagrams or T-O map, although spaces are left for both, including 5A (phases of the moon) and 6A (colours of the rainbow).

The codex includes Bede's *DTR* and *DT*.

Arévalo, *Isidoriana* 1:662; Beeson, *Isidor-Studien*, 67; Fontaine, *Traité*, 37; Fontaine, 'La Diffusion carolingienne', 116; Bischoff, *Katalog*, no. 6578. Online: *Bibliotheca Laureshamensis*.

110. VATICAN CITY, BAV, Regin. lat. 123, AD 1056, Ripoll; prepared by Abbot Oliva. Excerpt: ch. 13 only.

The codex includes Bede's *DNR* and *DT*.

Wilmart, *Codices Reginenses Latini* 1:287–92; Fontaine, *Traité*, 37.

111. VATICAN CITY, BAV, Regin. lat. 255, fols. 1r–22r, s. ix^{med}, Reims(?). *Isidorus de natura rerum*. Short recension. Diagrams 1–7 + 5A and another figure of a sphere enclosing a man extending a rod (fol. 11v). The *Epistle* of Sisebut follows (fols. 21v–22r).

Arévalo, *Isidoriana* 1:662; Beeson, *Isidor-Studien*, 45, 67; Wilmart, *Codices Reginenses Latini* 2:15–19; Fontaine, *Traité*, 37; Fontaine, 'La Diffusion carolingienne', 110–111; Mostert, *The Library of Fleury*, 260 (BF1362); Gorman, 'Diagrams', *stemma codicum*, 541; Bischoff, *Katalog*, no. 6647; Alberto, 'Sisebut's *Carmen*', 184.

112. VATICAN CITY, BAV, Regin. lat. 309, fol. 14v, s. x, St Denis. Excerpt: ch. 8.1 only.

The codex includes Bede's *DNR*.

Wilmart, *Codices Reginenses Latini* 2:160–74; Fontaine, *Traité*, 37.

113. VATICAN CITY, BAV, Regin. lat. 310, fols. 12r–51v, s. viii/ix, St-Denis(?). *Incipit rotarum liber isidori.* Short recension. Diagrams 1–7 + 5A. The *Epistle* of Sisebut is omitted.

Arévalo, *Isidoriana* 1:662; Beeson, *Isidor-Studien*, 24, 67; Wilmart, *Codices Reginenses Latini* 2:174–78; Fontaine, *Traité*, 37; Fontaine, 'La Diffusion carolingienne', 111; Mostert, *The Library of Fleury*, 262 (BF1377); Gorman, 'Diagrams', *stemma codicum*, 541; Bischoff, *Katalog* no. 6656.

114. VATICAN CITY, BAV, Regin. lat. 571, fol. 13, s. x. Excerpt: ch. 38.
Arévalo, *Isidoriana* 1:661; 2:321–23.

115. VATICAN CITY, BAV, Regin. lat. 1260, fols. 17r–43r, s. ix^{med}, Auxerre(?), Fleury(?). Lacks title. Short recension. The poem *De Ventis* (fol. 38) (see Appendix 4); the *Epistle* of Sisebut follows ch. 47(46) (fols. 43–44). Diagrams: 4 (fol. 24r), 6.

The codex includes Bede's *DNR* and *DT.*

Arévalo, *Isidoriana* 1:662; Baehrens, *Poetae Latini minores* 5:383; Beeson, *Isidor-Studien*, 67; Jones, *Bedae Pseudepigrapha*, 137; Fontaine, *Traité*, 37; Fontaine, 'La Diffusion carolingienne', 111; Mostert, *The Library of Fleury*, 279 (BF1482 or BF1483 or BF1484[?]); Eastwood, 'The Diagram of the Four Elements', 555; Gorman, 'Diagrams', *stemma codicum*, 541; Bischoff, *Katalog* no. 6776; Alberto, 'Sisebut's *Carmen*', 184.

116. VATICAN CITY, BAV, Regin. lat. 1573, s. xi, written by the scribes Gaufridus and Humbertus, Ferrières; prov. Fleury or Ferrières. *Liber astrorum caeli editus ab Isidoro.* Medium recension. T-O map (fol. 112v), with legends as in Appendix 5b.

Arévalo, *Isidoriana* 1:662; 2:333; Díaz y Díaz, *Index Scriptorum*, 32; Mostert, *The Library of Fleury*, 283–84 (BF1508).

117. VATICAN CITY, BAV, Urb. lat. 100, fols. 177r–191r, s. xv. *Ysidori de astronomia incipit feliciter.* Long recension (numbered as 49 chapters). Vegetius interpolation at end of ch. 48 (see Appendix 5a). Diagrams. T-O map. Fols. 190v–191r contain *De caelo*, a list of the heavens attributed to Isidore.

Arévalo, *Isidoriana* 1:662; 2:377–78; Biblioteca Apostolica Vaticana, *Catalogo Manoscritti.*

118. VATICAN CITY, BAV, Vat. lat. 214, s. xii. Fragment.
Fontaine, *Traité*, 38.

119. VATICAN CITY, BAV, Vat. lat. 5330, s. xi/xii. Long recension(?).
Diagram 7A (fol. 73r, mountains of Sicily).
Boese, *Die Lateinischen Handschriften*, 332 (s.v. Berlin, Ham. 689);
Biblioteca Apostolica Vaticana, *Catalogo Manoscritti*.

120. VENICE, Biblioteca Marciana Z. 497, fols. 191v–200v, s. xi[ex], Rome.
Epistula de natura rerum ad Sisebutum regem. The *Epistle* of Sisebut is
found on fols. 181v–182r, with the title, *Prologus Sisebuti regis Gothorum
de defectus soli* (!) *et lunae.*
Alberto, 'Sisebut's *Carmen*', 185–86.

121. VERDUN, Bibliothèque municipale 26, fols. 1r–69r, s. ix[2/3], St-Vanne.
Incipit liber de astra celi sancti. Long recension. Includes mystical addition
in ch. 1. Inserts *Hiemisphaeria* in ch. 12.3. Inserts *Aparctias* in ch. 37.1.
Adds formula of introduction for Diagram 7 in ch. 37 (see Appendix 2b).
Inserts 'Solinus' in ch. 40.1. Diagrams 1–4, 5A, 7 (5 and 6 lost due to
damage to MS). T-O map, with legends as in Appendix 5b.
Bischoff, *Katalog*, no. 7022. Online: *Manuscrits de Verdun.*

122. VERDUN, Bibliothèque municipale 27, s. ix[2/3], eastern France. *Isidori
liber Astracoelorum.* Long recension.
Catalogue général des départements 5:443; Fontaine, *Traité*, 37;
Bischoff, *Katalog*, no. 7023.

123. VIC, Arxiu I Biblioteca Episcopal 44 (36), fols. 1r–16v, AD 1064,
Spain. *Incipit liber astrologius* (!) *a beato ysidoro editus.* Short recension.
Omits *Hiemisphaeria* in ch. 12.3. Omits *Aparctias* in ch. 37.1. Omits
'Solinus' in ch. 40.1. Diagrams 1–3; 6–7. Space for Diagrams 4–5; no space
for 5A. No T-O map. The *Epistle* of Sisebut appears on fol. 16r–v (the
Epistle is not separated from the final chapter [*De monte ethna*] and gives
no indication of authorship).
Isidore's *De summo bono* follows on fol. 16v: *Incipiunt capitula libri
secundi* (= book 1 of *De summo bono*).
Gudiol, *Catàleg*, 61–63; Fontaine, *Traité*, 37; Mateu Ibars, *Colectánea
paleográfica* 1:474–76; cbk.

124. VIENNA, Österreichische Nationalbibliothek, Vindobonensis Palatinus 997, fol. 128r, s. x. Excerpt: ch. 14.
Tabulae codicum manu scriptorum 1:172; HMML.

125. WEIMAR, Herzogin Anna Amalia Bibliothek 414a, 4th fol. of a lost MS, s. viii^{ex}, probably England (Lowe)/Fulda (Fontaine). Fragment: long recension. Ch. 1.3 (*tabulas legis confregit ...*) to ch. 2.2 (*... crepusculum dicitur id est*). Fragment begins in middle of mystical addition in ch. 1.
Laistner, 'Fragment', 28–31; Lowe, *CLA* 9 (no. 1369); Fontaine, *Traité*, 34–35; Gneuss and Lapidge, *Anglo-Saxon Manuscripts*, no. 943.4; Gorman, 'Diagrams', 540; Martín, 'Isidorus Hisp.', 354. Fontaine's *W.*

126. WOLFENBÜTTEL, Herzog August Bibliothek 4640, fols. 33r–44v, 19r–20r, s. xii, Germany.
Kataloge der Herzog August Bibliothek Wolfenbüttel 9, 250; Fontaine, *Traité*, 38.

127. ZOFINGEN, Stadtbibliothek Pa 32, fols. 56r–75v, s. ix^1, St Gall.
Incipit liber ysidori de rerum natura ad syseputum regem. Long recension, redivided into 58 numbered chapters (Zofingen type, see Appendix 6a). Includes the mystical addition in ch. 1. Inserts *Hiemisphaeria* in ch. 12.3. Inserts *Aparctias* in ch. 37.1. Diagram 7 follows ch. 37.5. Inserts 'Solinus' in ch. 40.1. Vegetius interpolation (see Appendix 5a) + De Trinitate (*Etym.* 7.4) at end of ch. 48(47). Diagrams 1–5, 7 + 7A (mountains of Sicily). Omits diagram and text for 5A and 6. No T-O map.
Incorporated into text of Isidore's *Etymologiae* as the middle section of a supplementary book 4 in three parts under the comprehensive title *De libro rotarum* (fols. 50r–81v): *De astronomia* (*Etym.* 3.24–71 (Gasparotti and Guillaumin, *Etym.* 3.23–70)) + DNR + De temporibus (*Etym.* 5.28–39).
Fontaine, *Traité*, 37; Fontaine, 'La Diffusion de l'œuvre d'Isidore', 315–16; Gorman, 'Diagrams', *stemma codicum*, 541; von Büren, 'Isidore, Végèce et Titanus', 40–41. Online: *e-codices.*

The following manuscripts listed by Fontaine or others have been excluded or renumbered for the reasons given. An asterisk indicates that the manuscript displays a diagram from *DNR* with its inscribed legends only. Manuscripts that present Diagram 7 with the text of ch. 37.1–4 inscribed within the twelve segments of the winds are included in the main list.

BERLIN, Staatsbibliothek zu Berlin, Preussischer Kulturbesitz, lat. 641, s. x. Fontaine, *Traité*, 37, with an asterisk indicating only indirect and uncertain knowledge of the manuscript.

According to C. Thulin, *Die Handschriften des Corpus Agrimensorum Romanorum*, Abhandlungen der Königlich Preussischen Akademie der Wissenschaften, Philosophisch-Historisch Class, Fasc. 2 (Berlin: Verlag der Königlichen Akademie der Wissenschaften, 1911), 7, fols. 17r–257v of this manuscript contain: 'Isidori Origines, Ars Donati (Keil 4, 355–405), Glossare, Catos Disticha u. a. (Minuskel)'. No evidence of *DNR*.

BORDEAUX, Bibliothèque municipale 844, s. xii.

Díaz y Díaz, *Index Scriptorum*, 32. Not in Delpit, *Catalogue des manuscrits*, or *Catalogue général ... Départements*.

BRUSSELS, Bibliothèque Royale Albert Ier 1322, s. ix.

Fontaine, *Traité*, 37. This appears to be the same as BRUSSELS, Bibliothèque Royale Albert Ier 9311–9319 (1322), which Fontaine entered as a tenth-century manuscript.

*CAMBRIDGE, Corpus Christi College 391 ('Portiforium of St Wulfstan'), fol. 132r, s. xi$^{3/4}$, Worcester Cathedral Priory(?). Text legends of Diagram 6 (planets) only (no diagram).

Gameson, *Manuscripts*, no. 86; Gneuss and Lapidge, *Anglo-Saxon Manuscripts*, no. 104; Wallis, 'Calendar & Cloister'.

*CAMBRIDGE, University Library Kk.5.32, fol. 49r, s. xi/xii, England. Text legends of Diagram 6 (planets) only (no diagram) (cf. Cambridge, Corpus Christi College 391).

Gameson, *Manuscripts*, no. 44; Gneuss and Lapidge, *Anglo-Saxon Manuscripts*, no. 26; Wallis, 'Calendar & Cloister'.

CHARTRES, Bibliothèque municipale 63 (125), s. xi.

Fontaine, *Traité*, 37. Destroyed.

DURHAM, Cathedral Library, Hunter 100, fol. 1r, AD 1100–1128, Durham. Note on calendar, similar but not identical to ch. 4.5 and *Etym.* 16.18.11.

Gameson, *Manuscripts*, no. 278; Gneuss and Lapidge, *Anglo-Saxon Manuscripts*, no. 326; *ASMMF* 14:111–22. Microfiche: *ASMMF* 14.

ERLANGEN, Universitätsbibliothek 357, s. xii.

Fontaine, *Traité*, 38, with an asterisk indicating only indirect and uncertain knowledge of the manuscript. There is no record of Isidore in Erlangen 357 according to Fischer, *Handschriften*.

GRENOBLE, Bibliothèque municipale 347, s. xii.

Fontaine, *Traité*, 38, with an asterisk indicating only indirect and uncertain knowledge of the manuscript. Grenoble 347 in *Catalogue général* 7:128, is an unrelated manuscript of the seventeenth century.

*OXFORD, Bodleian Library, Auct. F.5.19, fol. 30v, s. xii[l]. Diagram 6 (planets) with legends only.

Wallis, 'Calendar & Cloister'.

*PARIS, Bibliothèque nationale, Nouv. acq. lat. 1615, fol. 181r, s. ix, Fleury. Diagram 6 (planets) with legends only. The codex includes Bede's *DTR*, *DNR*, and *DT.*

Beeson, *Isidor-Studien*, 60–61; Mostert, *The Library of Fleury*, 243 (BF1257 or BF1258(?)); Wallis, 'Calendar & Cloister'.

*PARIS, Bibliothèque nationale, Nouv. acq. lat. 1616, fol. 1v, s. ix[med], Fleury. Diagram 6 (planets) with legends only.

Mostert, *The Library of Fleury*, 243–44 (BF1259 or BF1260(?)); Wallis, 'Calendar & Cloister'.

ROUEN, Bibliothèque municipale 110, fragment, s. ix.

Fontaine, *Traité*, 37. According to the *Catalogue général*, vol. 1, *Rouen*, 110 (A377) is a biblical concordance of the thirteenth century. There are no other manuscripts of *DNR* in the *Catalogue*.

VIC, Arxiu I Biblioteca Episcopal 202, s. xii, Spain.

Isidore, *De summo bono* (no *DNR*).

Gudiol, *Catàleg*, 63–64; Fontaine, *Traité*, 38; Mateu Ibars, *Colectánea paleográfica* 1:476; cbk.

EDITIONS OF ISIDORE'S *DE NATURA RERUM*

ZAINER (Augsburg, 1472). The *editio princeps*, edited by Günther Zainer de Reutlingen. *Isidori iunioris, Hispalensis episcopi, Liber de responsione mundi et astrorum ordinatione ad Sesibutum regem* (cf. Edinburgh 123). Long recension (Zofingen type; see Appendix 6a): chs. 1–48 in 57 chapters, in the same arrangement and order as Zofingen P 32 (*q.v.*).[1] Diagrams: 1–7.

Fontaine, *Traité*, 141–42; Martín, 'Isidorus Hisp.', 362.

LA BIGNE (Paris, 1580). Complete edition of the works of Isidore, edited by Margerin de la Bigne. *DNR* is found on pp. 148–58. *Beati Isidori Iunioris Hispalensis archiepiscopi de natura rerum ad Sisebutum regem liber.* Running title: *de mundo.* Medium recension. No 'mystic addition' in ch. 1. The poem *De Ventis* (*Versiculi de supranominatis uentis*) (see Appendix 4) follows ch. 37. Diagrams 1–7.

Fontaine, *Traité*, 142.

GRIAL (Madrid, 1599). Complete edition, edited by Juan Grial. *DNR* is found in vol. 1, pp. 63–91. *De natura rerum liber I.* Short recension. Diagrams 1–7.

Fontaine, *Traité*, 142–43.

DU BREUL (Paris, 1601; repr. Cologne, 1617). Complete edition, edited by Jacques du Breul. *DNR* is found on pp. 354–73 (repr. pp. 246–59). *Beati Isidori Iunioris Hispalensis archiepiscopi de natura rerum ad Sisebutum regem liber.* Medium recension. Omits *Hiemisphaeria* in ch. 12.3. Omits *Aparctias* in ch. 37.1. Adds *rotatim* formula of introduction for Diagram 7 preceded by *Inde flatus* paragraph after ch. 37.4 (see Appendix 2c). The poem *De Ventis* (see Appendix 4) precedes ch. 38. Omits 'Solinus' in ch. 40.1. Inserts *Maeotis* in ch. 48(47).2. Diagrams 1–7. The *Epistle* of Sisebut is omitted. Close to Paris 15171.

Fontaine, *Traité*, 143. Online: *Gallica.*

---

1 Fontaine, *Traité*, 141. According to Fontaine, Zainer's MS was related to the Type II short-recension MS, Bamberg 61, from Monte Cassino. If so, this would be further evidence of an Italian connection at St Gall (where the Zofingen type probably originated).

ULLOA (Madrid, 1778). Edition of Grial, re-edited by B. Ulloa. Diagrams 1–7.

Fontaine, *Traité*, 143.

ARÉVALO (Rome, 1803). *S. Isidori Hispalensis Episcopi ... Opera Omnia*. Ed. Faustino Arévalo. 7 vols. Rome, 1797–1803. *DNR* in vol. 7, pp. 1–62. *S. Isidori Hispalensis episcopi de natura rerum ad Sisebutum regem liber.* Long recension, but without the mystical addition in ch. 1 (included in footnote). No *capitula*. Omits *Hiemisphaeria* in ch. 12.3 (included in footnote). Inserts *Aparctias* in ch. 37.1. Omits 'Solinus' in ch. 40.1 (included in footnote). Inserts *Maeotis* in ch. 48(47).2. Diagrams 1–7 (omits 5A). No T-O map. The *Epistle* of Sisebut is omitted.

Fontaine, *Traité*, 143–44; Martín, 'Isidorus Hisp.', 362. Online: HathiTrust.

BECKER (Berlin, 1857). *De natura rerum liber*, edited by Gustav Becker. Long recension. Diagrams 1–7. T-O map.

Fontaine, *Traité*, 144–45. Online: HathiTrust.

MIGNE (Paris, 1862). Reprint of Arévalo. *PL* 83:963–1018.

Fontaine, *Traité*, 144.

FONTAINE (Paris, 1960; repr. 2002). *Traité de la nature/De natura rerum*, edited with French translation by Jacques Fontaine. Long recension. Diagrams 1–7 (reproduced from Munich, CLM 14300). pp. 163–327. *Epistle* of Sisebut with French translation, pp. 328–35.

LABORDA (Madrid, 1996). Reprint of Migne (text of Arévalo), with Spanish translation, edited by Antonio Laborda.

PRINCIPLES GOVERNING THIS TRANSLATION

Our translation, the first in English of Isidore of Seville's *De natura rerum*, is based on the Latin text edited by Jacques Fontaine, under the title *Traité de la nature* (1960). For details, see Bibliography. Fontaine provides a facing-page translation into French. Isidore's text has been translated into Spanish by Antonio Laborda (see Editions, above). There is also an Italian translation, based on Fontaine's edition, by Francesco Trisoglio, *La natura delle cose* (Rome: Città Nuova, 2001).

In the text of our Introduction, we use the English title, *On the Nature of Things*; in the notes to the Introduction and the translation, and in the Inventory of Manuscripts, Commentaries, and Appendices, we use the abbreviation of the Latin title, *DNR*.

Isidore rarely quotes his sources absolutely word-for-word; he prefers to paraphrase, allowing for variations in wording and/or syntax and sometimes for additional comments of his own. We use italics to signal material clearly borrowed from the source indicated in the footnote, but not necessarily to imply verbatim copying. We do not include references to possible sources for material that he neither quotes nor paraphrases. For these, see Fontaine's edition.

Two-thirds of Isidore's biblical quotations are from the Vulgate text; the remainder are more or less close to the *Vetus Latina*, the Septuagint, or Isidore's source. See Fontaine, *Traité*, 13–14. When the quotation is from the Vulgate, we generally base our translation on the English of the Douay-Rheims version, modernized or altered slightly as appropriate. Quotations not from the Vulgate are flagged as 'not Vulgate'.

The Text

ISIDORE OF SEVILLE:
ON THE NATURE OF THINGS

Here begins
The Book of the Nature of Things
by Isidore, Lord Bishop of Seville,
addressed to King Sisebut. /**167**/

PREFACE:
ISIDORE, TO HIS LORD AND SON, SISEBUT

1. Although I am well aware that you are distinguished for the gift of eloquence and the flowering variety of your learning, nevertheless you are making an even more wide-ranging study, and you demand that I aid you in certain matters concerning the nature and causes of things.[1] And, after perusing the scholarly monuments of our predecessors, I hasten to satisfy your searching mind by explaining in some measure the reckoning of the days and the months, as well as the periods of the year and the alternation of the seasons, and the nature of the elements, and finally the courses of the sun and the moon and the properties of certain stars, that is to say, the signs of tempests and winds, and also the situation of the earth and the alternating tides of the sea.

2. I have addressed all these matters in a brief document, presenting them just as they were formulated by the scholars of antiquity and especially in the works of catholic authors. For to know the nature of these things is not superstitious knowledge, as long as they are investigated in accordance with sound and sensible teaching. Indeed, if they had no connection at all with inquiry into truth, that wise king would never have /169/ said: 'He has given me true knowledge of the things that are, that I may know the disposition of the heavens,[2] and the virtues of the elements ... the alterations of their courses, and the changes of seasons, the revolutions of the years, and the dispositions of the stars'.[3]

3. For this reason, I begin with the day, the creation of which appears nearly first in the order of visible things, and then take up in order the remaining issues about which I know that some pagan and Christian authors have expressed opinions, setting down on certain contested points both what they mean and what they say, so that their authority may inspire confidence in my statements.

Here the Preface ends.

---

1 See the comments of Fontaine, 'Isidoro de Sevilla', 38 and n. 56.
2 The received text of the Vulgate reads *orbis terrarum*, 'of the whole world'.
3 Wisd. 7:17–19.

LIST OF CHAPTERS

And here the list of chapters begins:

1 Days

2 Night

3 The Week

4 The Months

5 The Concordance of the Months

6 The Years

7 The Seasons

8 The Solstice and the Equinox

9 The World

10 The Five Circles of the World

11 The Parts of the World

12 Heaven and Its Name /**171**/

13 The Planets of Heaven

14 The Heavenly Waters

15 The Nature of the Sun

16 The Size of the Sun and the Moon

17 The Course of the Sun

18 The Light of the Moon

19 The Course of the Moon

20 The Eclipse of the Sun

21 The Eclipse of the Moon

22 The Course of the Stars

23 The Position of the Seven Wandering Stars

24 The Light of the Stars

25 The Fall of the Stars

26 The Names of the Stars

27 Whether the Stars have a Soul

28 Night

29 Thunder

30 Lightning

31 The Rainbow

32 Clouds

33 Rains

34 Snow

35 Hail

36 The Nature of the Winds

37 The Names of the Winds

38 Signs of Storms or Fair Weather

39 Pestilence

40 The Ocean's Tide

41 Why the Sea Does Not Grow in Size

42 Why the Sea has Bitter Waters /**173**/

43 The River Nile

44 The Names of the Sea and the Rivers

45 The Position of the Earth

46 Earthquake

47 Mount Etna

48 The Parts of the Earth

Here the list of chapters ends.

1 DAYS

1. Day is the presence of the sun between its rising and its setting. However, the word can be understood in two ways. Technically, it means the time from one sunrise to the next; loosely, *it refers to the time between the rising of the sun and its setting.*[1] The day [in the technical sense] is divided into two periods, daytime and night-time; a day consists of twenty-four hours, and each period, of twelve hours.

2. The three parts of the day, loosely speaking, are morning, midday, and evening. [In the technical sense] some people reckon the beginning of the day from the rising of the sun, others from its setting, still others from midnight. The Chaldeans put the beginning of the day at sunrise, calling that whole period one day. But the Egyptians take the start of the following day from nightfall. The Romans prefer to begin and end the day at midnight.[2]

3. *At the beginning of the works of God, the day took its commencement from light,*[3] */175/ to signify the fall of man; but now it takes it from darkness into light, in order that day not be darkened into night, but that night may be lightened into day, just as it was written that 'light began to shine from the darkness',*[4] *because freed from the darkness of his offences man attains to the light of faith and knowledge.*[5]

[The Mystical Addition][6]
Mystically, moreover, the day bears the image of the Law. For just as the splendour of the day illuminates the obscurity of the darkness, so likewise the Law, pointing out the way of life, dispels the darkness of errors, reveals the light of virtues, and, reproving the sins of evildoers, leads the good

1 Hyginus, *De astronomia* 4.19, ed. Ghislaine Viré (Stuttgart: Teubner, 1992), 157.

2 Isidore's distinction between the technical and colloquial use of the term 'day', and his comments on the customs of different peoples as to when the day begins, are replicated in *Etym.* 5.30.1 and 5.30.5.

3 Cf. Gen. 1:5. Sc. 'and ended it at darkness'. Augustine makes explicit what Isidore leaves to be understood.

4 2 Cor. 4:6 (not Vulgate).

5 Augustine, *Quaestiones Evangeliorum* 1.7 (*PL* 35:1325).

6 The mystical addition is found only, but not always, in MSS of the long recension. See Introduction, pp. 49–50.

to better things. And these are the feast days in the Old Law: *The day of unleavened bread, Passover,*[7] is the fourteenth day of the first month, when, after the old leaven had been thrown away, a lamb was sacrificed at the height of the full moon. *Pentecost is the day on which the Law was given to Moses on the summit of Mount Sinai, when the first loaves of the harvest were offered as showbread.*[8] The Sabbath is the day on which is celebrated a time of rest according to the Law, and when it was not permitted to gather manna in the desert. The day of the Neomenia is the celebration of the new moon. For the Jews always celebrated a feast day at the beginning of months, that is, on the first day of the moon. But they used to do this at the beginning of the month because a period of time ends with the disappearance of the moon, and begins again when the moon reappears. The day of the trumpets is the beginning of the seventh month when the Jews, celebrating the festival, used to play longer upon the trumpet and offer more sacrifices than in each of the other months.[9] *The day of the scenopegia is in the seventh month, and the fifteenth day of the month, and on this day the ancients celebrated the feast of the Tabernacles.*[10] *For scenopegia means 'tabernacles'.*[11]

The day of the fourth fast is in the month of July,[12] *on the seventeenth* /**177**/ day of the same month, when Moses, descending from the mountain, broke the tablets of the Law.[13] It was also on that day that Nebuchadnezzar first destroyed the walls of the city of Jerusalem.[14] The day of the fifth fast is in the month of August, when discord arose in the camp of the Hebrews on account of the spies which they sent to the Holy Land, and the consequence was that they wandered wearily through the desert for forty years and in

---

7 Jerome, *Commentaria in Zachariam* 1.1, ed. Marcus Adriaen, CCSL 76A (Turnhout: Brepols, 1970), 750.

8 Jerome, *Commentaria in Zachariam* 1.1 (Adriaen, CCSL 76A:750) (cf. Ex. 25:30) = *Etym.* 6.18.4. Isidore's *panis propositionis* is the Vulgate phrase; the Hebrew literally is 'bread of the Presence'. We borrow 'showbread' from the King James version of the Bible.

9 There is a buried reference here to Ps. 80:4(81:4): 'Blow up the trumpet on the new moon, on the noted day of your solemnity'. Cf. *Etym.* 6.18.11, where this verse is quoted.

10 Jerome, *Commentaria in Osee* 3.12.10, ed. Marcus Adriaen, CCSL 76 (Turnhout: Brepols, 1969), 137.

11 Jerome, *Commentaria in Zachariam* 1.1 (Adriaen, CCSL 76A:750).

12 Zech. 8:19. It is tempting to translate: 'the day of *the fast of the fourth month*', which is what Zechariah says, and probably what Isidore intends to be understood. July corresponds to the fourth month of the Jewish calendar (see n. 17, below).

13 Ex. 32:19. The biblical account does not mention a month or day.

14 2 Kings/4 Kings 25:1–7; Jerem. 39:1–8. According to Jeremiah, it was the fifth day of the fourth month.

those forty years they all perished in that waste.[15] *And furthermore in the same month the temple was cast down and burned and, as a reproach to a lost people, ploughed under by Nebuchadnezzar,*[16] *and much later by the Emperor Titus. The day of the seventh fast is in the seventh month, which is called October,*[17] *in which Gedaliah was killed and the rest of the people who were in Jerusalem were slain, according to what Jeremiah says.*[18] *The day of the tenth fast is in the tenth month, which we call January, when all the captives in Babylon learned that the temple had been destroyed in the fifth month, and they made lament and fasted.*[19] The blessed Jerome wrote these things in his commentaries on Zechariah.[20]

[End of the Mystical Addition]

In a prophetic sense, day signifies knowledge of the divine Law, and night the blindness of ignorance, according to the prophet Hosea, who says: 'I have made your mother similar to the night; my people have been made like one /**179**/ having no knowledge'.[21] *Likewise day sometimes signifies worldly prosperity, and night suffering.*[22]

4. Court [*fasti*] days are those on which judgement is spoken [*fatur*], that is, pronounced, just as non-court days are those on which it is not. Festival days are those on which a religious service takes place and people must abstain from lawsuits. Working days are the opposite of festival days, that is, they are without religious ceremony; festival days are similarly given over completely to rest and religious duty. Unlucky days are those that are also called common. Sidereal days are those on which the stars (*sidera*) advance and men are hindered from navigation.[23] There are thirty

15 Num. 14:33–36.

16 2 Kings/4 Kings 25:8–9; Jerem. 52:12–14.

17 Christian chroniclers equated Nisan, the first month of the Hebrew religious calendar, with April. Therefore, the seventh month would be October.

18 2 Kings/4 Kings 25:25; Jerem. 41:2–3.

19 Zech. 8:19.

20 Jerome, *Commentaria in Zachariam* 2.8.18–19 (Adriaen, CCSL 76A:819–821).

21 Hosea 4:5–6 (not Vulgate).

22 Augustine, *Enarrationes in Psalmos* 138.16, ed. Franco Gori. CSEL 95/4 (Vienna: Verlag der Österreichischen Akademie der Wissenschaften, 2002), 146.

23 Presumably, this is a reference to the belief that the rising or setting of certain stars or constellations is a sign of good or bad weather (cf. Isidore, *Etym.* 3.71.4–5 (Gasparotto and Guillaumin, 3.70.4–5)). The element of reflection or foresight is echoed in Isidore's *Differentiae* I (*Diferencias Libro I*, ed. Carmen Codoñer (Paris: Belles Lettres, 1992), 86): 'Sidera illa dicuntur quibus nauigantes considerant quod ad cursum dirigant consilium' ('*Sidera* are what sailors take into account (*considerant*) when they plot their course').

consecutive 'just' days.[24] Battle days are those on which it is lawful to attack an enemy in war. The Book of Kings testifies as to these, saying: 'at the time when kings go forth to war'.[25]

5. There are five intercalary days, which according to the Egyptians are left over after twelve months, and which begin with the ninth calends of September (24 August) and end with the fifth calends of the aforesaid month (28 August).[26] There are eleven days of epacts, which are added each year to the course of the moon. For while twelve lunar months have 354 days in a year, eleven days remain to make up the course of the solar year, and these the Egyptians called epacts,[27] /181/ because they must be added to find the day of the moon throughout the whole year. The solstitial days are those on which the sun comes to a standstill, and they are preceded or followed by the growing length of the days or the nights. The equinoctial days are those on which the day and the night unfold with hours of equal length.[28]

2 NIGHT

1. *Night is the absence of the sun from sunset until it returns again at sunrise.*[29] *And night, which we believe was given for the repose of the body and not for the performance of any work, is caused by the shadow of the earth.*[30] Night is understood in two ways in the Scriptures, that is, either it signifies the tribulation of persecution or the darkness of a blinded soul. Night [*nox*] derives from harming [*nocendo*], because it is harmful [*noceat*] to the eyes.[31]

---

24 A period of 30 days was allowed by law for the payment of debt (Lewis and Short).

25 2 Kings/2 Sam. 11:1.

26 This definition or explanation of intercalary days assumes that the Egyptians had a 365-day year and a calendar of twelve thirty-day months. Hence five additional days would be needed to bring the total days to 365. Isidore's August dates for the intercalation are based on the Julian calendar. For a fuller explanation, see below, *DNR* 4.7 and n. 69. In the *Etymologies*, written after *DNR*, Isidore explains that intercalary days 'are so-called because they are interposed so that the reckoning of the sun and the moon may be reconciled' (*Etym.* 6.17.28; trans. Barney, et al., *Etymologies*, 145).

27 One wonders whether 'Egyptians' is Isidore's shorthand way of referring to Alexandrian computists. 'Epacts' is a Greek word.

28 See ch. 8 and its Commentary for explanation of solstitial and equinoctial days.

29 Hyginus, *De astronomia* 4.19 (Viré, 157).

30 Ambrose, *Hexaemeron* 4.11, ed. Karl Schenkl, CSEL 32.1 (Vienna, 1897), 118.

31 = Isidore, *Etym.* 5.31.1.

2. There are seven parts of the night: dusk, eventide, the silent hours, the dead of night, cockcrow, and morning twilight.[32] The time that we say is doubtful, that is, between light and darkness, is called dusk [*crepusculum*], that is, dark [*creperum*].[33] Eventide [**uesper**um] takes its name from the star of that name which rises in the East.[34]

3. The silent hours [*conticinium*] are when everyone is hushed, for to cease speaking [*conticescere*] is to be silent.[35] The dead of night, that is, the unseasonable hour, is when nothing can be done and /183/ everything is quiet.[36] Cockcrow [*gallicinium*] is so-called on account of cocks [*galli*], the harbingers of light.[37] Morning twilight occurs between the departure of night and the advent of day.[38]

3 THE WEEK

1. Among the Greeks and the Romans the week completes its course in seven days. But among the Hebrews a week is seven years. Daniel makes this clear with respect to the seventy weeks.[39] The week consists of seven *feriae*. The *feria* takes its name from the verb *fari* ['to speak'] as if it were pronounced *faria*, because in the creation of the world God said 'let it be made' [*fiat*] on each separate day, and also because the day of the Sabbath is considered to have been a holiday [*feriatus*] from the beginning. Hence the day of the sun is called the first *feria*, because it is the first day after the day of rest [*feria*]. Similarly, the day of the moon is called the second *feria*, because it is the second day after the day of rest, that is, after the Sabbath, which is a holiday. And so the rest of the days take their names from a similar count.

---

32 = Isidore, *Etym*. 5.31.4. In *DNR*, Isidore gives seven words for the 'seven' parts of the night, but *crepusculum matutinum* is only one part, as is clear from the final sentence of the next paragraph, so he lists only six parts in all. There is no good explanation for this apparent contradiction. See Fontaine, *Traité*, n. 26 (pp. 339–40). In *Etym*. 5.31.4, the seven parts are: *uesper, crepusculum, conticinium, intempestum, gallicinium, matutinum*, and *diluculum*.

33 = *Etym*. 5.31.6–7.

34 i.e., **Vesper**, the evening star, by which Isidore probably understands Mars. For Isidore's identification of Vesper with Mars, see below, *DNR* 3.2, and *Etym*. 5.30.6.

35 = *Etym*. 5.31.8.

36 = *Etym*. 5.31.9.

37 = *Etym*. 5.31.11.

38 = *Etym*. 5.31.12.

39 Dan. 9:24.

2. However, among the Romans the days take their names from the planets, that is, the wandering stars. For the first day is called after the sun, which is the chief of all the heavenly bodies, just as this day is the head of all the other days.[40] The second day is called after the moon, which is nearest the sun both in brilliance and in size and borrows its light from the latter.[41] The third day is named for the star of Mars, which is called Vesper.[42] /185/

3. The fourth day is called after the star of Mercury, which some call the white circle.[43] The fifth day is named for the star of Jupiter, which they call Phaëthon.[44] The sixth day is named for the star of Venus, which they claim is Lucifer, and which of all the stars has the most light.[45] The seventh day is called after the star of Saturn, which, having been placed in the sixth[46] heaven, is said to complete its orbit in thirty years.[47]

4. And therefore the pagans gave the days the names of these seven stars, because they believed that they were to some extent formed from them, saying that they had[48] spirit from the sun, body from the moon, language and wisdom from Mercury, pleasure from Venus, fervour from Mars, temperance from Jupiter, and sluggishness from Saturn.[49] Such indeed is the foolishness of the pagans, who fashion such ridiculous fictions for themselves.[50]

---

40 = *Etym.* 5.30.5.

41 = *Etym.* 5.30.6. See also ch. 18, below.

42 = *Etym.* 5.30.6. See n. 34, above.

43 = *Etym.* 5.30.6; cf. Hyginus, *De astronomia* 2.42–43. The 'white circle' is a confused reference to the *circulus lacteus*, the Milky Way. The story is that Mercury laid the infant Hercules on the breast of Juno while she was asleep, so that he could suckle. When Juno woke up, she shoved baby Hercules away, spewing her milk across the heavens. See *Etym.* 5, ed. Yarza Urquiola and Andrés Santos, 90 n. 6.

44 = *Etym.* 5.30.7. Another unexplained confusion: Phaëthon was the son of Helios the sun god, and was sometimes used as an epithet for the sun.

45 = *Etym.* 5.30.7.

46 Possibly a blunder for 'seventh', or possibly deliberate. It is recorded in all the MSS, and repeated in the *Etymologies*. See Fontaine, *Traité*, n. 32 (p. 340), who suggests that Isidore may have wished to compare the spheres of the planets to the ages of the world. He also suggests that Isidore may have wanted to avoid identifying Saturn with the 'seventh heaven', i.e., the highest heaven reserved for spiritual beings; but we find no evidence that 'seventh heaven' was a proverbial term for the highest heaven in any ancient or biblical source; indeed, it appears to be of Jewish or Islamic origin.

47 = *Etym.* 5.30.7. The planets and their periods are also discussed in ch. 23, below.

48 Fontaine prints *ex aere ignem* after *habere*, a reading which is supported by only five of the 17 MSS he consulted. He omits the phrase in his French translation as we do in ours.

49 = *Etym.* 5.30.8.

50 = *Etym.* 5.30.8.

4 THE MONTHS

1. The month is the cycle and renewal of the lunar light, in other words, the period from new moon to new moon.[51] The form of the [cycle] is often understood as the course of this life, which advances like the moon by definite increments,[52] and comes to an end by equally precise reductions. *The ancients defined the month as the length of time it takes the moon to traverse the circle of the zodiac.*[53]

2. The pagans of antiquity gave some names to the months from their gods /187/, some from their characteristic conditions, and some from their number, beginning with March, because they observed the order of the new year from it. *They called this month March in honour of Romulus, because they believed he was the son of Mars. But they named April not from any name of their gods, but from its characteristic property, as though it were called 'Aperil', because then especially there is an opening up [**aperiatur**] from seeds into flower.*[54]

3. *Next the month of May is named for Maia, the mother of Mercury, whom they worship as a goddess,*[55] or for their ancestors [**maiores**].[56] *Then they name June after Juno,*[57] whom they assert was the sister and wife of Jupiter.[58] Others, however, have said that June [*Iunium*] is so-called for young persons [*iuniores*], like May for ancestors. Likewise, they named July for Julius Caesar, and August for Octavian Augustus. For they used to call July the fifth month (*Quintilis*) and August the sixth month (*Sextilis*), but their names were replaced by those of the Caesars Julius and Augustus.

---

51 In modern terms, the time between successive new moons (29.531 days) is the moon's synodic period. See Mark R. Chartrand, *National Audubon Society Field Guide to the Night Sky* (New York: Alfred A. Knopf, 1991), 635.

52 Fontaine, *Traité*, n. 37 (p. 341), proposes translating *mensis* as an adjective modifying *incrementis*: 'which advances by as it were "measured" increments', and sees in this a punning reference to the seven phases of the moon, the full development of an embryo of seven days, and the seven 'weeks' of years that lead to the onset of old age, as enumerated in Isidore's *Liber numerorum* 8.45–46 (ed. Guillaumin, 53–55).

53 Hyginus, *De astronomia* 4.19 (Viré, 157). In modern terms, this is the sidereal period of the moon – the length of time (27.322 days) it takes the moon to make a complete revolution of the earth with respect to the background of the stars. See Chartrand, *Field Guide to the Night Sky*, 635.

54 Augustine, *Contra Faustum* 18.5, ed. Joseph Zycha, CSEL 25.1 (Vienna, 1891), 494.

55 Augustine, *Contra Faustum* 18.5 (Zycha, CSEL 25.1:494).

56 = *Etym.* 5.33.8.

57 Augustine, *Contra Faustum* 18.5 (Zycha, CSEL 25.1:494).

58 Cf. Vergil, *Aeneid* 1.47, and Augustine, *DCD* 4.10.

4. Now September gets its name because it is the seventh from March, which is the beginning of spring [*uer*]. In a similar order according to number, October, November, and December took the name of rain [pl. < *imber*] and spring.[59] /189/ *Then they called January from the name of Janus,*[60] *but specifically January [Ianuarium] is so-called because it is the gateway [ianua] and beginning of the year.*[61] *They named February after the festival Februa, the sacred rites of the Luperci.*[62] Thus the ancient Latins used to reckon the cycle of the year as ten months. But the Romans added January, and Numa Pompilius[63] added February and divided the year into twelve months.

5. Most authorities, moreover, assert that Sancus, the king of the Sabines,[64] first divided the year into months, and instituted the ides, calends, and intercalary days. But in the books of the holy Scriptures the twelve months of the year are shown to have existed even before the Flood.

59 Fontaine, *Traité*, n. 40 (p. 341), provides an ingenious and satisfying explanation of these two obscure sentences, based on the Late Latin confusion of *b* and *v*, and Isidore's attempt to find a double etymological relationship between the suffix –*ber* and *uer* ('spring') as well as *imber* ('rain'). The association of the suffix with rain was part of the grammatical tradition, and is found, e.g., in Cassiodorus's *Variae*. What Isidore does that is new is to connect -*ber* with *uer* because the months once upon a time began with March '*qui est principium ueris*' and September through December are numbered with respect to March. In other words, September = 'seventh from spring' as well as 'seven-rainy'. As Fontaine observes, Isidore is really stretching things here!

60 Augustine, *Contra Faustum* 18.5 (Zycha, CSEL 25.1:494).

61 Jerome, *Commentaria in Zachariam* 2.8.18–19 (Adriaen, CCSL 76A:820).

62 Augustine, *Contra Faustum* 18.5 (Zycha, CSEL 25.1:494). The Luperci were priests of Lupercus, who was surnamed Februus. The festival of expiation was celebrated on the fifteenth of the month.

63 The second king of Rome (according to tradition, 715–673 BC).

64 For Sancus, see Augustine, *DCD* 18.19, and Fontaine, *Traité*, 341–42 n. 41. Sancus was a Sabine hero; Augustine claims he was the first king of the Sabines, and was divinized after death. No authority has been traced which links him to the calendar, however. Isidore may have confused Augustine's 'Sancus' with Cingius, a Roman jurist and grammarian of the late Republican or early Imperial period whose history of the calendar, *De fastis*, is quoted extensively (and with the author's name correctly rendered) in Macrobius's *Saturnalia* 1.12.2–8, 12.20–21. This section of the *Saturnalia* circulated on its own in the early medieval period under the title *Disputatio Chori* (or *Hori*) *et Praetextati*; it was well known in seventh-century Ireland, as well as to Bede. One late seventh-century Irish *computus* text, *De ratione conputandi*, identifies Macrobius's Cingius with Isidore's Sancus by claiming that 'Gingius' was the king of the Sabines who devised the division of the Roman months: Maura Walsh and Daíbhí Ó Cróinín (eds.), *Cummian's Letter De contro-versia Paschali and the De ratione conputandi* (Toronto: Pontifical Institute of Mediaeval Studies, 1988), 136.

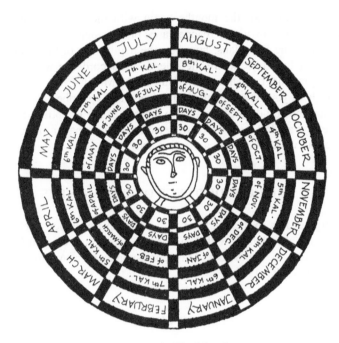

Diagram 1: The Months

So, for example, we read there: '*and the water decreased until the eleventh month; and in the eleventh month, the first day of the month, the tops of the mountains appeared*'.[65] *Thus the months were numbered then just as they are now; but they were delimited not by the calends, but by the beginning and end of the lunation.*[66]

6. The calends took their name from 'worshipping' [*colendo*]. For among the ancients the beginnings of the months were always a time for worship. The ides [*idus*] in turn took their name from 'days' [*dies*],[67] /191/ and the nones from the 'nundine' [*nundinae*, the ninth or market-day].[68]

65 Gen. 8:5 (not Vulgate), from Augustine, *De ciuitate Dei* 15.14, ed. Bernhard Dombart and Alphonse Kalb, CCSL 48 (Turnhout: Brepols, 1955), 473. In the Vulgate, it is the *tenth* month, which would not have suited Isidore's (or Augustine's) argument so well.

66 Augustine, *DCD* 15.14 (Dombart and Kalb, CCSL 48:473).

67 In *Etym.* 5.33.13, Isidore derives *idus* from 'eating' [*edendo*].

68 = *Etym.* 5.33.14.

Among the Latins all the months begin with the calends; among the Hebrews they begin with the reappearance of the moon.

7. Among the Egyptians, however, the beginnings of the months are fixed four or five days before the calends, in accordance with what the accompanying diagram shows [Diagram 1: The Months].[69]

From here you return to the 4th calends of September, and by such calculation the 360 days of the twelve months of the Egyptians are filled out. Five days are left over, which were called epagomenal or intercalary or added, about which there was a mention above.[70]

5 THE CONCORDANCE OF THE MONTHS

1. January accords in the length of its hours with December.[71] February's hours are equal in length with November's. March agrees with October. April is equal to September. May corresponds with August. June is comparable to July. /**193**/

6 THE YEARS

1. The year is the circuit of the sun and its return over a period of twelve months. Its name signifies figuratively the whole time of this life, as when Isaiah says: '[the Lord sent me] to preach the acceptable year of the Lord'.[72] *For the year in which the Lord preached was not the only one acceptable, but also the entirety of time, according to what the Apostle says: 'Behold, now is the acceptable time'.*[73] *Finally, he added the Day of Judgement as the*

---

69 See fig. 1 in Fontaine, *Traité*, facing p. 190. The diagram gives the first day of each of the twelve months of the Egyptians according to their dates on the Julian calendar. The first Egyptian month (Thoth) began on the fourth calends of September (29 August) in non-leap years (as Bede states in *DTR* 11 [Faith Wallis, *Bede: The Reckoning of Time*, TTH 29 (Liverpool: Liverpool University Press, 1999), 45] and Isidore probably intended the diagram to be oriented with September at the top. Each Egyptian month had 30 days. The last Egyptian month would therefore end on 23 August, and five intercalary days (six in the year preceding the Julian leap year) would be needed to fill the gap between it and the beginning of the next year (360 + 5 = 365 days).

70 See above, *DNR* 1.5.

71 See Commentary.

72 Isa. 61:2 [+ 58:5?] (not Vulgate).

73 2 Cor. 6:2.

end of this year, saying: '[the Lord sent me] *to preach the year of the Lord and the day of retribution'.*[74]

2. There are those who think that the year is called *annus* as if it were a ring [*anus*], that is, a circle. This is why finger-rings are called *anuli*, using the diminutive form. Some, like the Roman people, think the year begins with winter; others, like the Hebrews, with the spring equinox; others, like the Greeks, with the [summer] solstice; and still others, like the Egyptians, with autumn. The sages of this world have said that the word year refers sometimes to the civil year, sometimes to the natural year, and sometimes to the great year. The civil year is that which is ended by the cyclical return of a single star over twelve months.

3. The natural year is when the moon positions itself under the sun, so that, having been interposed halfway between the disk of the sun and our eyes, it darkens the entire disk, which is called an eclipse.[75] The reason for this was obscure for a long time, /**195**/ but it was explained by a certain Milesian philosopher.[76] It is said to be a great year when all the heavenly bodies, after the completion of a fixed number of temporal periods, come back to their original place and order. The ancients claimed that this year was completed or filled out after nineteen years.[77]

4. *The solstitial year is when the sun, after completing its circuit through all the signs* [of the zodiac], *returns to the one from which it took the beginning of its course.*[78] This is the solar or civil year, which is completed in 365 days. The common lunar year is the one that runs through twelve lunar months, or 354 days. The embolismic year is that which is assigned thirteen lunar months and 384 days.[79] In that year Easter day is delayed to

74 Jerome, *Commentaria in Isaiam* 17.61, ed. Marcus Adriaen. CCSL 73A (Turnhout: Brepols, 1963), 707–08: Luke 4:19 (cf. Isa. 61:1–2). Jerome borrows the words of Luke (and/or Isaiah) without acknowledgment; Isidore restores them to their source.

75 In the *Etymologies* 5.36.3, Isidore gives the three kinds of years as the lunar year of 30 days, the solstitial year of 12 months, and the great year. For a discussion of why Isidore chose to include the eclipse period here, and to call it a 'natural year', see Commentary.

76 The Milesian philosopher is Thales; see Augustine, *De ciuitate Dei* 8.2, ed. Bernhard Dombart and Alphonse Kalb, CCSL 47 (Turnhout: Brepols, 1955), 217. See Commentary for further discussion.

77 Isidore does not give a figure for the term of the great year in the *Etymologies*.

78 Ambrose, *Hex.* 4.5.24 (Schenkl, CSEL 32.1:131). This definition of a 'solstitial year' might tempt one to suppose that Isidore (and Ambrose?) intended to refer technically to the solar year of 365¼ days in contradistinction to the everyday language of the common year or *annus communis* (365 days) and leap year or *annus bissextilis* (366 days). But the phrasing of the next sentence appears to rule this out.

79 = *Etym.* 6.17.21–22.

a later date. The leap year is the one to which is added the sum total of one day acquired over a period of four years from the reckoning of a quarter of a day per year. The Jubilee year is a year of remission, which is set at the end of seven weeks of years, that is, forty-nine years, in which, according to the Law, the trumpets were sounded, and property formerly held reverted to all [their previous owners].[80]

5. The Olympiad among the Greeks is the fourth year[81] after the Olympic competition, which takes place after the passage of four years. The end of this year is chosen as the time of this competition on account of the course of the sun over a period of four years, and because in the quadrennium one full day is formed by taking from each of the years the extra three[82] hours. At this time they used to send messengers around the city-states to announce that people should gather together not only from all parts, but also from every class, age, and sex.

6. The lustrum is a period of five years among the Romans. It is called a lustrum because in the Republic the census was conducted every five years, and after its completion, and after a sacrifice had been performed, the city of Rome was purified [*lustrare*]. The Romans created the indictions, which, advancing through successive years up to the fifteenth, return again to the beginning of the first year.

7. The era was instituted in the time of Caesar Augustus. It was called an era [*aera*] because the whole world promised to render tribute [*aes, aeris*] to the state.[83] The era takes its start from the day of the calends of January. In the course of the moon, the leap-year day [*bissextus*] is added between the 6th [*sextus*] nones of March [2 March] and the day before the calends of January [31 December].[84] The year of the Egyptians without the

---

80 = *Etym.* 5.37.3.

81 'Olympiad' usually refers to the period of four years between celebrations of the Olympic games, just as Isidore defines it in *Etym.* 5.37.1. Here, for whatever reason, he seems to restrict its meaning to the fourth year only of that quadrennial period.

82 The manuscript evidence unanimously supports *trium*. Isidore apparently imagines three hours of daylight being left over after 365 days, which after four years will add up to one extra day (as opposed to night). Or he nodded, or *ui* in the archetype was miscopied as *iii*. The point is (or should be) that six hours are left uncounted for at the end of each 365-day year. Bede comments on this error in *DTR* 39.

83 = *Etym.* 5.36.4.

84 = *Etym.* 6.17.27. In the *Etymologies*, Isidore adds the phrase *atque inde detrahitur*, 'and is then subtracted'. This strange sentence seems to be an attempt to find a way to apply the bissextile day of the Julian calendar to a lunar calendar (see Bede, *DTR* 41, on incorporating the leap year day into lunar reckoning). In the Julian calendar, the bissextile day is the sixth

bissextile day begins on the 4th calends of September [29 August], but with the bissextile day it begins on the 3rd calends of the aforesaid month [30 August]. /199/

7 THE SEASONS[85]

1. According to Ambrose: *the seasons are an alternating succession of changes,*[86] *by means of which the sun, by the fixed dimension of its course, adorns the circuit of the year* with *a variety that is devoid of confusion.*[87] *For the seasons derive from the motion of the heavenly bodies. Hence, God, when he set them in place, said: 'and let them be for signs and for seasons and for days and for years',*[88] *that is, they are in constantly changing movement; one of them advances earlier, another later, because they cannot be together at the same time.*[89] According to the Hebrews, a season is a whole year, as expressed in that passage in Daniel: 'a season, and seasons, and half a season',[90] signifying by season a year, by seasons two years, and by half a season six months.[91]

2. *According to the Latins, however, four seasons are assigned to one year: winter, spring, summer, and autumn. It is winter when the sun tarries in the southern regions. Then the sun is furthest away, and the earth grows rigid and hardened with frost, and the lengths of the night are more extended than those of the day. This is the reason that an excessive quantity*

calends of March (24 February), which day is 'repeated', i.e., there are two twenty-fourths of February.

85 Isidore's title for this chapter is *De temporibus*, literally 'On times'. English cannot express the ambiguity of the Latin *tempus*, which may refer to time in general, or to a particular period of time, like a season. It is clear that Isidore's topic in this chapter is the seasons of the year, and we have translated *tempus* accordingly (as does Fontaine in his French version). But Augustine's concern in the passage quoted by Isidore in the first paragraph is with the nature of time itself, not with the seasons. See also below (n. 91).

86 Ambrose, *Hex.* 4.5.21 (Schenkl, CSEL 32.1:121).

87 Clement/Rufinus, *Recognitiones* 8.21.2 (ed. Rehm, p. 230.10).

88 Gen. 1:14 (from Augustine, not Vulgate).

89 Augustine, *DCD* 12.16 (Dombart and Kalb, CCSL 48:371).

90 Dan. 12:7.

91 Cf. Jerome, *Commentaria in Danielem* 12.7, ed. F. Glorie, CCSL 75A (Turnhout: Brepols, 1964), 940. Jerome states explicitly that the phrase in Daniel means three and a half years. The Latin, '*tempus et tempora et dimidium tempus*', can also be translated as 'time, times and a half time' (as in the King James Bible), thus illustrating the ambiguity of the term *tempus* noted above (n. 85).

of snow and rain is hurled abroad by the winter winds. It is spring when the sun, withdrawing from the southern regions, returns above the earth and makes the length of the day and night equal and restores a moderate temperature to the air, and, nourishing everything, stimulates birth upon birth, so that the earth sprouts afresh and seeds released by ploughing /201/ spring to life, and the offspring of all the races which are on the lands and in the waters is propagated by yearly breeding.[92]

3. *It is summer when the sun ascends into the north and extends the duration of the day, but contracts and shortens the nights. And so the more the sun by constant intimacy embraces and mingles with this air, the more it heats the air itself; and, with the moisture drawn out of it,*[93] *the earth crumbles into dust*[94] *and makes the seeds grow and drives as it were the young sap to ripen the fruit of the orchards. Then while the sun blazes with summer heat, it makes shadows in the south shorter because it lights that region from high above.*[95]

It is autumn when the sun, descending again from the height of the sky, diminishes the strength of its fires and, after little by little slackening and laying aside its heat, produces cooler conditions,[96] with a subsequent maelstrom of winds and storm of tempests and violence of lightning and claps of thunder.

4. Since we have described the changes of seasons by means of particular distinctions according to the definitions of our predecessors, let us now explain how these same seasons are joined together by their natural linkages.[97] For, in fact, spring is composed of moisture and fire, summer of fire and dryness, autumn of dryness and cold, and winter of cold and /203/ and moisture. Hence also the seasons [*tempora*] are named from the

---

92 Ambrose, *Hex.* 4.5.21 (Schenkl, CSEL 32.1:127–28).
93 Ambrose, *Hex.* 4.5.21 (Schenkl, CSEL 32.1:128).
94 Ambrose, *Hex.* 2.3.12 (Schenkl, CSEL 32.1:52).
95 Ambrose, *Hex.* 4.5.21 (Schenkl, CSEL 32.1:128).
96 Ambrose, *Hex.* 4.5.23 (Schenkl, CSEL 32.1:131).
97 Fontaine emends the reading of the MSS, *circulis*, to *uinculis*. For his justification, cf. *Traité*, n. 62 (p. 344). The radical emendation, which we accept, improves the sense, as a cycle cannot join anything. But it remains to explain why *uinculis* was banalized to *circulis*. This change involves only two letters, and it is not difficult imagine palaeographical scenarios that could lead to their metamorphosis; but, whether it was an accidental or a deliberate alteration, *circulis* had to have made sufficient sense to enter the text tradition at a very early stage, and totally extinguish the original reading. One plausible theory, hinted at by Fontaine in his note, would point to the influence of the diagram itself, where the *linkages* between the elements (i.e., the shared paired qualities) are denoted by semi-*circles*.

Diagram 2: The Seasons

due proportion [**temperamentum**] of their mutual interaction.[98] This is the figure of this interaction [Diagram 2: The Seasons].[99]

5. These are the beginnings of the seasons: spring begins on the 8th calends of March [22 February], and lasts for 91 days; summer begins on the 9th calends of June [24 May], lasting 91 days; autumn takes its beginning on the 10th calends of September [23 August], lasting 93 days; and winter begins on the 7th calends of December [25 November], lasting 90 days. Hence there are 365 days in the revolving year. This also accords with the natural difference of the seasons.

6. On the other hand, according to allegory winter is understood as temporal tribulation, when the tempests and storms of the world are oppressive. Summer is the persecution of the faith, when doctrine is desiccated by the aridity of falsehood. But spring is the peace or the

98 = *Etym.* 5.35.1.
99 See fig. 2 in Fontaine, *Traité*, facing p. 202.

renewal of faith, when after winter's /205/ tribulation the tranquillity of the Church returns, when 'the month of the new [grain]',[100] that is, the Pasch of the lamb, is celebrated, and when the earth is adorned with flowers, that is, when the Church is adorned with the company of the saints.

7. Next, a recapitulation of what is said above. The year is regulated by the cycle of the sun and the months. The seasons unfold by a succession of changes. The month is produced by the waxing and waning of the moon. The week is bounded by a period of seven days. The day and the night are renewed by the alternating successions of recurring light and darkness. The hour is made up of certain intervals and moments.[101]

8 THE SOLSTICE AND THE EQUINOX

1. There are two solstices: the first is the winter solstice on the 8th calends of January [25 December], when the sun stands still[102] and the days lengthen; the second is the summer solstice on the 8th calends of July [24 June], when the sun comes to a halt and the nights lengthen. Opposite to these are the two equinoxes: the one is the spring equinox on the 8th calends of April [25 March], after which the days lengthen; the other is the autumnal equinox on the 8th calends of October [24 September], when the days shorten.

2. The solstice [*solistitium*, cl., *solstitium*] gets its name from the sun's standing still [*solis statio*], as it were, while the equinox gets its name because at that time the day and the night [*nox*] come back to an equality of twelve hours /207/ of equal durations. Moreover, the summer solstice is

---

100 *Mensis nouorum*: so in Exod. 23:15 and 34:18. See Calvin B. Kendall, *Bede: On Genesis*, TTH 48 (Liverpool: Liverpool University Press, 2008), 197 n. 307. The month is Nisan, the first month of the Hebrew calendar, some part of which, according to Bede, always falls in April: cf. Bede, *DTR* 11.

101 Isidore leaves units of time shorter than an hour undefined, both here and in *Etym.* 5.29.1. Bede, in contrast, refers to precise units of a quarter of an hour (a *punctus*, or 15 minutes), a tenth of an hour (a *minutum*, or 6 minutes), a fifteenth of an hour (a *pars*, or 4 minutes), and a fortieth of an hour (a *momentum*, or 1½ minutes). See Bede, *DTR* 3 (Wallis, *Reckoning*, 14–16) and Bede, *DT* 1 (Kendall/Wallis, *On Times*, 107). These divisions and their quantities were apparently the creation of early medieval Irish computists; see Maura Walsh and Dáibhí Ó Cróinín, *Cummian's Letter De controversia Paschali and the De ratione conputandi* (Toronto: Pontifical Institute of Mediaeval Studies, 1988), 123–24. Hence Isidore's use of the term 'moment' (*momentum*) is non-technical.

102 The rising and setting sun moves south on the horizon until the winter solstice when it appears to 'stand still' and then to begin moving north (and similarly with respect to the summer solstice).

called the 'lamp' because from that day the lamp of the sun takes on greater brightness and pours out the excessive heat of the summer coming on.

9 THE WORLD[103]

1. The world is the entire universe, which consists of heaven and earth. The apostle Paul says of it: 'for the fashion of this world passes away'.[104] In a mystical sense, the world properly signifies the human being, because just as the former is compounded out of four elements, so also the latter is made up out of four humours mixed together in one temperament.

2. Hence the ancients considered man to be closely connected with the structure[105] of the world, since in Greek the world is called the 'cosmos' and a human being is called the 'microcosm (*micros cosmos*)', that is, the lesser world,[106] although by world Scripture sometimes also means sinners, of whom it is said: 'and the world knew him not'.[107]

3. The design of the world is described as follows. Just as the world is elevated towards the north, so it is tilted down towards the south.[108] The eastern region is its head and as it were face, while the northern part is the last. For it has four parts. The first part of the world is the eastern; the second is the southern; the third is the western; and the last /**209**/ and the extreme is the northern, of which Vergil spoke as follows:

> Around which are extended the dark extremes
> On the right and the left, congealed with ice.[109]

103 The Latin term *mundus* may refer to the universe, as it does here, or to the terrestrial world, as it does in subsequent chapters. We translate it as 'world' throughout.

104 1 Cor. 7:31.

105 *fabrica*: This term, with its architectural connotations, was particularly favoured by Isidore to denote the universe. See Marek Hermann, 'Die astronomischen Metaphern in Isidors von Sevilla *Origines* und *De natura rerum*', *Analecta Cracoviensia* 38–39 (2006–2007): 443.

106 '[M]an is called ... the lesser world': cf. Solinus, *Collectanea* 1.93–94 (ed. Mommsen, 26). Solinus's term is *minorem mundum*. Isidore is the first Latin writer to use the transcribed Greek word *microcosmos* (cf. *Etym.* 3.23.2): Fontaine, 'Isidore de Séville et l'astrologie', 283 n. 5; *Thesaurus linguae latinae* v. 8 (Leipzig: Teubner, 1936–1976) pt. 1, col. 932.

107 John 1:10.

108 Isidore is referring to the north celestial pole (around which the heavens revolve) as being above the horizon (in the northern hemisphere) and the south celestial pole as being below. Cf. Vergil, *Georgics* 1.240–43. See Commentary.

109 Vergil, *Georgics* 1.235–36. The two extremes of which Vergil speaks are the Arctic

And Lucan:

> Thus lies the last part of the world, which
> The snowy zone and perpetual winter oppress.[110]

10 THE FIVE CIRCLES OF THE WORLD

1. In marking out the boundaries of the world, the philosophers say there are five circles, which the Greeks call parallels, that is, zones, into which the earth is divided. Vergil mentions these in the *Georgics*, saying:

> Five zones comprise the heavens.[111]

But let us imagine them in the likeness of our right hand, so that the thumb is the Arctic circle, uninhabitable because of the cold; the second finger is the summer[112] circle, temperate and habitable; the middle finger is the equatorial[113] circle, torrid and uninhabitable; the fourth finger is the winter[114] circle, temperate and habitable; the little finger is the Antarctic[115] circle, cold and uninhabitable.

2. The first of these circles is the northern; the second, the solstitial; the third, the equinoctial; the fourth, the winter; and the fifth, the southern. Concerning these, Varro speaks as follows: /**211**/

> The celestial orb is girdled by five zones
> And wintry cold devastates the ones at both ends and heats the middle.
> Thus the lands between the extremes and the middle are cultivated,
> Which the force of the sun never destroys by its strong fire.[116]

and Antarctic zones (projected onto the heavens). See below, ch. 10.

110 Lucan, *The Civil War* 4.106–07.

111 Vergil, *Georgics* 1.233.

112 Isidore's term is *therinos*, i.e., θερινὸς [τροπικὸς], 'summer circle'; cf. *Etym.* 3.44.2 (Gasparotto and Guillaumin, 3.43.2).

113 *Isemerinos*, i.e., ἰσημερινός, 'equatorial [circle]'; cf. *Etym.* 3.44.3, and ed. of Gasparotto and Guillaumin, 3.43.3, p. 106 n. 262.

114 Xeimerinos, i.e., χειμερινός, 'winter [circle]'. In *Etym.* 3.44.3 (Gasparotto and Guillaumin, 3.43.3) and 13.6.5, the fourth circle is *antarcticos*.

115 *Antarcticos*. In *Etym.* 3.44.4 (Gasparotto and Guillaumin, 3.43.4) and 13.6.6, the fifth circle is *xeimerinos*. On the inversion of the fourth and fifth circles in the *Etymologies*, see Commentary.

116 Varro Atacinus, fragment. This is one of several passages from different authors quoted by Isidore that are known only from Isidore's citation of them in *DNR*. See below (n. 329).

Diagram 3: The Circles of the World

A figure such as this distinguishes the divisions of these circles [Diagram 3: The Circles of the World].[117]

3. The equinoctial circle is uninhabitable because *the sun, as it courses through the middle of the heavens, creates heat in that region, so that not only do crops not grow there on account of the parched earth, but men are not able to live there on account of the excessive heat. But on the other hand the northern and southern circles,* which are associated with each other,[118] *are not inhabited because they are placed far from the course of the sun,*[119] *and are laid waste by the excessive cold of the heavens and the icy blasts of the winds.*[120]

117 See fig. 3 in Fontaine, *Traité*, facing p. 210.
118 This statement is explained visually by the wheel, which arranges the five circles in a circle, with the first and fifth circles touching each other. See Commentary.
119 Hyginus, *De astronomia* 1.8 (Viré, 11–12).
120 Ambrose, *Hex.* 2.4.16 (Schenkl, CSEL 32.1:56).

4. The solstitial circle that is located in the east between the northern and summer circle[121] and the one that is placed in the west between the summer and southern circle *are temperate because they /213/ have cold from one circle and heat from the other.*[122] Of these, Vergil says:

> Between these [outer zones] and the middle one, two zones have been granted
> As a gift of the gods to suffering mortals.[123]

But the people who are closest to the summer circle are the Ethiopians who are scorched by the excessive heat.[124]

11 THE PARTS OF THE WORLD

1. There are four parts[125] of the world: fire, air, water, and earth. This is their nature: fire is thin, sharp, and mobile; air is mobile, sharp, and thick; water is thick, blunt, and mobile; and earth is thick, blunt, and immobile. These are also mixed together as follows. Earth, which is indeed thick, blunt, and immobile, combines with the thickness and bluntness of water. Then water is united with the thickness and mobility of air. In turn air combines with fire by reason of their common properties of sharpness and mobility.[126] However, earth and fire are separated from each other, but they are connected by the two middle parts, water and air.[127] Therefore, lest

---

121 'Summer circle' in this section may refer to the equatorial circle; see Commentary on ch. 10.4.

122 Hyginus, *De astronomia* 1.8 (Viré, 13).

123 Vergil, *Georgics* 1.237–38.

124 Hyginus, *De astronomia* 1.8 (Viré, 12–13).

125 In the chapter title and in this first sentence, Isidore speaks of *partes* rather than the more usual *elementa*, which he employs below and in the corresponding passage in the *Etymologies* (cf. Bede, *DNR* 4, '*De elementis*'). Perhaps his use of *partes* here is intended to call to mind the descending hierarchy of the cosmic regions of fire, air, water, and earth as well as their status as elementary particles. *Pars* in Latin can mean either a region or a constituent.

126 'These are also ... sharpness and mobility': cf. *Etym.* 13.3.2, where Isidore states, 'Indeed, [the elements] are said to be joined thus among themselves with a certain natural logic, now returning to their origin, from fire to earth, now from earth to fire, since fire ends in air, and air is condensed into water, and water thickens into earth; and in turn, earth is loosened into water, water rarefied into air, and air thinned out into fire' (trans. Barney, et al., *Etymologies*, 272).

127 Isidore is pointing out that this schema of the elements is a hierarchy, not a circle. See Commentary.

Diagram 4: The Elements

these bewildering matters not be understood, I have represented them by the adjoining picture [Diagram 4: The Elements].[128]

2. On the other hand, this is how Saint Ambrose distinguishes these same elements according to the qualities with which they are mixed together by a certain kinship of nature. *Earth*, he says, *is dry and cold, water is cold and moist, air is hot and moist, and fire is hot and dry. Thus the separate elements are intermingled by these linkable qualities. For earth, although it is dry and cold, is linked to water by the affinity of their cold quality. In turn, water is linked to air by moisture, because air is moist. And water with, so to speak, its two arms of cold and moisture seems to embrace earth with the one and air with the other, earth with the cold one and air with the moist one.*[129]

---

128 See fig. 4 in Fontaine, *Traité*, facing p. 212. Fontaine prints the inscription of earth in the figure before that of water, which violates the logic of the statement above. The order, fire, air, water, earth, is clear in the figure in St Gall 238 (p. 337).

129 Ambrose, *Hex.* 3.4.18 (Schenkl, CSEL 32.1:71–72).

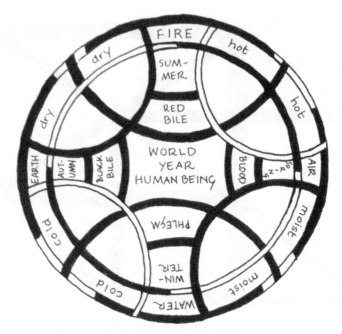

Diagram 5: The Macrocosm and Microcosm

3. *Likewise, air itself, intermediate between two naturally contending elements, that is, between water and fire, connects each of these elements to itself, since it is joined to water by moisture and to fire by heat. Fire also, since it is hot and dry, is connected by heat to air, but by its dryness is closely associated with earth, and so by this circuit, like a kind of dance in the round, the elements gather themselves together in a harmonious union. /217/ Hence, what in Latin are called elements are called* cena[130] ['common origins'] *in Greek, because they combine with one another and harmonize.*[131] The figure of the appended circle reveals the precise combination of the elements [Diagram 5: The Macrocosm and Microcosm].[132]

---

130 Fontaine's emendation of MSS *oena* (*era, eona*). See his discussion of the emendation and its assumed meaning, *Traité*, n. 74 (p. 347).

131 Ambrose, *Hex.* 3.4.18 (Schenkl, CSEL 32.1:72).

132 See fig. 5 in Fontaine, *Traité*, facing p. 216.

12 HEAVEN

1. *Heaven in a spiritual sense is the Church, which gleams in the night of this life with the virtues of the saints as if with the brightness of the stars.*[133] In the plural, moreover, all the saints and angels are understood by the term heaven, just as we must also take the heavens as the prophets and the apostles, of whom it is written: 'The heavens show forth the glory of God',[134] especially since these heavens proclaimed to the world the coming of Christ and his death, and likewise these proclaimed his resurrection and glory.

2. And Saint Ambrose in the books that he wrote on the creation of the world speaks in this way about the term heaven: *the Greek word for heaven is* uranus; *but by Latin speakers it is called heaven [caelum] because, having the lights of the stars imprinted upon it like signs, it is said to be engraved [caelatum], in the same way as silver, which gleams with signs in relief, is said to be engraved.*[135] For Scripture also proves that its nature /**219**/ is exquisitely fine, saying: 'that he shall make heaven firm like smoke'.[136]

3. Its parts are the vault, the axis, the clime, the pivots, the domes, the poles, and the hemispheres. The vault is that by which heaven is enclosed. Hence Ennius says:

Hardly to fill with terrors the entire vault of heaven.[137]

The axis is the straight line that extends through the middle of the ball of the sphere.[138] The clime is a region or part of heaven, such as the eastern clime and the southern clime. The pivots are the farthest parts of the axis. The domes are the extremities of heaven. *The poles are the points of the heavenly circles upon which the sphere principally bears. One of them facing to the north is called Boreus; the other set at the opposite end of*

---

133 Gregory, *Moralia in Iob* 32.15.25, ed. Marcus Adriaen, CCSL 143B (Turnhout: Brepols, 1985), 1648.

134 Psalm 18:2/19:1.

135 Ambrose, *Hex.* 2.4.15 (Schenkl, CSEL 32.1:54).

136 Isa. 51:6 (not Vulgate) (Ambrose, *Hex.* 1.6.21: Schenkl, CSEL 32.1:17).

137 Ennius, *Annales* 557, trans. E.H. Warmington, in *Remains of Old Latin in Three Volumes*, vol. 1, *Ennius and Caecilius* (Cambridge, MA: Harvard University Press/London: Heinemann, 1935).

138 = *Etym.* 3.36 (Gasparotto and Guillaumin, 3.35); 13.5.3. Isidore's comparison of the sphere to a ball (*pila*) is original. See Hermann, 'Die astronomischen Metaphern', 447–48.

the earth is called Austronotius.[139] **There are two hemispheres, one of which is above the earth and the other below.**[140]

4. Philosophers think that heaven turns once in the day and night from east to west. *Furthermore, they have stated that it is round, revolving, and fiery.*[141] *Its sphere is said to have been placed upon the waters, in order that it might revolve in them and that they may moderate its heat.*[142] *And they affirm that the sphere has neither a beginning nor an end because on account of its rotundity, like a circle, it is not easy to perceive where it starts or where it leaves off.*[143] *Indeed, they say* /**221**/ *that it is compacted equally and entirely and in like fashion oriented in all directions and separated at equal distances from the centre of the earth, and that because of its very equality it is so stable that this equality concentrated on all sides does not permit it to slope towards any part, and that, borne aloft, it is not held up by any support.*[144]

5. In the course of treating the perfection of this sphere or circle with many proofs, Plato shows that the Artificer of the world is rational. First, because the zodiac, which is drawn from the angles formed from five lines, consists of one line.[145] Second, because it is without a beginning and without an end. Third, because it is constructed beginning from a point. Again, because it derives its motion from itself. Next, because it lacks any indication of angles and because it incorporates in itself all the signs of the stars and because it possesses perfect motion, since the six other motions are subject to error: forward, back, to the right and to the left, upwards and

---

139 Hyginus, *De astronomia* 1.4 (Viré, 5).

140 Cf. Cassiodorus, *Cassiodori Senatoris Institutiones* 2.7.2, ed. R.A.B. Mynors (Oxford: Oxford University Press, 1937), 154; *Etym.* 3.43.1 (Gasparotto and Guillaumin, 3.42.1). For this 'Isidorian' interpolation, a feature associated primarily with the long recension, see Introduction, pp. 46–47.

141 Ambrose, *Hex.* 1.1.4 (Schenkl, CSEL 32.1:4).

142 Ambrose, *Hex.* 2.3.12 (Schenkl, CSEL 32.1:50); Clement/Rufinus, *Recognitiones* 8.21.2 (ed. Rehm, 229.24–25). This is a doctrine of the Stoics. See Fontaine, *Isidore et la culture*, 478–79; Smyth, *Understanding the Universe*, 92–93.

143 Ambrose, *Hex.* 1.3.10 (Schenkl, CSEL 32.1:9); Clement/Rufinus, *Recognitiones* 8.21.2 (ed. Rehm, 229.23–26).

144 Clement/Rufinus, *Recognitiones* 8.21.2 (ed. Rehm, 229.23–26).

145 = *Etym.* 3.45 (Gasparotto and Guillaumin, 3.44). Fontaine's translation ('tracé à partir des angles *qu'il forme* avec cinq lignes') adds a clause which postulates that it is the zodiac which creates the angles, not the lines. See Commentary.

downwards.[146] Finally, because it necessarily follows that this line cannot be drawn beyond the circle.[147]

6. As we have stated, there are two axes on which heaven revolves. One is Boreus, which we call the northern axis: here are the Bears, that is, /223/ the seven pole stars, which are always visible to us. Opposite to this is Notius, which is called the southern axis. This is the one, as Cicero says, which is concealed by the earth, and which is called *aphanes* [invisible] by the Greeks. But the pole is said to move with such great rapidity that, unless the stars ran against its headlong course, it would cause the destruction of the world. For they say that its rapid whirl is tempered by the course of the stars; hence Lucan:

> The stars, which alone moderate the flight of heaven
> And rush to meet it [i.e., the pole], potent against its opposite course ... [148]

13 THE SEVEN PLANETS OF HEAVEN AND THEIR REVOLUTIONS

1. Saint Ambrose in his book the *Hexaemeron* speaks out plainly, saying: we read in David: 'Praise him, you heavens of heavens'.[149] *There is a controversy whether there is one heaven or more, as some assert that there are many, while others claim that there are no others except the one. But worldly philosophers have maintained that there are seven heavens, that is, planets, in a harmonious motion of spheres. They say that all of them are linked together by their orbits, which, being connected with and as it were nested inside each other, they believe are whirled backwards and carried in a contrary motion to the other stars.*[150] Furthermore, 'the heavens of heavens' /225/ are also found in Christian books,[151] *and the apostle Paul knows that he had been caught up to the third heaven.*

146 For the seven motions, cf. Plato, *Timaeus* 34A and 43B, where the motions are actually enumerated.

147 Presumably, the line of the zodiac; see Commentary.

148 Lucan, *The Civil War* 10.199–200. Isidore alters the final phrase of Lucan's line 200 from *diuersa potentia prima* to *diuerso potentia cursu*. Fontaine, *Traité*, n. 84 (p. 347), raises the possibility that this phrase, which does not scan, is Isidore's prose comment on Lucan's text.

149 Ps. 148:4 (Ambrose, *Hex.* 2.2.6: Schenkl, CSEL 32.1:45).

150 Ambrose, *Hex.* 2.2.5–6 (Schenkl, CSEL 32.1:44–45).

151 3 Kings/1 Kings 8:27; 2 Para. 2:6; 2 Para. 6:18; Ecclesiasticus 16:18; Ps. 148:4.

But concerning their number, let not the rash opinion of humanity be presumptuous.[152]

2. *God did not make the heavens formless or confused, but distinct in accordance with a rational plan in a particular order. For he stretched out the heaven of the superior circle, after setting it apart with its own proper boundary and assembling it with equal distances in every direction, and he placed in it the powers of the spiritual creatures. God, the artificer of the world, tempered the nature of this heaven with waters, lest the conflagration of the superior fire should set the inferior elements ablaze. Then he established the circle of the inferior heaven not with a uniform, but with a complex motion, calling it the firmament*[153] *on account of its support of the superior waters.*[154]

14 THE WATERS THAT ARE ABOVE THE HEAVENS

1. This is the opinion of Ambrose: *the worldly philosophers declare that there cannot be waters above the heavens,*[155] saying that heaven is fiery and the nature of water cannot be harmonized with it. *They add as well that the sphere of heaven is round and revolving and burning and that waters cannot possibly remain in that revolving circuit. For it is necessary that they flow and slide down when that sphere rotates from top to bottom,* /227/ *and because of this they say that the waters cannot possibly stay in place, since the axis of heaven, twisting itself by its violent motion, would pour them out in the course of its revolution.*[156]

2. *But in the end these men cease to rave and in confusion admit*[157] *that one who could create everything from nothing could also stabilize the nature of the waters in heaven with the solidity of ice.*[158] *For since they*

---

152 Hilary of Poitiers, *Tractatus super psalmos* 135.10, ed. Jean Doignon, CCSL 61B (Turnhout: Brepols, 2009), 167–68 (Paul = 2 Cor. 12:2).

153 The reference is to Gen. 1:7–8: 'And God made the firmament and separated the waters which were under the firmament from the waters which were above the firmament ... And God called the firmament heaven'.

154 Hilary, *Tractatus super psalmos* 135.9 (Doignon, CCSL 61B:167).

155 Ambrose, *Hex.* 2.3.9 (Schenkl, CSEL 32.1:47).

156 Ambrose, *Hex.* 2.3.9 (Schenkl, CSEL 32.1:47).

157 Ambrose, *Hex.* 2.3.9 (Schenkl, CSEL 32.1:47).

158 Augustine, *De Genesi ad litteram* 2.5, ed. Joseph Zycha, CSEL 28.1 (Vienna, 1894), 39. See Commentary.

themselves also say that the sphere which glitters with the burning stars revolves, is it not the case that divine Providence necessarily saw to it that water would flow within as well as above the sphere of heaven, in order to temper the fires of the burning axis?[159]

15 THE NATURE OF THE SUN

1. These are the words of Ambrose in his book, the *Hexaemeron*: *the philosophers, he says, deny that the sun is in its nature hot, because it is white, not glowing red or orange like fire; and therefore that its nature is not fiery. If it has any heat, they say that it happened because of the excessive rapidity of its revolution. They think that this must be said, therefore, lest the sun seem to consume moisture, because it does not have natural heat, by which moisture is commonly either diminished or exhausted. But they accomplish nothing when they propose such things, because it makes no difference* /**229**/ *whether someone has heat naturally, or has it from a disease or from some other cause.*[160]

2. *But we believe that just as the sun has the power of giving off light, so also it has the power of vaporizing. For the sun is fiery. Indeed, fire not only gives off light but also burns.*[161] Moreover some say that the fire of the sun is fed by water and receives the power of light and heat from this contrary element.[162] *Hence we frequently see the sun full of moisture and dropping dew: in this it manifestly gives proof that it has taken in the element of water for the sake of tempering itself.*[163]

3. This is pertinent as far as concerns the sun's nature. But indeed according to spiritual understanding the sun is Christ, just as it is written in Malachi: 'but to you who believe, the sun of justice shall arise, and health in his wings'.[164] Christ is rightly understood to be called the sun, because, having arisen, he set[165] according to the flesh, and according to the spirit after setting he arose again. Likewise, the sun gives light

---

159 Ambrose, *Hex.* 2.3.12 (Schenkl, CSEL 32.1:50).
160 Ambrose, *Hex.* 2.3.14 (Schenkl, CSEL 32.1:53).
161 Ambrose, *Hex.* 4.3.9 (Schenkl, CSEL 32.1:116).
162 = *Etym.* 3.49 (Gasparotto and Guillaumin, 3.48).
163 Ambrose, *Hex.* 2.3.13 (Schenkl, CSEL 32.1:53).
164 Malachi 4:2 (not Vulgate).
165 The verb *occidere* can be used of the sun ('to set') and of a person ('to die'). The double entendre is partially obscured in translation.

and burns and in overcast weather warms the healthy, but sets ablaze the feverous with the burning of redoubled heat. So also Christ illuminates believers with the invigorating spirit of faith, while he will burn those who deny him with the heat of eternal fire. /231/

16 THE SIZE OF THE SUN AND THE MOON

1. Again in the same work the same doctor bears witness in this way: the rays of the sun are not nearer to one person, or farther away from another. Similarly, the sphere of the moon is equal in size for all. The sun is alike both to Indians and to Britons; it is seen when it rises at the same moment by both, nor when it inclines towards setting does it appear smaller to Orientals, nor is it thought to be lesser to Occidentals than to Orientals when it rises. 'As far as the east is from the west', [the psalmist] says,[166] so great is the distance the one from the other. But the sun is not more or less distant from either; to neither is it closer or more remote.[167]

2. *Let it not disturb anyone that it seems as if the sun's disk is a cubit wide when it rises; but it is necessary to consider how great the distance is between the sun and the earth, which the weakness and diseased condition so to speak of our sight is scarcely able to grasp.*[168] But the philosophers describe it as being fractionally larger than the earth.[169]

3. *They say that the moon is smaller than the sun; for everything which is close seems to us larger, and a visible object diminishes with the remoteness of its location. Therefore, what we see when we look at the moon is that it is very near, and not that it is larger than the sun. And for that reason, although the sun is higher by far* /233/ *than the moon, and nevertheless seems larger to us, if it should ever come close to us, it would be much larger still.*[170]

---

166 Ps. 102:2 (103:12) (Ambrose).
167 Ambrose, *Hex.* 4.6.25 (Schenkl, CSEL 32.1:132–33).
168 Ambrose, *Hex.* 4.6.26 (Schenkl, CSEL 32.1:133).
169 Cf. Cassiodorus, *Institutiones* 2.7.2 (ed. Mynors, 155).
170 Hyginus, *De astronomia* 4.14 (Viré, 154). See also *Etym.* 3.50 (Gasparotto and Guillaumin, 3.49).

17 THE COURSE OF THE SUN

1. *The ancients Aratus and Hyginus say that the sun moves of its own accord and that it does not revolve with the world, remaining in one place. For if it remained fixed, it would necessarily set in the same place and rise in the same place from which it had risen the day before, just as the other signs of the stars rise and set. Besides, if it were so, it would follow that all the days and nights would be equal, and however long the present day was, it would always be exactly as long in the future.*[171]
2. *Night also would always remain equal for the same reason, but since we realize that the days are unequal, and we see that the sun will set in one place [tomorrow] and that it set in another place yesterday, and because it sets and rises in different places, philosophers think that it definitely does not revolve with the world, remaining itself fixed, but that it moves of its own accord.*[172] *After dipping its burning wheel in the ocean, it returns by ways unknown to us to the place from which it had emerged, and, with the completion of the night's revolution, it quickly bursts out again from its place.*[173] *For it proceeds on an oblique and uneven line through the south to the north and so returns to the east.*[174] *And in wintertime /235/ it runs through the southern region, but in summer it neighbours the north.*[175] *But when it runs through the south, it is nearer to the earth; when it is close to the north, it is raised up on high.*[176]
3. God ordained diverse places and seasons for the sun's course, lest, by always lingering in the same places, it destroy them with its daily heat. *But, as Clement says, it takes diverse courses, in order that the temperature of the air may be regulated in accordance with the rhythm of the seasons, and that the order of their changes and alterations may be preserved.*[177] *For as the sun ascends to the higher regions, it tempers the spring; but when it reaches the height of heaven, it kindles summer heat. Declining again, it restores the moderate temperature of autumn; and when it returns to its*

---

171 Hyginus, *De astronomia* 4.13 (Viré, 145).

172 Hyginus, *De astronomia* 4.13 (Viré, 145).

173 Jerome, *Commentarius in Ecclesiasten* 1.5, ed. Marcus Adriaen, CCSL 72 (Turnhout: Brepols, 1959), 254.

174 Jerome, *Commentarius in Ecclesiasten* 1.6 (Adriaen, CCSL 72:255).

175 Jerome, *Commentarius in Ecclesiasten* 1.6 (Adriaen, CCSL 72:254).

176 Jerome, *Commentarius in Ecclesiasten* 1.6 (Adriaen, CCSL 72:255). When Jerome says 'nearer to the earth', he means 'closer to the horizon'.

177 Clement/Rufinus, *Recognitiones* 8.45.2 (ed. Rehm, 245.17–19).

lower orbit, it bequeaths to us from the icy structure of heaven the severity of our wintry cold.[178]

4. *The hours derive from the sun; day is created from it when it arises; night is also formed from it when it sets; the months and years are reckoned from it; the changes of the seasons derive from it, and although it is a good* /237/ *servant, to be thanked for moderating the changes of the seasons, nevertheless when by the will of God a scourge is inflicted upon mortals, it glows more fiercely, and burns the world with more furious flames,*[179] *and the air is unsettled, and affliction of men and corruption is cast upon the earth, and plague is decreed upon living things and a pestilential year upon mortals everywhere.*[180]

5. *As to the fact that the rising sun takes its course through the south, that is, the meridional region, and, after having descended through the southern region, travels invisibly, returning to its starting place, truly this world was created in the likeness of the Church, in which the Lord Jesus Christ, the eternal sun, traverses his own region – hence they call it the* meridianum[181] *– but he does not arise for the north, that is, for the hostile region, just as, when he comes on Judgement Day, these people will say: 'the light of justice has not shined on us, and the sun has not risen upon us'.*[182] *So also it is written: 'but for those fearing the Lord, the sun of justice arises, and health in his wings'.*[183] *Indeed, it is night at midday for the wicked, just as we read: 'while they await the light, darkness falls upon them; while they await brightness, they have walked in the dark night'.*[184]

178 Clement/Rufinus, *Recognitiones* 8.45.6 (ed. Rehm, 246.5–9).

179 Clement/Rufinus, *Recognitiones* 8.45.5, 8.46.1 (ed. Rehm 246.3–5, 10–13).

180 Clement/Rufinus, *Recognitiones* 8.45.3 (ed. Rehm 245.20–22). Cf. ch. 39, where the notion of scourging or smiting (*correptio*) is likewise connected with pestilence. The word we have rendered here as 'plague' (*plaga*) usually denotes a blow, stroke or wound, but by Late Antiquity had extended its meaning to an outbreak of deadly disease; hence the reference to 'corruption' of the atmosphere, the putative cause of epidemics. *Plaga* need not refer to the disease now called 'plague', caused by infection with *Yersinia pestis*.

181 Fontaine, *Traité*, n. 104 (p. 350), explains this obscure phrase as perhaps being a distant echo of an allegorical interpretation of *meridianum*, signifying 'the region of the eternal sun'.

182 Wisdom 5:6.

183 Tyconius, *Liber Regularum* 7, ed. F. Crawford Burkitt, *The Book of Rules of Tyconius* (Cambridge: Cambridge University Press, 1894), 73–74, quoting Malachi 4:2 (not Vulgate).

184 Isaiah 59:9 (not Vulgate).

18 THE LIGHT OF THE MOON

1. Saint Augustine makes an assertion in his exposition of the tenth Psalm: *the question is,* he says, *where the moon gets its light. Two opinions are passed down, but it appears doubtful that anyone can know* /**239**/ *which of them is true. For some say that it has its own light, and that one part of its sphere beams with light, while the other is dark, and as it moves in its orbit, the part that shines is turned little by little towards the earth, so that it can be seen by us; and for that reason it first gleams with light in the shape of a crescent.*[185]

2. *For if you fashion a ball half white and half dark and then hold before your eyes the part that is dark, you will not see any white; as you begin to turn the white part little by little towards your eyes, first you will see as it were a crescent of white; then gradually it will enlarge, until the entire white part is before your eyes and none of the darkness of the other half is visible. But if you rotate it anew little by little, the darkness begins to appear and the whiteness to diminish, until again it goes back to a crescent, and thus all the white is turned away from your eyes, and again the dark part alone can be seen. They say that this is what happens when the light of the moon is seen to increase up to the fifteenth day, and then to diminish and revert again to a crescent up to the thirtieth day, until absolutely no light appears in it.*[186]

3. *Others on the contrary say that the moon does not shine with its own light, but receives its light from the sun.*[187] *For indeed the sun is placed above it. Hence it happens that,* /**241**/ *when it is under the sun,*[188] *its upper half is bright, but its lower, which is turned towards the earth, is dark; but when it begins to move away from the sun, it will be illuminated as well on the side which is turned towards the earth, beginning with the crescent. Thus, little by little, as it moves farther away from the sun, all the lower part is illuminated, until the fifteenth day of the moon comes to pass. After the middle of the month, when it begins to approach the sun from the other semicircle, the more it is illuminated on the upper half, the more it is unable*

---

185 Augustine, *Enarrationes in Psalmos* 10.3, ed. Clemens Weidmann, CSEL 93/1A (Vienna: Verlag der Österreichischen Akademie der Wissenschaften, 2003), 221.

186 Augustine, *En. in Ps.* 10.3 (Weidmann, CSEL 93/1A:221–22).

187 Augustine, *En. in Ps.* 10.3 (Weidmann, CSEL 93/1A:222).

188 i.e., when the moon is in conjunction with the sun.

to capture the rays of the sun on the side which it turns towards the earth, and therefore it seems to diminish.[189]

4. *It is plainly evident and easily known by anyone giving it attention that the moon does not increase in our sight except by receding from the sun, and that it does not diminish except by approaching the sun from the other side* [of its orbit].[190] Therefore it takes its light from the sun and, when it is beneath it, it is always slender; but when it recedes farther from it, it enlarges and fills out its circumference. *Indeed, if it employed its own light, it would necessarily always be equal and not become thin on the thirtieth day; and if it employed its own light, it would never be eclipsed.*[191] /**243**/

5. Otherwise, insofar as pertains to the mystical sense, *the moon is an image of this world, because just as it weakens in the course of completing its monthly revolutions, so the world running towards the completion of times stumbles with its daily weaknesses.*[192] Indeed, the moon in accordance with the inconstancy of its basic nature wanes in its diverse courses so that it may wax, and waxes that it may wane. But it exhibits this reversal of its starry nature in its alternating changes in order to teach us that men will die after birth and that they will live after death. And when it grows old, it proclaims the death of our bodies; when it increases, it indicates the eternity of our souls.

6. But sometimes as well the Church is signified by the moon, because just as it is illuminated by the sun, so the Church is illuminated by Christ. Indeed, just as the moon wanes and waxes, so the Church has her waning and waxing. *For she has frequently grown by means of her weaknesses and has deserved to be enlarged by them, during times when she is diminished by persecutions and crowned by the martyrdom of the confessors.*[193] Likewise *just as the moon abounds in dew and governs watery substances,*[194] so also the Church abounds in baptism and governs preaching. And /**245**/ as all streams of water also increase with the waxing moon, and diminish as it diminishes,[195] so we understand in the same way the Church, in whose

---

189 Augustine, *En. in Ps.* 10.3 (Weidmann, CSEL 93/1A:222–23); cf. *Epistola* 55.7 (Daur, CCSL 31, 239).
190 Augustine, *Epistola* 55.7 (Daur, CCSL 31, 240).
191 Hyginus, *De astronomia* 4.14 (Viré, 149–50).
192 Gregory, *Moralia* 34.14.25 (Adriaen, CCSL 143B:1750).
193 Ambrose, *Hex.* 4.8.32 (Schenkl, CSEL 32.1:138).
194 Ambrose, *Hex.* 4.7.29 (Schenkl, CSEL 32.1:134).
195 Cf. Dracontius, *De laudibus Dei* 1, lines 733–37 (ed. Vollmer, 64).

increase we benefit along with her; but when she suffers persecution and diminishes, we also suffer and diminish with her.

7. Also, just as the moon has seven shapes, so the Church has the same number of graces of merits. The moon's first form is two-horned, thus [Diagram 5A: The Phases of the Moon]:

Its second is thin-sliced,[196] thus:

Its third is half-sized, thus:

Its fourth is full-sized, thus:

Its fifth is again half of its larger size, thus:

Its sixth is again thin-sliced, thus:

Its seventh is two-horned, thus:[197]

The distribution of the gifts which are conferred by the Holy Spirit upon the whole Church also consists of the same number.[198] On day seven and a half and day twenty-two and a half of its orbit, the moon is a half circle. The other moons are proportional.

Diagram 5A: The Phases of the Moon

19 THE COURSE OF THE MOON

1. *It is necessary,* Hyginus says, *because of its different risings and settings, that the moon move rather than stand still, a fact which can more readily /247/ be understood about it than about the sun. Since it takes its light from the sun and thus seems to us to shine, there is no doubt that it moves rather than stands still.*[199] Moreover, the moon, which is close to the

---

196 Fontaine translates 'en forme de croissant', which clearly is what is meant.
197 See Commentary.
198 The seven gifts of the Holy Spirit are wisdom, understanding, counsel, fortitude, knowledge, godliness, and fear of the Lord (Isa. 11:2–3).
199 Hyginus, *De astronomia* 4.14 (Viré, 149). What Hyginus means is that the moon moves with its proper motion, like the sun, and is not passively moved by the diurnal rotation of the outermost sphere of the fixed stars. The relatively rapid displacement of the moon against the

earth,[200] revolves in a more constricted orbit, and the journey that the sun accomplishes in 365 days, it makes in 30 days. Hence the ancients reckoned months from the moon, and years from the course of the sun.

2. *Therefore the moon, accomplishing its course in twelve repetitions of thirty days, completes the year according to the Hebrews by means of the addition of several days; according to the Romans* [the year is completed] *by proclamation of the addition of one intercalary day once every fourth year.*[201] *Likewise by its waxing and waning,*[202] in accordance with a wonderful arrangement of Providence, everything that is born is nourished and grows. *For the elements are in sympathy with its disappearance; and those things which were exhausted are filled out by its advance, like the brain of sea-creatures, since sea-urchins and oysters are said to be found plumper at the time of the moon's waxing.*[203]

20 THE ECLIPSE OF THE SUN

1. *Wise men say that the sun orbits higher* [than the moon], *and that the moon is very near to the earth. Therefore, when the latter* /249/ *arrives underneath* [the former] *in the same sign or path in which the sun travels,*[204] it passes in front of the sun and plunges its disk into darkness. This only happens at the time of the new moon.[205] *For then the moon is in the same part of the sign in which the sun travels, and therefore comes closest to it, and seems by its opposition to hide the sun's light from our eyes. It is just as if someone were to hold up an open hand to his eyes – the closer he held*

backdrop of the fixed stars, as well as its daily alteration of phases, makes this more obvious to the casual observer than is the case with the sun's motion.

200 Cf. Hyginus, *De astronomia* 4.14 (Viré, 152).

201 Ambrose, *Hex.* 4.5.24 (Schenkl, CSEL 32.1:131). Twelve lunar months of 29.5 days (not 30, as in Ambrose) totals 354 days, which falls short of a solar year by 11 days. An intercalary month is inserted every three years or so to bring the Jewish lunar months back into phase with the seasons. Ambrose compares this to the leap year day which compensates for the discrepancy between the Roman year of 365 days and the length of the astronomical solar year (as then calculated), namely 365.25 days.

202 Augustine, *DCD* 5.6 (Dombart and Kalb, CCSL 47:133–34). But Augustine's point is rather different: that the waxing and waning of the moon affects the tides and the variable size of certain sea-creatures. See next sentence.

203 Ambrose, *Hex.* 4.7.29 (Schenkl, CSEL 32.1:134–35); Augustine, *DCD* 5.6 (Dombart and Kalb, CCSL 47:134) ['sea-urchins'].

204 Hyginus, *De astronomia* 4.14 (Viré, 150).

205 Cf. Hyginus, *De astronomia* 4.14 (Viré, 150).

it, the less he would be able to see; but the farther out it went, the more
everything could be seen by him.[206]

2. *In the same way when the moon reaches the place and path of the*
sun, then it seems to be very near to it and thus to shut off its rays from
our eyes, so that the sun's light cannot shoot out. However, when the moon
moves away from that place, then the sun shoots out its light and transmits
it to our eyes.[207] Therefore the moon blocks the sun since the earth is on the
opposite side of the moon. When the lights of either of these bodies do not
reach the earth, they are said to be eclipsed. But others say that an eclipse
of the sun happens, if the hole of the air, through which the sun pours its
rays, is contracted or shut off by some exhalation. These are the views of
the natural philosophers and wise men of the world.

3. Additionally, our doctors have said that the mystery of this eclipse was
completed mystically in Christ, at the time when, there being a rupture in
the regular course /**251**/ of eternal law, the elements, thrown into unnatural
confusion, lost their order, and the true Sun, bristling at the crime of the
sacrilegious conspiracy, the darkness of error having penetrated the Jewish
people, hid himself for a little while in death and, after being deposed from
the cross, darkened himself, concealed in a sepulchre, until on the third
day, more exalted than ever, he displayed the power of his brightness to
this world, that is, to the nations, and gleaming like the sun in his might, he
illuminated the darkness of a darkened age.[208]

21 THE ECLIPSE OF THE MOON

1. The moon does not disappear, but is overshadowed, and it does not
experience a diminution of its body, but undergoes a loss of light because
of the intervention of the overshadowing earth. For the philosophers
maintain that it does not have its own light, but is illuminated by the sun.
And because the moon is separated from the sun by such a measure that
if a straight line be extended through the middle of the earth, it will touch
the sun beneath the earth and the moon above the earth; and because the
earth's shadow extends as far as the lunar orbit, it sometimes happens that

---

206 Hyginus, *De astronomia* 4.14 (Viré, 150).
207 Hyginus, *De astronomia* 4.14 (Viré, 150).
208 Fontaine, *Traité*, 13, remarks on the eloquence and grandeur of this long periodic
sentence and its sources in Origen and Jerome.

the rays of the sun do not reach the moon, when the mass or shadow of the earth intervenes.[209] /**253**/

2. The moon undergoes this event on the fifteenth day just until the moment when it passes the centre and shadow of the blocking earth and sees the sun or is seen by the sun. It is certain, therefore, that the moon receives its light from the rays of the sun and that it loses its light when the sight of the sun is blocked by the earth. For the Stoics say that the whole earth is surrounded by mountains, by whose shadow the moon is said suddenly to disappear. Hence Lucan says:

> Now when Phoebe reflected her brother with her entire orb,
> Struck of a sudden by earth's shadow, she turned pale.[210]

3. Allegorically, the eclipse of the moon signifies the persecutions of the Church, when, because of the slaughter of the martyrs and the effusion of blood, that eclipse and darkening as it were, she seems almost to display the moon's bloody face, so that the weak are terrified of the name of Christian.[211] But just as the moon reappears with a clear light, so much so that it seems to have suffered no loss, so likewise the Church, after shedding her blood for Christ through the testimony of the martyrs, shines again with a greater brightness of faith, and, adorned with worthier light, spreads herself more broadly throughout the whole world. /**255**/

22 THE COURSE OF THE STARS

1. *In truth, the stars revolve with the world;*[212] *they do not roam about while the world stands still.*[213] For, with the exception of those that are called planets, that is, wandering stars, which move in irregular courses, all the rest, which are called *aplanes* ['fixed'],[214] revolve with the world, fixed in one place. And for that reason the planets, that is, the wandering stars,

---

209 Hyginus, *De astronomia* 4.14 (Viré, 151).

210 Lucan, *The Civil War* 1.538–39.

211 Augustine, *En. in Ps.* 10.4 (Weidmann, CSEL 93/1A:224).

212 Here, as frequently elsewhere, *mundus* refers to the celestial sphere, which revolves with the fixed stars and planets around the stationary earth. On Isidore's usage of this word, see Commentary, ch. 9.

213 Hyginus, *De astronomia*, preface (Viré, 2); 4.8 (Viré, 135–36).

214 Cf. Calcidius, *Timaeus a Calcidio translatus commentarioque instructus* 69, ed. Jan Hendrik Waszink, 2nd edn. (London: Warburg Institute, 1975), 116.

are so called because they roam with varied motion in different directions through the whole world.[215]

2. The stars, however, move amongst each other in different parts [of the sky]. For some run higher, some lower. *Hence, those that are nearer to the earth seem to us as if they are larger than those that revolve about the heavens, for a visible object diminishes with the remoteness of its location.*[216] *Hence it happens that, by reason of the wide separation of their different orbits, some return more swiftly, others more slowly, to the beginning of their course.*[217]

3. *For some stars,*[218] *having risen more quickly, are thought to set later; and some, having risen later than others, reach their setting sooner; and still others arise together but do not set together.*[219] *But all the stars return in their own time to their proper course.*[220] Stars[221] hindered by the rays of the sun become irregular or retrograde or stationary, just as the poet likewise mentions, when he says: /257/

> The sun divides the periods of time,
> It alternates day with night, and by its powerful rays
> It prevents the planets[222] from moving, and brings
> their wandering courses to a standstill.[223]

23 THE POSITION OF THE SEVEN WANDERING STARS[224]

1. In the circuit of the seven celestial orbits, the moon comes first, situated in the circular course of the lower sphere; it is positioned nearest to the earth, in order more easily to display its light to us at night. Next, the star Mercury is located in the second circle, equal to the sun in swiftness, but

---

215 = *Etym.* 3.67 (Gasparotto and Guillaumin, 3.66) (cf. Lucan, *The Civil War* 1.643).

216 Hyginus, *De astronomia* 4.14 (Viré, 154); = above, *DNR* 16.3.

217 Hyginus, *De astronomia*, preface (Viré, 2).

218 Isidore substitutes *sidera* for Hyginus's *signa*. Presumably, therefore, in this paragraph he intends *sidera* in the classical sense of 'constellations'. But his usage is inconsistent. See below (n. 221).

219 Hyginus, *De astronomia*, preface (Viré, 2–3).

220 Hyginus, *De astronomia* 4.14 (Viré, 155). See Commentary.

221 Here, *sidera* refers to the planets.

222 Lucan's word is *astra*, by which he clearly means 'planets'.

223 = *Etym.* 3.66.3 (Gasparotto and Guillaumin, 3.65.3) (Lucan, *The Civil War* 10.201–03).

224 On the order of the planets, see Fontaine, *Isidore et la culture*, 512–15.

possessed, so the philosophers say, of a kind of contrary force.[225] In the third circle is the orbit of [the star] Lucifer, which is called Venus by the pagans, because of the five [wandering] stars it has the most light.[226] For, as we have said above, this star also makes a shadow just like the sun and the moon.[227]

2. The orbit of the sun is located in the fourth circle. It is placed in the middle precisely because it is more luminous than all the other stars, so that it may furnish light to those above as well as to those beneath. It is placed in this way by divine reason, because everything noble ought to be in the middle. Next, they say that the star Vesper, which these people assign to Mars, is located in the fifth circle.[228] In the sixth circle is placed the star Phaëton, which they call /259/ Jupiter. Then, in the highest heavens, that is, at the pinnacle of the world, is positioned the star Saturn. While it certainly occupies the highest heavens and is loftier than all the rest, nevertheless it is said to be of a frigid nature, as Vergil attests:

To where the frigid star Saturn withdraws.[229]

3. These are called the wandering stars, not because they themselves wander, but because they make us go astray. These stars are called 'planets' in Greek. Indeed, when it comes to the sun and the moon, their rising and setting are common knowledge on account of the fact that the sun and moon move on a straight course. But the others retrograde or become irregular, that is, when they add and retract small parts [of their course]. When they only retract, they are said to be retrograde; and when they come to a standstill, they 'make station'.

---

225 Cf. Plato, Timaeus 38D (Cicero, *Timaeus* 9.29). What exactly Isidore means by 'contrary force' is not evident. He may be referring to Mercury's retrograde motion; see Commentary.

226 Isidore probably meant to say both that the planet Lucifer is so-called because of the five stars it has the most light and that it is called Venus by the pagans. In the *Etymologies* he clearly states: '*Lucifer dictus eo quod inter omnia sidera plus lucem ferat*', 'Lucifer is so-called because of all the stars it carries the most light' (*Etym.* 3.71.18 (Gasparotto and Guillaumin, 3.70.18)).

227 There is no previous reference to this phenomenon in *DNR*, nor is it found in the *Etymologies*. The claim is made by Pliny (*NH* 2.6.37), whose work was apparently unavailable to Isidore when he wrote this passage, and is found also in Martianus Capella, *De nuptiis Philologiae et Mercurii* 8.883, ed. James Willis (Leipzig: Teubner, 1983), 335. See Fontaine, *Isidore et la culture*, 513 and n. 3.

228 Isidore likewise gives the name Vesper to Mars in *DNR* 3. See above (n. 34), and Fontaine, *Isidore et la culture*, 514. 'These people' (*illi*) are presumably the pagans.

229 *Georgics* 1.336 (*receptet*, Vergil; *receptat*, Isidore).

Diagram 6: The Planets

4. These are the years of the individual stars that are contained in the appended sphere.[230] After these years are completed, the stars retrace their orbit from the same signs and regions. The moon is said to complete its orbit in eight years; Mercury in twenty years; Lucifer in nine years; the sun in nineteen years; Vesper in fifteen years; Phaëton in twelve years; and Saturn in thirty years. /**261**/ The appended figure shows the position of these orbits and stars [Diagram 6: The Planets].[231]

---

230 i.e., Diagram 6.
231 See fig. 6 in Fontaine, *Traité*, facing p. 260.

24 THE LIGHT OF THE STARS

1. They say that stars do not have their own light, but are illuminated by the sun,[232] and that they never depart from the heavens, but are hidden by the coming of the sun. All the stars are obscured by the rising sun; they do not fall. *For as soon as the sun has sent ahead the signs of its rising, all the fires of the stars disappear beneath the brilliance of its light,*[233] so that aside from the fire of the sun no star's splendour may be seen. This is also the reason it is called the sun [*sol*] because, when all the stars are obscured, it appears alone [*solus*]. Nor is this fact about the sun astonishing, since when the moon is full and gleams throughout the night it is also the case that many stars do not shine. *Moreover, the eclipse of the sun proves that the stars are in the heavens during the day, because, when the sun has been obscured by the interposed orb of the moon, the stars are seen very clearly in the heavens.*[234] /263/

2. The stars, according to mystical allegory, are understood to be holy men, about whom it was said: 'Who tells the number of the stars'.[235] For just as all the stars are illuminated by the sun, so the saints are made illustrious by Christ with the glory of the kingdom of heaven. And just as the stars are dimmed by reason of the brightness of the sun and the very great power of its light,[236] so too all the splendour of the saints is obscured to a certain extent in comparison with the glory of Christ. And just as the stars differ among themselves in brightness,[237] so the diversity of the just follows from the difference of their merits.

25 THE FALL OF THE STARS

1. It is a false and common belief that stars fall at night, although we know that sparks which have fallen from the ether travel through the heavens, are carried by the winds, and imitate the light of a wandering star; but the stars remain immobile and fixed in the heavens.

---

232 = *Etym.* 3.61 (Gasparotto and Guillaumin, 3.60).
233 Ambrose, *Hex.* 4.6.27 (Schenkl, CSEL 32.1:133).
234 Jerome, *Commentaria in Isaiam* 6.13.10, ed. Marcus Adriaen, CCSL 73 (Turnhout: Brepols, 1963), 230.
235 Ps. 146:4 (147:4).
236 Cf. Hyginus, *De astronomia* 4.13 (Viré, 146).
237 Cf. 1 Cor. 15:41.

2. For this is what the poet says:

Often too you will see stars fall headlong from the heavens
With the threatening wind, and through the darkness of the night
Long trails of flames grow white behind ...[238]

And again:

... the falling stars[239]
Trail scattered furrows through the high atmosphere, /265/
And even the stars which remain fixed
Around the highest poles ...[240]

But these poets voluntarily aligned themselves with the popular belief. However, the philosophers, whose business it is to examine the system of the world, assert the things that are reported above.

26 THE NAMES OF THE STARS

1. We read in Job that the Lord says: 'Are you able to join together the shining stars, the Pleiades, or can you stop the turning about of Arcturus? Can you bring forth Lucifer in its time, and make Vesper rise upon the children of the earth?'[241] And again elsewhere: 'Who makes Arcturus and Orion and the Hyades'.[242] When we read these names of the stars in the Scriptures, let us not give approval to the meaningless nonsense of the poets, who, in accordance with their false beliefs, have imposed these names on the stars from the names of men or of other creatures. For wise men of the pagans gave names to certain stars, just as they did to the weekdays.

2. As to the fact that holy Scripture uses these same names, it does not on that account approve the worthless fables that go with them, but, making from visible things the symbols of invisible things,[243] it employs these names, which are widely known, so that people will recognize them, in order that whatever it signifies that is unknown may become known more easily to human understanding through that which is known. /267/

---

238 Vergil, *Georgics* 1.365–67.
239 Translating *cadentia sidera* as 'meteors' (as does Duff in the Loeb edition) obscures the problem that Isidore and his authorities faced.
240 Lucan, *The Civil War* 5.561–64.
241 Job 38:31–32.
242 Job 9:9.
243 Cf. Romans 1:20.

3. Arcturus. *Arcturus is what the Latins call Septentrion,*[244] *which, gleaming with the rays of seven stars, revolves about itself in a circle,* and which for that reason is called the Wain because it turns like a wagon *and at one moment it lifts three stars to the highest point, at another it brings four stars down. And, located on the axis of the heavens, it turns continually and never dips below the horizon, but, while it revolves about itself, the night also comes to an end.*[245]

4. By this Arcturus, that is, Septentrion, we understand the Church, gleaming with a sevenfold virtue. For just as Arcturus always declines and is lifted up again on the axis of the heavens, so the Church is humbled by various misfortunes, but soon rising again is lifted up by hope and virtues; and just as the Wain is composed of three and four stars, so also the Church is brought to perfection by faith in the Trinity and by the operation of the four cardinal virtues.[246] For man is justified by faith and by works.

5. Boötes[247] is the star which follows the Wain, that is, Septentrion. It is also called by the Ancients Arctophylax ['bear-keeper'] or the Lesser Bear [*Arctos*]: hence some have called it Septentrion. Those who are navigators in particular observe it. Lucan says of it: /**269**/

... where at night swift Boötes, ...[248] because after it rises, it quickly sets.[249]

6. The Pleiades are many stars yoked together. We also call them the Cluster from its multitude of stars. *There are said to be seven of them also, but no one is able to see more than six.*[250] *They arise from the east and, as the brightness of the day approaches, the arrangement of its stars spreads out.*[251] *The Pleiades derive their name from their plurality, because the*

244 *Septentrio*, literally, 'the seven ploughing-oxen' = the Great Bear or Big Dipper. In his more detailed discussion in *Etym.* 3.71.6–9 (Gasparotto and Guillaumin, 3.70.6–9), Isidore distinguishes Arcturus in the constellation Boötes from the Great Bear. Here, however, he reproduces Gregory the Great's mistake.

245 Gregory, *Moralia* 9.11.13, ed. Marcus Adriaen, CCSL 143 (Turnhout: Brepols, 1979), 465; 29.31.72 (Adriaen, CCSL 143B:1484).

246 Cf. Gregory, *Moralia* 29.31.72 (Adriaen, CCSL 143B:1484).

247 Literally, 'the ploughman'.

248 Lucan, *The Civil War* 3.252.

249 This misinformation derives from Isidore's misreading of a scholium on Lucan; see Fontaine, *Isidore et la culture*, 521. Homer, *Odyssey* 5.272, says the opposite – that Boötes is 'slow to set', and Latin sources such as Germanicus's translation of Aratus reproduce this.

250 Hyginus, *De astronomia* 2.21 (Viré, 64).

251 Gregory, *Moralia* 29.31.73 (Adriaen, CCSL 143B:1485). Isidore omits Gregory's likening of the Pleides cluster that rises in the east to the grace of the New Testament, which explains why the arrangement of its stars becomes clearer with the approach of day.

Greeks call plurality apo ton pliston.[252] *The Latins call them Vergiliae, because they arise after spring [ver], and they are more frequently referred to than the other constellations because summer is signified by their rising and winter by their setting, information which is not conveyed by the other signs at all.*[253]

7. *Because there are seven of them and they shine brilliantly, all the saints gleaming with the sevenfold virtue of the Spirit are signified by these stars; and because they are near together, but do not touch, they signify the preachers of God who are close together in charity but separated in time.*[254]

8. There is the star Orion. It takes its name Orion from the sword.[255] Hence the Latins also call it Jugula,[256] as the constellation seems armed and terrible from the light of its stars. It is very difficult to miss, /**271**/ because, however untrained the eyes, it nevertheless attracts them to itself by the splendour of its brightness. *Indeed, these Orions arise in the dead of winter and by their rising call forth rains and storms and disturb the seas and the lands.*[257]

9. *The Orions signify martyrs.*[258] For just as they arise in the heavens in wintertime, so martyrs appear in the Church in time of persecution. When the Orions appear, the sea is disturbed, and the land; and when martyrs appear, the hearts of earthly men and of infidels are tossed by the tempest.[259]

10. There is Lucifer, the shining star, which seems to be the greatest and brightest of all. For this star also makes a shadow, just as the sun and the moon do.[260] Accordingly, it precedes the rising sun and, announcing the morning, it scatters the darkness of night with the brilliance of its light. It symbolizes Christ, who, like Lucifer, is brought forth by the mystery of the

252 Gregory, *Moralia* 29.31.67 (Adriaen, CCSL 143B:1481).
253 Hyginus, *De astronomia* 2.21 (Viré, 66).
254 Gregory, *Moralia* 29.31.68 (Adriaen CCSL 143B:1481–82).
255 This apparent non-sequitur is partially explained by the next sentence. In *Etym.* 3.71.10–11 (Gasparotto and Guillaumin, 3.70.10–11), Isidore derives Orion from urine (*urina*), and mentions the sword in connection with Jugula, which he describes as he does here. In Roman mythology, Orion was a hunter who was transformed into a constellation by Diana.
256 Jugula is related to *iugulum*, 'throat' with the auxiliary sense of slaughter.
257 Gregory, *Moralia* 9.11.14 (Adriaen, CCSL 143:465–66).
258 Gregory, *Moralia* 9.11.14 (Adriaen, CCSL 143:466).
259 'For just as ... by the tempest'; cf. Gregory, *Moralia* 9.11.14 (Adriaen, CCSL 143:466).
260 See above (n. 227).

Incarnation, and through whom the light of faith is made known like the day that is to come.

11. Lucifer is divided into two [senses]:[261] thus, part of it is holy, as the Lord says of himself and the Church in the Apocalypse: 'I am the root and stock of David, the bright and morning star',[262] and, again: /273/ 'he who overcomes, I will give him the morning star'.[263] But the other part of Lucifer is known to be the devil, of whom it is written: 'How you are fallen from heaven, O Lucifer, who did rise in the morning?'[264] He is also the one who says that he will place his seat in heaven above the stars of God, and falling from heaven is crushed.[265]

12. Vesper is the western star bringing in the night.[266] It follows the setting sun and precedes the approaching darkness. It represents the type of the Antichrist, who, like Vesper, 'arises upon the children of the earth', as Job says,[267] so that he darkens the hearts of the carnal by the approach of blind night. This happens by the authority of God, because those who have refused to believe in Christ have deserved to receive the Antichrist.

13. A comet [*cometes*] is a star which spreads out, so to speak, a mane [*comae*] of light, from itself.[268] When it appears, it is said either to signify a change of reigns or the onset of wars and plagues. Prudentius says of it:

The melancholy comet perishes.[269]

And Lucan:

… and the comet, changing reigns on earth.[270]

/275/And Vergil:

… nor did fearful comets burn so often.[271]

Astrologers say that all the wandering stars become comets at certain

---

261 The bivalent symbolism of Lucifer is a classic example of the *sensus in bono et malo* of medieval allegorical interpretation. See Kendall, *Bede: On Genesis*, 18.
262 Apoc. 22:16.
263 Apoc. 2:26; 28 (not Vulgate).
264 Isa. 14:12.
265 Cf. Isa. 14:13–19.
266 = *Etym.* 3.71.19 (Gasparotto and Guillaumin, 3.70.19).
267 Job 38:32 (not Vulgate).
268 = *Etym.* 3.71.16 (Gasparotto and Guillaumin, 3.70.16).
269 Prudentius, *Liber Cathemerinon* 12.21. *Intercidit*, Isidore; *intercidat*, Prudentius.
270 Lucan, *The Civil War* 1.529.
271 Vergil, *Georgics* 1.488.

times,[272] and that in accordance with the motions of each they portend good or bad fortune.

14. There is the star Sirius, which is commonly called the Dog Star. It is called Sirius on account of the brilliance of its flame,[273] because its brightness is such that it seems to shine more brightly than the other stars.[274] At its rising it burns the world with the excessive intensity of its heat and destroys crops with its warmth. Sometimes also it inflicts bodies with disease, corrupting the air with the glowing heat of its fire. The 'dog days' get their name from this star,[275] because these days burn with more heat than the entire summer season.

27 WHETHER THE STARS HAVE A SOUL

1. *It is often asked,* Saint Augustine says, *whether the sun, moon, and stars are only bodies or whether they have their own spirits directing them; and, if they have, whether they are also inspirited by them in such a way that they become living beings, just as the flesh of living creatures is animated by the souls of living creatures, or whether they are [directed] by their presence alone, without any intermingling [of spirit].*[276] /277/ And, while without a soul there cannot be motion in any body, it cannot be readily understood whether the stars, which move with such order and system that their course is absolutely never in any way impeded, are animate and rational beings.

2. *When Solomon says of the sun, 'the spirit advances, wheeling around in a gyre, and returns to its circuits',*[277] *he shows that the sun itself is a spirit, and that it is a living being and breathes and is active and in its course completes its annual orbits, just as the poet also says:*

Meanwhile the sun revolves around the great year;

---

272 Cf. Seneca, *Natural Questions* 7.4.1, though it is more likely that Isidore found this distant echo of Babylonian astronomical theory in a scholium on Lucan: Fontaine, *Isidore et la culture,* 525.

273 Sirius derives from the Greek word for 'scorching'.

274 Hyginus, *De astronomia* 2.35 (Viré, 83–84); = *Etym.* 3.71.15 (Gasparotto and Guillaumin, 3.70.15). In the *Etymologies,* Isidore substitutes *Canis* for *Sirius* in this sentence, and implies that *Canis* is so-called from 'the brilliance [*candorem*] of its flame'.

275 = *Etym.* 3.71.14 (Gasparotto and Guillaumin, 3.70.14).

276 Augustine, *DGAL* 2.18 (Zycha, CSEL 28.1:62).

277 Eccles. 1:6 (not Vulgate).

And elsewhere:

> A spirit within nourishes the shining globe
> Of the moon and the Titanian stars.[278]

Therefore, if the bodies of the stars have souls, we must ask what they will become in the resurrection.

28 NIGHT[279]

1. I have read Ambrose saying in the book of the *Hexaemeron: there is a question how the shadow of the earth takes over the space of air, making night for us, when the sun retreats from us and hides the day, when it illuminates the lower regions of the northern heavens.*[280] *For every body makes a shadow and naturally the shadow adheres to the body, so that /279/ even painters try to depict the shadows of bodies before they paint in colour, and maintain that it is the duty of art not to neglect the principles of nature.*[281]

2. Therefore, just as during the day whenever any figure of a man or a tree appears with one side facing the sun, a shadow remains on the side where the light is blocked, so also, as the day wanes,[282] and the sun reaches the place where it is said to set, it is separated from us by the bulk of the mountains there,[283] and thus the air in the northern part is darkened by the interposition of the earth, to such an extent that this very shadow of the earth makes night for us.[284]

278　Jerome, *Commentarius in Ecclesiasten* 1.6 (Adriaen, CCSL 72:255) (quoting Vergil, *Aeneid* 3.284 and 6.725–26). The 'Titanian stars' refer to the sun (Titan) and the stars.

279　This is the second of two chapters on 'night' (see Commentary).

280　Ambrose, *Hex.* 6.2.8 (ed. Karl Schenkl, CSEL 32.1:209). Ambrose, as Bede understood (see *DTR* 7, ed. Jones, CCSL 123B:298; Wallis, *Reckoning*, 30), is discussing how far the earth's shadow extends into space. Isidore's paraphrase of Ambrose somewhat obscures the point. By 'lower regions' (*inferiora*), Isidore may intend the reader to understand the opposite side of the world where it is day when it is night here, though this is far from clear.

281　Ambrose, *Hex.* 4.3.11 (Schenkl, CSEL 32.1:118).

282　Ambrose, *Hex.* 4.3.11 (Schenkl, CSEL 32.1:118).

283　Hyginus, *De astronomia* 4.9 (Viré, 136). See Commentary.

284　Ambrose, *Hex.* 4.3.11 (Schenkl, CSEL 32.1:118). Isidore misunderstands Ambrose when he inserts 'in the northern part'.

29 THUNDER

1. Thunder is produced by the clashing of clouds. For, the blasts of winds conceived in the bosom of the clouds are whirled about there, and when they are about to break out they clash violently, and burst forth by the innate power of their mobility somewhere, and resound with a great roar,[285] and the sound of their clashing is transmitted to our ears like four-horse teams bursting out from their stalls.[286]

2. In another sense, however, thunder is the supernal rebuke of the divine voice or the clear preaching of the saints, which rings with a loud cry /281/ through all the lands in the ears of the faithful, in order that the world, having been warned, can recognize its sin.

30 LIGHTNING

1. *Those who investigate the natural causes of things say that lightning is produced by the collision and rubbing together of clouds, like hard flints from which fire escapes when you strike them together,*[287] or just as fire is produced when you rub wood with wood. Hence, Papinius says:

And as often as lightning flashed from clashing winds.[288]

For this reason, therefore, when clouds clash together, lightning is immediately produced.

2. Then thunder follows; it is emitted simultaneously with lightning, even if it is slower to sound, and the brilliance of the light precedes the shock. Its sound penetrates the ears more slowly than the flash of lightning penetrates the eyes; it is just like an axe felling a tree at a distance, the blow of which you see before the sound reaches your ears.[289] Therefore lightning is produced by the clashing of clouds. For lightning has never flashed from a clear sky. Hence Vergil says:

---

285 Cf. Ambrose, *Hex.* 2.4.16 (Schenkl, CSEL 32.1:56).

286 The image is from chariot races in the Roman circus.

287 Jerome, *Homiliae XIV in Jeremiam* 5 (*PL* 25, 629). For detailed analysis of Isidore's sources in this chapter, and the traditions they drew on, see Gasparotto, *Isidoro e Lucrezio*, ch. 2.

288 Cf. (Publius Papinius) Statius, *Thebaid* 1.354.

289 Ch. 30.2 up to this point is a paraphrase of Lucretius, *De rerum natura* 6.164–72.

Never did more lightning fall from a clear
Sky.[290]

3. Lightning is made from cloud, rain, and wind. For when wind is violently agitated in the clouds, it is heated so that it is set on fire. /**283**/ Then, as has been said above, lightning and thunder are emitted together. But the former is seen more quickly because it is brilliant, whereas the latter reaches the ears more slowly. After a lightning strike, the violence of the winds breaks forth, and issuing towards the earth, they thus send forth the storm's fury, which they stirred up when they were shut up in the clouds.

4. Lucretius says that lightning consists of minute particles, and therefore that it is capable of penetration.[291] And wherever lightning strikes, it emits the door of sulphur. Vergil says:

... and far and wide the region all around reeks with sulphur;[292]

and Lucan adds:

And the wicked sword reeked with sulphur from the sky.[293]

5. In lightning, the miracles of the saints are signified, gleaming and reaching to the inmost parts of the heart by the brilliance of their marvels and powers. Lofty places feel the violence of winds and lightning more than lower land. Hence, Horace says:

... and the lightning
Strikes the summits of the mountains.[294]

But very high places are sheltered from storms, like Olympus, which on account of its height feels neither the violence of winds nor of lightning, because it overtops the clouds.[295] /**285**/

---

290 Vergil, *Georgics* 1.487–88.
291 Cf. Lucretius, *De rerum natura* 2.381–87; 6.225–27. The 'minute particles' in Lucretius's account are particles of fire. For further discussion, see Commentary.
292 Vergil, *Aeneid* 2.698.
293 Lucan, *The Civil War* 7.160.
294 Horace, *Odes* 2.10.11–12.
295 Cf. Lucan, *The Civil War* 2.269–71.

31 THE RAINBOW

1. Clement, the Roman bishop and martyr, writes thus: this is the way the rainbow is formed in the air from the image of the sun. When the sun shines directly opposite thinning clouds, and, spreading out [its light], impresses its rays in a direct line onto the cloudy moisture, it produces a reflection of its splendour in the clouds from which the flashing brilliance forms the likeness of a bow. For just as wax which has been impressed takes the image of a signet ring, so the clouds taking on a shape from the round form of the sun opposite them make a circle and depict the image of a bow. However, this is only seen when the clouds of the sky thin out. For when the clouds come together again and thicken, the form of the bow immediately dissipates. Indeed, in a thick mass of clouds the bow embraces the air in an incomplete curve. Finally, without the sun and clouds a bow never appears, because its figure is formed from the impression of the rays of the sun.[296]

2. The rainbow has four colours, and acquires its appearance from all the elements: for from heaven it takes its fiery colour, from water its purple colour, from air the colour white, and from earth the colour black.[297] Moreover, the rainbow, because it shines brightly in the clouds from the sun, signifies the glory of Christ gleaming in the prophets and the doctors [of the Church].[298] Others have said that /287/ two judgements are signified by its two colours, that is, watery and fiery: the one by which the wicked perished in the flood in the past, the other by which sinners are to be burned in hell in the future.[299]

32 CLOUDS

1. It should be noted that this visible air,[300] according to the book of Job, is compressed so that it is heaped together, and having been heaped together

---

296 Cf. Clement/Rufinus, *Recognitiones* 8.42.3–7 (ed. Rehm, 243.3–19).

297 See Commentary.

298 Cf. Augustine, *Contra Faustum* 12.22 (Zycha, CSEL 25.1:350).

299 Cf. Gregory, *Homiliae in Hiezechihelem* 1.8.29, ed. Marcus Adriaen, CCSL 142 (Turnhout: Brepols, 1971), 119. The reference is to Noah's flood and the rainbow which God put in the sky as a sign that he would not again inflict that watery judgement upon the wicked (Gen. 9:13–15) and to the fires of the Last Judgement awaiting the wicked in the future (Apoc. 20:14–15). Bede offers a moving elaboration of Gregory and Isidore's allegorization in the second book of his commentary on Genesis (Kendall, *Bede: On Genesis*, 208).

300 What Isidore means by 'visible air' (which reading is overwhelmingly supported

it is turned into clouds. Job puts it this way: 'the air on a sudden shall be thickened into clouds, and the wind shall pass and drive them away'.[301] And Vergil says:

The winds arise, and the air is compressed into clouds.[302]

The clouds are understood to signify holy preachers, who pour out the rain of the divine word upon believers.[303]

2. This thin and empty air signifies the vacant and inconstant minds of men, which nevertheless when condensed turns into clouds, because when the souls of infidels are gathered together out of their inane vanity they are made solid by faith. And just as rain clouds are made from empty air, so holy preachers are collected from the vanity of the world to faith. They are called clouds [*nubes*] because they cover the ether.[304] Hence brides [*nuptae*] are so-called, because they veil their faces.[305] And hence also Neptune, because he covers the earth with cloud and sea.[306] /**289**/

33 RAINS

1. We read in the prophet Amos: 'who calls the waters of the sea, and pours them out upon the face of the earth'.[307] *For the very bitter waters of the sea are suspended in a fine mist by the heat of the air, like a medicinal cupping-glass, which draws up moisture and blood by the heat of its upper dome.*[308]

---

by the MSS) can probably best be understood in terms of his statement in *Etym.* 13.7.1: 'The term [air] refers partly to earthy and partly to heavenly material. That which is very fine, where winds and tempests cannot exist, makes up the celestial part, but that which is more turbulent, which takes on bodily substance with exhalations of moisture, is defined as earthy' (trans. Barney, et al., *Etymologies*, 273). See also, Fontaine, *Traité*, n. 173 (p. 357).

301 Job 37:21.

302 Vergil, *Aeneid* 5.20. The received text of Vergil reads 'cloud' (*nubem*).

303 Cf. Gregory, *Moralia* 17.26.36, ed. Marcus Adriaen, CCSL 143A (Turnhout: Brepols, 1979), 871. Gregory explains that 'holy preachers' are the apostles.

304 = Isidore, *Etym.* 9.7.10; 13.7.2 (where, instead of 'ether', Isidore uses the word *caelum*, 'heaven'). For 'ether', see above, ch. 25.1, and Commentary thereon. Isidore's substitution of 'heaven' for 'ether' in the *Etymologies* may suggest that he had refined his understanding of ether, and wanted clearly to restrict it to the super-celestial realm.

305 = Isidore, *Etym.* 9.7.10; 13.7.2.

306 = Isidore, *Etym.* 13.7.2.

307 Amos 5:8; 9:6.

308 Jerome, *Commentaria in Amos* 2.5.8 (Adriaen, CCSL 76 (1969):281). A cupping glass is semi-spherical, like the dome of the heavens; see Commentary.

And so by a similar process the waters of the sea, suspended in the air in very fine mists, little by little are condensed, and being heated there by the fire of the sun, they are converted into the sweet taste [i.e., fresh water] of the rains.

2. Then, as the cloud grows heavier, the waters they contain, now squeezed out by the force of the winds, now released by the heat of the sun, are sprinkled upon the face of the earth. Therefore, the waters of the sea are snatched by the clouds and are returned again by them to the earth. But, as we have said, in order that the waters in the rains may be sweet, they are heated by the fire of the sun. Others, however, say that not only are clouds increased by the waters of the sea, but also that little clouds grow from vapours exhaled from the land, and that as the vapours are thickened and condensed the clouds rise higher, and, as they begin to sink again, they pour out the rains.

3. Clouds, as we have already said, signify the apostles and teachers. The rains from the clouds, therefore, are the eloquent words of the apostles, which come as it were drop by drop, that is, in the form of definitive statements, but which infuse /**291**/ the fruitfulness of religious instruction very abundantly.[309]

34 SNOW

1. Ambrose says *that very often waters which are congealed by icy blasts of the winds solidify into snow, and that when the air is burst open the snow falls.*[310]

35 HAIL

1. The coagulation of hail occurs in a similar way. The waters of the clouds are drawn together and harden into ice by the rigour of the winds. Then the ice itself, partly crumbled into fragments by the crashing of the winds, partly loosened by the heat of the sun, falls piecemeal to the earth. As to the

---

309 Cf. Gregory, *Moralia* 20.2.5 (Adriaen, CCSL 143A:1005). Fontaine, *Isidore et la culture*, 562 notes the 'subtlety' here of Isidore's very concrete application of Gregory's allegorization of texts like Ps.142.6 (143.6), 'my soul is as earth without water unto you', to the physical phenomenon of rainfall.

310 Ambrose, *Hex.* 2.4.16 (Schenkl, CSEL 32.1:56).

fact that hail appears to be round, this is caused by the heat of the sun and the delay imposed by the restraining air as the fragments descend through the long distance from the clouds to the ground.

2. Allegorically, hail is the hardness of infidelity, frozen by the heavy lethargy of malice,[311] whereas snow stands for the unbelievers, who are exceedingly cold and sluggish,[312] and cast down into the depths by a heavy lethargy of spirit.[313] Likewise, in another interpretation, the snows are men frigid with respect to love, who, even though they /293/ are shining with the purity of baptism, do not burn with the spirit of charity.[314]

36 THE WINDS

1. *Wind is air stirred up and agitated,*[315] as Lucretius demonstrates:

For wind occurs when air is roused by being agitated.[316]

Even in a place of total calm and free from all winds, this can be proved with a small fan, with which we stir up the air and feel a breeze as we drive away the flies. When this happens as a result of some less obvious motion of heavenly or terrestrial bodies over a great expanse of the world, it is called wind, and from the diverse regions of the sky it is assigned names equally diverse.[317]

2. Some say that winds are born from air because air is born from waters. But Clement says that it is *because lofty mountains are found in certain regions, and the air, which has been as it were compressed and contracted, is by the ordinance of God driven and forced out from them as winds, by whose breath not only does the bud blossom into fruits but the summer's heat is tempered, when the fiery Pleiades are kindled by the heat of the sun.*[318] /295/

3. Sometimes, moreover, the winds are understood to be the spirits of

---

311 Cf. Gregory, *Moralia* 29.20.37 (Adriaen, CCSL 143B:1459).

312 Cf. Augustine, *Enarrationes in Psalmos* 147.23, ed. Franco Gori, CSEL 95/5 (Vienna: Verlag der Österreichischen Akademie der Wissenschaften, 2005), 234.

313 Cf. Augustine, *En. in Ps.* 147.23 (Gori, CSEL 95/5:233).

314 Cf. Augustine, *En. in Ps.* 147.23 (Gori, CSEL 95/5:233).

315 Augustine, *De quantitate animae* 4.6, ed. Wolfgang Hörmann, CSEL 89 (Vienna: Hölder-Pichler-Tempsky, 1986), 138.

316 Lucretius, *De rerum natura* 6.685.

317 Augustine, *De quantitate animae* 4.6 (Hörmann, CSEL 89:138).

318 Clement/Rufinus, *Recognitiones* 8.23.1–2 (ed. Rehm, 230.15–19).

the angels, who are sent by God's secretariat[319] throughout the whole world for the salvation of the human race. *Also the winds are sometimes* taken to be *inciting spirits, because by the breath of their evil tempting they kindle the hearts of the wicked to earthly desires, according to what is written: 'a burning wind shall take him up'.*[320]

37 THE NAMES OF THE WINDS

1. The first cardinal wind is Septentrio [N], which is cold and snowy; it blows straight from the north pole and generates arid chills and dry clouds. **It is also called Aparctias.**[321] Then there is Circius [NNW], which is also called Thrascias; thundering forth from the right of Septentrio,[322] it produces snow and coagulations of hail. [From the left there is] the wind Aquilo [NNE], which is also called Boreas. Blowing from on high, cold and dry and without rain, it does not disperse the clouds, but constricts them; hence and not without reason it represents the devil, because it constricts the hearts of the heathen with the cold of iniquity.

2. The second cardinal wind is Subsolanus [E], which is also called Apeliotes; it thunders forth from the east and is temperate. Then there is Vulturnus [ENE], which is also called Caecias, to the right of Subsolanus; it evaporates and dries up everything. Eurus [ESE], coming from the left side of Subsolanus, bedews the east with clouds. /**297**/

3. Auster [S] of the southern region, which is also called Notus, is the third cardinal wind; it blows from low down, moist, hot and filled with lightning, generating big clouds and very abundant rains, and destroying flowers. Euroauster [SSE], a hot wind, thunders forth from the right of Auster. Euronotus [SSW], a temperate wind, and hot, blows from the left of Auster.

4. The fourth cardinal wind is Zephyrus [W], which is also called Favonius; it blows from due west. With its delightful return, this wind

---

319 The image of the secretariat is borrowed from high imperial administration. See Fontaine, *Traité*, n. 178 (p. 357). However, Isidore was probably thinking of Psalm 103:4 (104:4): 'you who make your angels spirits, and your ministers a burning fire', here and in the following sentence, which he took from Gregory.

320 Gregory, *Moralia* 18.20.32 (ed. Adriaen, CCSL 143A:906) (Job 27:21).

321 This phrase is an 'Isidorian' interpolation, for which, see Introduction, pp. 46–47.

322 Personifications of the winds are depicted facing inwards around the circumference of a compass rose. Hence, a wind to the west of the north wind would be on its right, etc.

Diagram 7: The Winds

softens winter's rigor, and brings forth flowers. Then there is Africus [WSW], which is also called Lips, which thunders forth on the right side of Zephyrus. It generates tempests and rain-storms, and causes the collision of clouds and claps of thunder and the sight of frequent lightning flashes and the shock of lightning strikes. Next there is Corus [WNW], which is also called Argestes, blowing from the left side of Favonius; when it blows there are clouds in the east, but fair weather in India [Diagram 7: The Winds].[323]

5. Tranquillus[324] gives particular names to certain local breezes. These include the Syrus in Syria, the Carbasus in Cilicia, the Thracidas in the Propontis, the Sciron in Attica, in Galicia /**299**/ the Circius, in Spain the

323 The wind diagram is inserted at this point in Type 2 recensions. See fig. 7 in Fontaine, *Traité*, facing p. 296. For the discrepancies between Isidore's text and the inscriptions, see Commentary.

324 C. Suetonius Tranquillus (*c*.AD 69–*c*.140), author of the *Lives of the Twelve Caesars* and *The Meadows* (for which, see n. 329, below).

wind of the Sucro.[325] There are besides innumerable winds which take their names from rivers or lakes or regions. Besides these there are two that are found everywhere, exhalations rather than winds, the breeze and the sea-breeze.[326]

38 SIGNS OF STORMS OR FAIR WEATHER

1. *The tempest is the whirlwind of divine judgement,*[327] as the prophet says: 'God is in the tempest, and his ways are in the whirlwind'.[328] But fair weather is the joy of eternal light. In *The Meadows*, Tranquillus remarks as follows about the signs of storms for sailors: *a change for the worse in the weather is to be expected when during a nocturnal voyage the water sparkles around the oars and rudders.*[329] When squid[330] or flying fish fly, or when dolphins reveal themselves entirely in their leaps and strike the water with their tails, it signals a change of wind towards the south.[331] For the wind always rises from the direction towards which they head. It is no wonder that dumb animals forecast the weather although they are beneath the waves. For the waters are always churned up by the motion of an incipient breeze. The inhabitants of the waves are the first to feel this alteration of the sea. And so because of this impulse they struggle either from fear lest they be carried onto the shore, or from instinct lest the surge /**301**/ precipitate their destruction once they have turned in flight. What

---

325 The river Sucro, now the Júcar, in eastern Spain.

326 = *Etym.* 13.11.16. The wind diagram is inserted here in Type 1 recensions. For the two insertion points, see Introduction, pp. 34–35.

327 Gregory, *Moralia* 18.19.31 (Adriaen, CCSL 143B:905).

328 Nahum 1:3 (not Vulgate?).

329 Suetonius, *C. Suetoni Tranquilli praeter Caesarum libros reliquiae* 152, ed. August Reifferscheid (Leipzig, 1860), 233–34. *The Meadows* (*Prata*, or *De variis rebus*), is a largely lost encyclopaedic work of Suetonius that dealt with various aspects of natural history. Many of the fragments published by Reifferscheid are drawn from Isidore's text in *DNR*. Fontaine's scepticism about Reifferscheid's attribution of one particular passage in the *Etymologies* to Suetonius has general application: 'L'éditeur y reproduit simplement le texte d'ISID. *Orig.* 3,30, sans autre motif que sa conviction subjective qu'Isidore a utilisé le *Pratum* suétonien comme "source unique" païenne' (Fontaine, *Isidore et la culture*, 471 n. 1).

330 It is well known that several species of squid can propel themselves out of the water, e.g., the species of squid sold in Spanish markets even today as *voladores*. Cf. Pliny, *NH* 18.87.361.

331 Cf. *Etym.* 12.6.11. On the lore of dolphins as indicators of weather, see Commentary.

then? Do dolphins alone fear this violence? No indeed, other fish do as well. But only dolphins are seen, because they leap up out of the water.

2. Likewise, Varro says that it is a sign of a storm when there is lightning from the direction of Aquilo [NNE] and thunder from the direction of Eurus [ESE].[332] And Nigidius says: if the moon has black spots at the top of the crescent in the first part of the month, there will be (he says) rain; if it has spots on that same crescent area in the middle of the month, at the time when it is full, there will be fair weather.[333] Certainly, if it is red like gold, it forecasts winds. For wind is generated from the thickness of the air; when the sun and the moon are overspread by this thickness they become red. Similarly, if its horns are covered by a foggy mist, there will be a storm.

3. Moreover, Aratus says: if the northern horn of the moon is straighter, Aquilo [NNE] is imminent. Likewise, if the southern horn is rather more upright,[334] Notus [S] is imminent.[335] And the fourth day of the moon is a very certain indicator of future winds. Hence Vergil says:

But if at the fourth rising (for this is a very certain sign).[336]

4. Likewise, the same Vergil says: if the sun is spotted at its rising and hidden under a cloud, or if only half of it appears, there will be /303/ showers.[337] Similarly, Varro says: if the sun at its rising seems flattened, so that it shines in the middle and shoots out rays partly to the south and partly to the north, it forecasts wet and windy weather.[338]

5. The same author also says: if the sun is red, the day will be nearly perfect; if it looks pale, it signals storms.[339] Nigidius too says: if a pale sun sets into black clouds, it forecasts a north wind.[340] Likewise, the Lord says in the Gospel: 'If in the evening the sky is red, in the morning it will be fair weather. If in the morning the gloomy sky is red, there will be a storm;'[341] 'and when the south wind blows, there will be heat'.[342]

332 Varro, fragment.
333 Nigidius, fragment.
334 See Commentary.
335 Cf. Aratus, *Phaenomena*, ed. and trans. Douglas Kidd (Cambridge: Cambridge University Press, 1997), 794–95.
336 Vergil, *Georgics* 1.432.
337 Cf. Vergil, *Georgics* 1.441–43.
338 Varro, fragment.
339 Varro, fragment.
340 Nigidius, fragment.
341 Matt. 16:2–3 (not Vulgate).
342 Luke 12:55 (not Vulgate).

39 PESTILENCE

1. Pestilence is a disease spreading widely and infecting by its contagion whatever it touches. This sickness does not have a period of time during which either life or death may be expected, but a sudden faintness comes followed immediately by death. Some have stated what the cause of this pestilence is: *when plague [plaga] smites the earth because of mankind's sins,*[343] then from some cause, that is, either the force of drought or of heat or an excess of rain, the air is corrupted. And thus, with the tempered balance of the natural order disturbed, the elements are infected, and a corruption of the air and a pestilential breeze occurs, and /**305**/ a ruinous defect of corruption takes place in men and other living things.

Hence, Vergil says:

When the sky's expanse was corrupted and a piteous plague
Came upon the trees and the crops.[344]

2. Others say that *many plague-bearing seeds of things* [6.1093] are borne in the air and suspended and transported *into foreign* climatic regions either by the winds or *the clouds* [6.1099].[345] Then, wherever they are carried, either they fall throughout the region and corrupt all plant life resulting in the death of animals, or *they remain suspended in the air and, as breathing we inhale the breezes, we at the same time also absorb them into our body* [6.1128–30], and then, *enfeebled* by illness *the body* [6.1157] expires either *from foul sores* [6.1200] or from a sudden stroke. For just as the bodies *of those coming from afar* can *be disturbed by the strangeness of the climate or of the waters* [6.1103–05], with the result that they become diseased, so corrupt air coming from other regions strikes the body *with a sudden scourge* [6.1125] and unexpectedly extinguishes life.

---

343 Clement/Rufinus, *Recognitiones* 8.45.3 (ed. Rehm, 245.20–21). See n. 180 above.
344 Vergil, *Aeneid* 3.138–39.
345 Ch. 39.2 is unparalleled in *DNR* for Isidore's profound and often verbatim debt in this paragraph to Lucretius's lengthy analysis of the cause of disease in Book 6 of *De rerum natura*. We indicate verbatim borrowings and near echoes with the relevant line numbers from the poem.

40 THE OCEAN

1. Why the ocean turns back upon itself in alternating tides. *The philosophers, as Solinus puts it, say that the world, like a living creature, is formed from four elements and is moved by a certain breath, and just as reciprocal respirations take place in our bodies,* /307/ *so*[346] *some say that in the depths of the ocean there are certain passages for the breath of the winds, the nostrils of the world as it were, through which blasts of air, emitted or drawn back, by their accession and recession first blow out the seas with their exhalation, and then draw them back with their inhalation.*[347] But others are of the opinion that the ocean rises as the moon becomes fuller, *and it is as though it were drawn back by certain respirations of the moon and by the push and pull of the same planet it returns again to its wonted measure.*[348]

2. Others also say that the stars are nourished by the ocean's waters and that the sun draws up the water from the ocean by its fires and pours it around all the stars to temper them because they are fiery.[349] Then they say that, when the sun draws up the waters, it raises the ocean. But whether the waters are raised by the breath of the winds, or increase with the course of the moon, or subside when the sun draws them off, God alone knows, whose work the world is, and the entire plan of the world is known only to him.

3. *The size of the ocean is said to be beyond compare*[350] *and its width impassable. The disciple of the apostles, Clement, seems to indicate this when he says: the ocean is impassable to men, and those worlds which are beyond it are also unreachable.*[351] Moreover, the philosophers say /309/ that there is no land beyond the ocean, but that the sea and likewise the land beneath it are enclosed by the dense air of the clouds alone. Therefore, Lucan says:

---

346 The sentence up to this point, which is a feature associated primarily with the long recension, is an 'Isidorian' interpolation, for which, see Introduction, pp. 46–47.

347 Solinus, *Collectanea* 23.20–21 (ed. Mommsen, 107). In *Etym.* 13.15.1, Isidore attributes the tides to surface winds.

348 Ambrose, *Hex.* 4.7.30 (Schenkl, CSEL 32.1:136). The connection between the tides and the phases of the moon was first established by Posidonius (c.135–c.51/50 BC). See Stahl, *Roman Science*, 48.

349 Isidore here makes an oblique reference to the classical secular observation that the sun is the principal agent of evaporation. See chapter 41.1, below. For the notion that the fire of the sun is fed by water, see ch. 15.2; on the notion that celestial waters cool the extreme heat of the sphere of the stars, see ch. 14.

350 Augustine, *DGAL* 4.34.54 (Zycha, CSEL 28.1:135).

351 Rufinus, *Origen, De principiis* 2.3.6 (PG 11:194).

When the sea overwhelmed the peoples, when Thetis,[352]
Encompassed by the sky, would not let any shore be occupied.
Then also such a great mass of sea would have reached to the stars,
If the ruler of the gods above had not held down the waters by the clouds.[353]

41 WHY THE SEA DOES NOT GROW IN SIZE

1. Bishop Clement says that the reason why the sea does not become greater and does not increase at all with so great an abundance of rivers is *because salt water naturally consumes the fresh water that flows into it,*[354] and therefore it happens that, howsoever great the quantity of water that it receives, that same salty element of the sea nevertheless drinks it up. Add to this besides that which the winds bear away and the warmth and heat of the sun absorb. Finally, we see that lakes and many ponds are dried up within a short space of time by blasts of winds and the heat of the sun. Moreover, Solomon says: 'unto the place from whence the rivers come, they return'.[355] /**311**/

2. From this we understand that the sea does not increase: namely because the waters, after they have returned through certain hidden passages of the depths, flow back to their sources and run again in their accustomed course in their own rivers. The sea, therefore, was made to receive the courses of all the rivers. *Although its depth is varied, nevertheless the even smoothness of its surface is unbroken. Hence, it is thought to be called 'smooth and level' [aequor], because its surface is smooth [aequalis].*[356] But the natural philosophers say that the sea is deeper than the land is high.[357]

---

352 *Tethys*, Lucan; *Thetis*, Isidore. *Tethys*, with a long first syllable, was a sea-goddess, the wife of Oceanus; *Thetis*, with a short first syllable, was a sea-nymph, the mother of Achilles. It is not clear whether Isidore was aware of the distinction between the two or of the quantity of their first syllables (the hexameter requires that the syllable be long). See Fontaine, *Traité*, n. 196 (p. 360).

353 Lucan, *The Civil War* 5.623–26.

354 Clement/Rufinus, *Recognitiones* 8.24.2 (ed. Rehm, 231.9–10).

355 Eccles. 1:7 (Isidore perhaps assumes that his readership will recall the first half of this verse: 'All the rivers run into the sea, yet the sea does not overflow').

356 Ambrose, *Hex.* 3.2.8 (Schenkl, CSEL 32.1:64); = *Etym.* 13.14.2.

357 Fontaine translates *altius* as 'plus élevée', but to say that the sea is higher than the land makes little sense. It seems better to take it in the sense of 'deeper' when applied to the sea (see *altitudo*, 'depth', in the previous sentence) and in the sense 'high' (implied) when referring to the land.

42 WHY THE SEA HAS BITTER WATERS

1. Again, it is the doctor Ambrose who stated for our instruction: *the ancients say that the sea has salty and bitter waters because the waters that flow into it from the various rivers are consumed by the heat of the sun and the blasts of the winds; and they also say that as much water is consumed by daily heating as is carried into the sea every day from all the courses of the rivers. They claim that this phenomenon happens because of the sun, which absorbs whatever is pure and light, but leaves whatever is heavy and earthy, and this is what is bitter and undrinkable.*[358] /313/

43 THE RIVER NILE

1. Egypt is always sunny because of the heat of the atmosphere; it never receives clouds or rain. In the summer season, the region is inundated by the river Nile, which people use in place of rain. This river arises in the south-east. The Etesian winds[359] blow from the direction of Zephyrus, that is, from the west, and have a fixed duration. They spring up in the month of May. At first their breezes are light, but they increase from day to day.

2. These winds blow from the sixth to the tenth hour. Therefore, as the wind pushes back the water and heaps of sand obstruct the mouths of the river through which it flows into the sea, the surging flood of the Nile swells and is forced back, and so the bursting waters are propelled towards the south. When the waters pile up, the Nile bursts out over Egypt; but when the Etesian winds quieten down and the heaps of sand are broken through, the river returns once more to its channel.

44(–) THE NAMES OF THE SEA AND THE RIVERS[360]

1. In *The Meadows*, Tranquillus makes this assertion: the external sea is the Ocean; the internal sea is the sea which flows out of the Ocean; the Upper and the Lower Sea are the seas by which Italy is washed. /315/ Of

---

358 Ambrose, *Hex.* 2.3.14 (Schenkl, CSEL 32.1:53–54).
359 An annual, steady, summer wind of the eastern Mediterranean.
360 Ch. 44(–) is found always and only in MSS of the long recension. See Introduction, pp. 47–49.

the latter two, the Upper Sea is also called the Adriatic and the Lower, the Tuscan.[361]

2. A strait [*fretum*] is a narrow, practically boiling [*feruens*] sea, like the Straits of Sicily and Cadiz.[362] Estuaries are all the places through which the sea alternately first rushes in and then recedes.[363] The 'high' sea is, strictly speaking, the deep sea. Shoals are places in the sea where one can stand, which Vergil calls '*brevia*' and the Greeks call '*brachia*'.[364]

3. The greater recesses of the sea are called 'gulfs', like the Caspian, the Arabian, and the Indian;[365] lesser ones are called 'bays', like the bay of Paestum, the bay of Amyclae, and others similar to them. A swell [*flustra*] is the motion of the undulating [*fluctuantis*] sea in the absence of a storm.

Naevius in *The Punic War* says:

Loaded merchant ships were lying at anchor in the swells,[366]

as if to say: 'in the open sea'.

4. Moles are huge projections which extend into the sea. Pacuvius says of these:

All retreats blocked by a mole washed by the sea, bays ...[367]

A groundswell is a sea surge not yet flecked with foam, of which Atta in a *togata*[368] says:

Before the people they make groundswells by their discord.[369]

And Augustus says: *we came to Naples in spite of a groundswell.*[370] /**317**/

5. The shore [*litus*] is whatever is washed [*adluitur*] by water.[371] A stream is every body of water which flows at least moderately. A torrent

---

361 In *Etym.* 13.16.7, Isidore explains that the 'Upper' and 'Lower' seas probably derive their names 'from their position with respect to the sky ... because the east is upper and the west is lower' (trans. Barney, et al., *Etymologies*, 278).

362 = *Etym.* 13.18.2. The Straits of Cadiz = the Straits of Gibraltar.

363 = *Etym.* 13.18.1.

364 = (in part) *Etym.* 13.18.6.

365 = *Etym.* 13.17.1. In the *Etymologies*, Isidore specifies that these are gulfs of the ocean, not of the Mediterranean (the Caspian, of course, is not a gulf, but an inland sea).

366 Gnaeus Naevius, fragment.

367 Marcus Pacuvius, fragment.

368 *Comoediae togatae* were plays on Roman themes.

369 Titus Quinctius Atta, fragment.

370 Augustus, letter, fragment.

371 = *Etym.* 14.8.41.

is a river that is swollen from rain; it becomes parched [*torrescit*], that is, dries up, with drought, about which Pacuvius says:

The torrent is parched by the fiery heat.[372]

Mouths are the outlets of rivers into the sea.[373] Cascades are chutes of water, like those that are in the river Anio where its pitch is the steepest.

45(44) THE POSITION OF THE EARTH

1. As to how the earth, which rests upon the air, is thought to remain poised in balanced equilibrium, Ambrose says: *concerning the nature or position of the earth let it suffice to be known, according to the book of Job, that [God] suspended the earth in the void.*[374] Similarly, the philosophers also suppose that the earth is held up by thick air and hangs immobilized like a sponge because of its mass, *so that it is buoyed up by an equal motion on one side and the other,*[375] as by *wings beating like oars on every side,*[376] and it is unable to be tipped in any direction.

2. Nevertheless, whether it is held up by the thickness of the air or hangs above the water (because it is written: 'who established the earth above the waters'),[377] or *how the soft air is able to sustain so great a terrestrial mass, /319/ or, if such an immense weight is above the waters, how it is not submerged[378] or how it maintains an equal balance, so that, being weighed down, it does not incline to one side, this is not permitted to any mortal to know nor is anyone allowed to investigate so great a perfection of the divine art, just so long as it is known that the earth remains fixed by the law of the majesty of God either above the waters or above the clouds.*[379] For 'who is able to declare his works?',[380] says Solomon, or 'who has inquired

---

372 Pacuvius, fragment; = *Etym.* 13.21.2.
373 = *Etym.* 14.8.43.
374 Ambrose, *Hex.* 1.6.22 (Schenkl, CSEL 32.1:18); Job 26:7 (not Vulgate: paraphrased from wording in Ambrose).
375 Ambrose, *Hex.* 1.6.22 (Schenkl, CSEL 32.1:18).
376 Ambrose, *Hex.* 1.7.25 (Schenkl, CSEL 32.1:24), quoting Vergil, *Aeneid* 1.301.
377 Ps. 135:6 (136:6) (not Vulgate).
378 Ambrose, *Hex.* 1.6.22 (Schenkl, CSEL 32.1:18).
379 Ambrose, *Hex.* 1.6.22 (Schenkl, CSEL 32.1:18–19).
380 Ecclesiasticus 18:2.

into[381] the power of his majesty?'[382] Therefore, that which is hidden from mortal nature must be left to divine power.

46(45) EARTHQUAKE

1. Wise men say that the earth is like a sponge and that the wind when it has arisen swirls and passes through its cavities. And when this is so far advanced that the earth is unable to control it, the wind sends its roarings and rumblings in all directions. Then, when the earth is no longer able to keep it in, as it seeks a path of escape, the earth either shakes or splits open in order for the wind to blow out. Thereupon they say that an earthquake happens when a confined wind shakes all things to their foundations.

2. Hence Sallust says: *the winds rushed through the cavities of the earth; some mountains and hills, blasted apart, tumbled down.*[383] Therefore, an earthquake arises either because of a breath of wind through the cavities of the earth (as we have said) /**321**/ or because of the subsidence of lower parts and the movement of water. For this is what Lucan says also:

… with the earth splitting open,

… the Alps shook with unusual motions.[384]

3. Moreover, an earthquake happens frequently where there are cavities in the earth into which the winds enter and make the earth quake. For where the soil is sandy or the earth is tightly packed an earthquake does not occur. The earthquake refers to the Last Judgement, when sinners and earthly mortals will be convulsed after being struck by the breath of the mouth of God.[385] Also the earthquake is the conversion of earthly mortals

---

381 *Inuestigauit*, Isidore; *inuestigabit*, Vulgate, Fontaine. Fontaine, *Traité*, n. 205 (pp. 360–61), adopts the received Vulgate reading, despite the unanimity of the MSS of *DNR* against it. But *inuestigauit* is the form found in two Vulgate Spanish MSS (Cava, Archivio della Badia 1 [14], s. ix²; Madrid, Bibl. Nac., Vitr. 13–1 [Tol. 2–1], s. x), as well as the Northumbrian Amiatinus (Florence, Bibl. Mediceo-Laurenz., Amiatino I, s. viiiⁱⁿ), and an Austrian MS (Salzburg, Stiftsarchiv St Peter a. IX. 16, s. viii). See *Biblia Sacra Iuxta Vulgatam Versionem*, ed. Robert Weber. 4th edn, rev. Roger Gryson (Stuttgart: Deutsche Bibelgesellschaft, 1994), apparatus. We believe this justifies retaining the manuscript form.
382 Ecclesiasticus 18:3.
383 Sallust, *Histories*, fragment; = *Etym.* 14.1.2.
384 Vergil, *Georgics* 1.479 and 1.475 (not Lucan).
385 Cf. 2 Thess. 2:8.

to the faith. Hence it is written: 'his feet stood and the earth shook',[386] undoubtedly for the sake of faith.[387]

47(46) MOUNT ETNA

1. Justinus wrote this about Mount Etna in his book of *Histories*: *the earth of Sicily is thin and brittle, and so penetrable through certain cavities and openings that it lies almost completely open to the blasts of the winds. Also the substance of the soil itself is of a nature for generating and nourishing fires; indeed, it is said to be layered on the inside with sulphur and pitch. This is the cause of the fact that, as the air struggles against the fire within, the earth frequently and in many places belches sometimes flames, sometimes vapours, and sometimes smoke.*

2. *This is the reason, finally, that the fire of Mount Etna endures through so many ages,* /323/ *and, whenever the wind has pressed more violently through the vents of the caverns, masses of sand and stones are discharged. The Aeolian Islands are also always ablaze,*[388] *as though that fire were fed by their very waters. For so great a fire would never otherwise have been able to endure for so many ages in so narrow a place, unless it were fed by the nutriment of moisture.*

3. *Hence, therefore, this has been the cause of the fables of Scylla and Charybdis, of reports of barking, of incredible phantoms of monstrosity, while sailors, terrified by the great whirlpools of the sundered sea, imagine that the waves, which the swirl of the sucking surge beats together, are barking. The same cause creates the perpetual fires of Mount Etna as well. For that convergence of waters draws the air trapped within it into the lowest depth and holds it compressed there until, after being diffused through the vents of the earth, it kindles the nutriment of fire.*[389]

4. This is beyond doubt a symbol of hell, whose perpetual fire will spew out flames to punish sinners who will be tormented forever and ever. For

386 Isidore's source for this composite quotation (cf. Zech. 14:4; also Ps. 131:7(132:7) + Ps. 17:8(18:8)) is Gregory, *Homiliae in Hiezechihelem* 1.10.28 (Adriaen, CCSL 142:158), where Gregory employs it in the same context.

387 That is, the Old Testament reading has both its literal sense and a spiritual meaning for Christians: in this case, an earthquake is also a sign of conversion.

388 The Lipari Islands, off the north coast of Sicily.

389 Ch. 47 to this point is drawn from Marcus Junianus Justinus's *Epitome* of Pompeius Trogus's *Historiae Philippicae* 4.1.

just as these mountains have endured for such a long time and continue to endure /**325**/ right up to the present with raging flames that can never be extinguished, so that fire for torturing the bodies of the damned will never have an end.[390]

48(47) THE PARTS OF THE EARTH[391]

1. Now we will determine the position of the earth and explain systematically in what places the sea seems to have spread. *The earth, as Hyginus testifies, which is located in the middle of the universe, being separated by an equal distance from all its parts, occupies the centre. The ocean, spread over the region of the circumference of the globe, bathes the borders of nearly the whole orb. And accordingly the constellations when they set are thought to sink into it.*[392]

2. *The territory of the earth is divided into three parts, of which one is called Europe, another Asia, and the third Africa. Consequently, Europe is separated from Africa by the [Mediterranean] sea, beginning at the outer limits of the Ocean and the Pillars of Hercules. The mouth of the river Nile, which is called the Canopic mouth, determines the boundary between Asia and Libya together with Egypt.* **The river Don divides Asia from Europe; it thrusts itself in two branches into the marsh which is called Maeotis.**[393] *And Asia,* as the very blessed Augustine says, *extends from the south through the east to the north. Europe extends from the north to the west, and then Africa extends from the west to the south.*[394] /**327**/

3. In consequence, two parts, Europe and Africa, seem to occupy half of the world; Asia alone occupies the other half. But these two parts are formed because all the water which bathes the lands enters from the Ocean between them, and this water makes our Great Sea.[395] The geographers

---

390 Cf. Matt. 25:41, Apoc. 20:9–10.

391 Ch. 48 is found always and only in MSS of the medium and long recensions. See Introduction, pp. 42, 45.

392 Hyginus, *De astronomia* 1.9 (Viré, 11).

393 Hyginus, *De astronomia* 1.9 (Viré, 11). The sentence in **bold** is found in MSS of the medium recension only. See Introduction, p. 45. Maeotis is the Sea of Azov.

394 Augustine, *DCD* 16.17 (Dombart and Kalb, CCSL 48:521).

395 Augustine, *DCD* 16.17 (Dombart and Kalb, CCSL 48:521). The Great Sea is the Mediterranean.

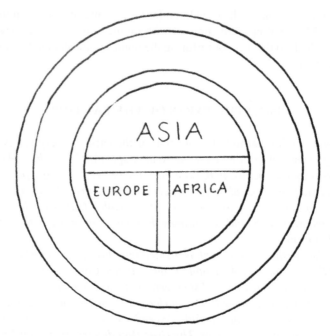

T-O Map: The World

have estimated the circumference of the whole earth to be one hundred and eighty thousand stades.[396] [T-O Map.][397]

---

396 Ambrose, *Hex.* 6.2.7. (Schenkl, CSEL 32.1:208).
397 A T-O map, which Fontaine does not reproduce, accompanies most MSS of the medium and long recensions. See Introduction, pp. 29–30. In some later MSS, the three continents are accompanied by extensive legends, for which, see Appendix 5b.

Commentary

PREFACE

The preface takes the form of a letter to King Sisebut, whom Isidore addresses as a king endowed with both the natural gift of eloquence and the acquired accomplishments of culture.[1] But Sisebut is extending his range into 'certain matters concerning the nature and causes of things' – nature and cause being interchangeable terms – and the implication is that *DNR* is the response to a request for guidance.

The order of the book is then laid out: units of time, the nature of the elements, celestial bodies, meteorology, the earth, and 'the alternating tides of the sea'. The underlying theme is a classic one in ancient natural philosophy – as old as the pre-Socratic physiologoi: the dialogue between order and change. Isidore, however, does not address the fundamental philosophical problem of change; rather, he exhibits change-within-order as the succession of times and seasons, the transformation of elements into one another, the movements of the sun, moon and stars, winds and weather, and the ambivalent nature of the earth as both stable and shaken by earthquakes and subterranean fires.

Isidore claims to be presenting extracts taken verbatim from 'the scholars of antiquity and especially ... the works of catholic authors'. The latter are especially important, because he wishes to deflect criticism that 'the nature of these things' is 'superstitious' knowledge (*superstitiosae scientiae*). Isidore claims that such knowledge is not *superstitiosa* as long as it is considered in accordance with teaching (*doctrina*) that is sound sana and sensible (*sobria*) – a phrase possibly lifted from Augustine, Epistle 204.7, where he speaks of *iudicio doctrinae sobriae*.

---

1 Fontaine, *Traité*, 337.

What exactly did Isidore mean by *superstitiosa*? Its classical meanings include 'irrational credulity' about cause and effect or 'excessive dread of the gods'; it implied immoderate and inappropriate scrupulousness about omens or apotropaic rites. It is tempting to think that Isidore, like Lucretius, is writing natural history to allay ignorant fears that natural phenomena were the intentional acts of divine beings. That he contrasts *superstitiosa* with *sobria* suggests that he grasped the ancient connotation of 'superstition' with intemperate religious behaviour.

However, *superstitio* underwent a significant change in the hands of Christian writers of the Patristic age. Origen and Eusebius in particular turned the tables on pagans who derided Christians as adherents of a bizarre and alien cult, by applying the term to pagan beliefs in the existence of the gods and daimones.[2] Hence Isidore's meaning may be that his learning (e.g., about the names of the planets, or the days of the week) is free of the taint of pagan religion.

Whatever he meant by *superstitiosa*, Isidore was explicitly concerned to defend his project from 'fundamentalist' critics, and this is worth dwelling on. One reason why he may have been so neutral regarding the origin of his information, so generally conciliatory in his attitude to pagan learning, and so interested in presenting classical cosmology as unimpeachably Christian, was the trend in sixth-century thought that consciously amplified the opposition between classical cosmography and a biblical understanding of the world.[3] But what ultimately justifies the study of the subjects covered in *DNR* is Sisebut himself. As a king, Sisebut can lay claim to the model of Solomonic wisdom represented in Isidore's quotation from Wisdom 7:17–19. This model encompasses the cosmos: 'the disposition of the heavens, and the virtues of the elements … the alterations of their courses, and the changes of seasons, the revolutions of the years, and disposition of the stars'. Oddly, given that *DNR* includes a hemerology, Isidore omits from the quotation 'the beginning and ending and midst of the times', but possibly he dropped the reference because his treatise will not deal with history. He also does not extend the catalogue of Solomonic wisdom to verse 20: 'the natures of living creatures and rage of wild beasts, the force of winds and reasonings of men, the diversities of plants and the virtues of roots'. In short, Isidore tailors his biblical justification to the

---

2 Dale B. Martin, *Inventing Superstition: From the Hippocratics to the Christians* (Cambridge MA: Harvard University Press, 2004).

3 See discussion of Cosmas Indicopleustes in Introduction, pp. 13–14 and 21.

outlines of a Lucretian 'nature of things', not vice versa. But he is also, at one stroke, evoking both the Bible and royal authority for his project.

Finally, Isidore gestures towards the order of creation in Genesis by saying that he will begin with the day. Of course, the first thing to be created after heaven and earth was light, which God subsequently called day, but there seems to be no precedent before Grosseteste for talking about light itself. In his *Etymologies* (13.10.14), Isidore only discusses light *qua* light in relation to the rainbow; light is a substance in itself (*ipsa substantia*), distinct from its brightness (*lumen*). Perhaps because light was not discussed in classical cosmology or meteorology, Isidore chose not to integrate it into his explanation of the day.

CHAPTER 1 DAYS

In the first chapter of *DNR* Isidore establishes a pattern which he will follow throughout the hemerological section of his book: (1) definition and differentiation; (2) parts of the day; (3) time-keeping customs and terms of ancient peoples; (4) different species of days.

The chapter on the day exposes the overlap in Isidore's mind between hemerology (the lore of divisions of time) and cosmology. A day is both a uniform unit of time-measurement and an event in the natural world – 'the presence of the sun between its rising and setting'. As Fontaine observes, this ambivalence is rooted in Isidore's awareness of the Genesis narrative, where the first three days (in the sense of units of time) unfold before the sun is created on the fourth day.[4] Hence the distinction between *propria* (technical, exact) and *abusiua* (casual, loose, colloquial) uses of the term 'day' also refers to primacy in the order of creation: the day as a unit of time preceded the day as a physical phenomenon. A 'technical' or hemerological day can thus begin at sunrise, noon, sunset, or midnight, according to regional or ethnic custom.

The paragraph (1.3) beginning 'Mystically' [*Mystice*] has long been considered a non-Isidorian interpolation into the text. This is a view which we challenge. To begin with, it parallels the unquestionably authentic closing sentence of this section, which begins with a cognate adverb *Prophetice* ('In a prophetic sense'). The content of the two paragraphs is

---

4 Fontaine, *Traité*, 7 n. 2; Hermann, 'Zwischen heidnischer und christlicher Kosmologie', 317.

similar; in both cases, the day represents the law. The mystical addition is an example of an 'exegetical' expansion.[5]

In the mystical addition Isidore lists the days of the ancient Jewish calendar that had particularly close associations with the Law: Passover; Pentecost, which commemorated the giving of the Law on Sinai; the Sabbath, the first day which God legislated as holy; the new moon; the seventh month, with its feast of Tabernacles. These are all times of rejoicing, but the next paragraph lists fasts associated with darker events: 'the fourth fast' in July, marking the disobedience of the Israelites and Moses's destruction of the tablets of the Law; the 'fifth fast' in August, marking the beginning of the forty years' wandering of the Jews, and the parallel destructions of Jerusalem by Nebuchadnezzar and Titus; the 'seventh fast' in October, coinciding with the slaughter of the Jews remaining in Jerusalem; and the 'tenth fast' that commemorated the arrival in Babylon of news of the destruction of the Temple. These parallels of rejoicing and mourning are picked up in the next paragraph ('In a prophetic sense'), a fact which further supports an argument for the authenticity of the mystical addition.

Isidore now shifts over to secular time-reckoning, explaining the terms used to distinguish days in the Roman calendar. In the Roman calendar, days were distinguished by their character in public law as *comitalis* (assembly days), *fastus* (days of public business), and *nefastus* (days when public business was forbidden). Of these Isidore only mentions *fasti*, but adds the common term *feriae* (festival holidays), as well as the more recherché or learned *profesti* (working days) and *proeliares* (days on which battle can be joined).[6] Isidore's sources are, as is typical, hard to pin down, but there are parallels in antiquarian treatises like Macrobius' *Saturnalia* 1.16 and Aulus Gellius, as well as in works of *grammatica* like Varro's *De lingua latina*. But Isidore innovates by linking the term *proeliares* to a phrase from the book of Kings.

Chapter 1 ends with a double handful of miscellaneous adjectives that can qualify the word 'day'. The first set refers to days that are inserted (intercalary) and days that are dropped out or left over (epacts); the second comprises the solstitial days, when the day (loosely speaking) seems to

---

5 See Introduction, p. 50.
6 On the character of days in the Roman calendar, see Bonnie J. Blackburn and Leofranc Holford-Strevens (eds.), *The Oxford Companion to the Year* (Oxford: Oxford University Press, 1999), 674–75.

remain the same length for a considerable time, and the equinoctial days, when the (technical) day is divided into equal portions of day and night.

CHAPTER 2 NIGHT

The Genesis account refers to the six 'days' of creation, beginning with the initial act of the creation of light. While the night is created along with the day by the separation of light and darkness, night itself is not treated as a kind of time; the day as a span of time is formed of 'evening and morning'. Night is only a cosmological phenomenon – the absence of sunlight between sunset and sunrise.[7] Isidore explains its cause (which is also its nature), namely the shadow cast by the earth as the sun rotates around it. While God made night as a time of rest, its meaning in Scripture is more sinister – persecution or spiritual blindness. This introduces a long chain of etymologies – a feature largely absent from chapter 1, and relatively rare in *DNR* as a whole. Unlike the precise and numbered hours of the day, the divisions of the night are indefinite in inception and duration. The divisions represent indeterminately bounded degrees of darkness, from dusk (*crepusculum*) to morning twilight (*matutinum*). Even their exact number is undefined: Isidore says there are seven, but lists only six (he corrects this in *Etym.* 5.31.4).

The identity of the evening star that defines eventide is uncertain, because Isidore says that it rises in the east, and Venus as the evening star would never rise in the east at sunset (though Mars could). In *Etym.* 5.31.5, Isidore states, 'Eventide [*uesperum*] is named for the western star [*Vesper*], which follows the setting sun and precedes the subsequent darkness', implicitly correcting the confused astronomy he gives here. Pliny's *Natural History* was almost certainly not available to Isidore at the time when he was composing these sections. If it had been, he would have learned that 'Vesper' was the Romans' name for Venus when it shone after sunset, and that 'Lucifer' was their name for Venus when it rose as the morning star in the east before dawn (*NH* 2.6.36).

7 See also the Commentary on ch. 28.

CHAPTER 3 THE WEEK

Isidore's statement that the Greeks and Romans have a seven-day week, while for the Hebrews a week is seven years, is perplexing, since the seven-day week with its Sabbath rest is inscribed in the account of creation in Hebrew scripture.[8] It seems probable that Isidore meant that the Jews also use the term 'week' to denote a period of seven years, as the remainder of chapter 3.1 makes plain. The terse reference to the seventy prophetic weeks of Daniel 9:24 inspired Bede to dilate on this in *The Reckoning of Time* 9, 'The Seventy Prophetic Weeks'.[9] For Bede, however, the seventy weeks became an exegetical and chronological puzzle of how to reconcile 70 × 7 years with the chronology of the period from the restoration of the Temple to the Incarnation.

Isidore introduces the Roman planetary names for weekdays in a neutral tone, reserving his scorn for the astrological anthropology that assigned elements of human nature or character traits to different planets. Where he found this precise arrangement of qualities is unknown: Servius' commentary on the *Aeneid* 11.51[10] assigns spirit to the Sun and body to the Moon, but the remaining planets are responsible for a different set of qualities from those enumerated by Isidore.

CHAPTER 4 THE MONTHS

In this chapter, Isidore distinguishes three types of month: the arbitrary and variable months of the Roman calendar, the lunar months of the Jewish calendar, and the mathematically equal months of the Egyptian calendar. However, he says nothing about the lengths of the Roman months, concentrating instead on an antiquarian and etymological exposition of their names. The subject of the names of the Roman months proved very popular

8 The ancient Romans used both a seven-day planetary week and a nine-day (nundinal) market week. The Judeo-Christian seven-day week with Sunday holiday was legislated for the Roman Empire by Constantine the Great, but the old planetary names for the weekdays persist in many European languages including English; see Daryn Lehoux, *Astronomy, Weather, and Calendars in the Ancient World*, 12–13 and Blackburn and Holford-Strevens, *Oxford Companion to the Year*, 566–68.

9 See commentary in Wallis, *Reckoning*, 279–80 and also Bede, *On Ezra and Nehemiah* 3, trans. Scott DeGregorio, TTH 47 (Liverpool: Liverpool University Press, 2006), 159–60.

10 See Yarza Urquiola and Andrés Santos's edition of book 5 of the *Etymologies*, p. 92 n. 2.

in the Middle Ages; Bede took it up in *DTR* 12 and *DT* 6, and it was from the Carolingian period onward frequently included on the pages of calendars.[11] Isidore's only traceable sources are Patristic, but this section (ch. 4.2–4) is paralleled by Macrobius' history of the Roman calendar in *Saturnalia* 1.12–15; indeed, Bede used the *Saturnalia* to flesh out Isidore's account.

Neither classical Antiquity nor the Middle Ages had a universally agreed upon month and day for the beginning of the new year. For the Romans, 1 January was the New Year from at least 153 BC; it was the day the consuls took office.[12] Antiquarian writers, however, preserved the memory of the earlier March beginning of the year. Since Isidore seems (from his content) to be drawing on an antiquarian source or sources, this is probably why he begins with March. The Church wanted to unhook the beginning of the year from 1 January, with its games and pagan auguries, while still keeping as close as possible to the Roman calendar. Gaul and Spain opted for Epiphany (6 January). Here and in the *Etymologies*, Isidore identifies both January and March as the first month. But, whereas in *DNR* he begins his enumeration of the months with March, in the *Etymologies* he begins with January.

Isidore comments that the division of the year into twelve months greatly pre-dates the innovation of 'Sancus, king of the Sabines', because it is found in the Old Testament. This is typical of his consistent policy of emphasizing the precedence of biblical over Classical tradition.[13]

While the Roman calendar seems to divide the months arbitrarily, and the Jewish calendar begins each month with the new moon, the Egyptian calendar comprises 12 months of exactly 30 days each, with five intercalary days added at the end to approximate the solar year. In the Pharaonic period, there was no leap year, so the calendar slowly drifted with respect to the seasons. When Augustus annexed Egypt, he instituted a reform initially planned by Ptolemy III in 238 BC to rectify the drift by adding a sixth intercalary day every four years. By Isidore's time, the Egyptian

11 Arno Borst's synthetic recreation of the 'Carolingian Imperial Calendar', based on a detailed comparison on dozens of ninth- to eleventh-century manuscripts, contains for each month of the year an appendix of additions (*Anhänge*) to the basic calendar-martyrology page. These additions include Isidore's histories of the Roman month names; Arno Borst, *Die karolingishe Reichskalender und seine Überlieferung bis ins 12. Jahrhundert,* MGH: Libri memoriales 2 (Hanover: Hahnsche Buchhandlung, 2001), 3 vols.

12 See Blackburn and Holford-Strevens, *Oxford Companion to the Year*, 6–7.

13 The assertion of the temporal priority of Christian over pagan wisdom and custom was a feature of Christian apologetics since the second century. See discussion of the Christian world-chronicle tradition in Kendall/Wallis, 25–30.

new year's day on 1 Thoth fell on 29 August (4th calends of September) on every year except the one preceding the Julian leap year, when it began on 30 August.[14]

To illustrate this, Isidore introduces the first of the many circular diagrams that gave *DNR* its alternate title *Liber rotarum*: the concordance wheel of Roman and Egyptian months. Isidore refers to the diagram somewhat unusually as a *formula*: his more common term is *figura*.[15] In this case, however, the *figura* is essentially a reference table presented in circular form.[16]

The *rota* of the months probably goes back to an ancient Greek hemerology, as conveyed in school manuals of Greek and Roman astronomy such as Hyginus, whose lost final book may have provided Isidore with his immediate model.[17] It takes the form of a wheel with 12 segments and 5 concentric rings. The data, reading from the periphery to the centre, are: (1) the name of the Roman month; (2) the date expressed in terms of the Roman month, when the Egyptian month begins; (3) the name of the Roman month; (4) '*diebus*' ('days'), and (5) 'XXX' ('30'). Isidore follows this diagram with a short explanation of the five intercalary days that round out the Egyptian year; he does not mention the sixth day added every fourth year. This information appears, however, in chapter 6.7. The repetition of the name of the Roman month is odd; it is possible that the ancestor of the *rota* contained the names of the Egyptian rather than the Roman months in (1), which would provide a true concordance. Bede in *DTR* 11 presents the same material in prose, but adds the names of the Egyptian months that correspond to the Roman ones.

Our 'translation' of this diagram, based on Fontaine's rendering of the illustration in Munich 14300, includes the human head which the artist placed in the hub of the wheel. Similar and equally enigmatic busts are found in the diagram of the zones in chapter 10, the *rota* of the planets in chapter 23, and the wind-rose in chapter 37. Other early manuscripts of *DNR*

---

14 Blackburn and Holford-Strevens, *Oxford Companion to the Year*, 708–10.

15 Four of the seven diagrams in *DNR* are identified by Isidore as *figurae*: the *rota* of the year, seasons and qualities (ch. 7.4); the schema of the five circles of the world (ch. 10.2); the *rota* of the macrocosm and microcosm (ch. 11.3); and the *rota* of the planets (ch. 23.4). The *figura solida* of the elements (ch. 11.1) is so labelled on the diagram, though Isidore's text refers to it as a *pictura*. The wind-rose (ch. 37.4) is not described by Isidore.

16 On the typology of diagrams, with particular reference to Isidore's *figurae*, see Wallis, 'What a Medieval Diagram Shows? A Case Study of Computus'.

17 Fontaine, *Traité*, 16; Fontaine, *Isidore: Genèse*, 298.

introduce human heads into the diagram of the elements (ch. 11). Bernard Teyssèdre suggests that the human heads constitute a moral or spiritual 'exegesis' of the diagrammatic material; this receives some support from the diagram of the months in Paris 6400 G, fol. 117v, showing a human figure from hips up, facing front, arms raised in *orans* posture.[18]

CHAPTER 5 THE CONCORDANCE OF THE MONTHS

The information conveyed in this brief chapter often appears in medieval *computus* manuscripts under the rubric *horologium* or *sundial*, because it describes the differences and similarities between different months of the year with respect to the lengths of daylight. The comparison is fairly coarse: December and January, which fall on either side of the winter solstice, are said to have equivalent amounts of daylight, as do June and July, which flank the summer solstice. Daylight increases from January to June at the same rate as it decreases from June to December; hence February and November have equivalent daylight, as do March and October, and so forth.

This quantum of daylight could be expressed either as the length of the shadow cast by the gnomon at noon, or the number of equal hours of daylight and night-time. Medieval calendars frequently contained a note at the foot of each month-page giving either or both pieces of information for each month, e.g., in January 'the night has 16 hours, the day 8 and [a gnomon of] 9 feet [casts a shadow] 16 feet long'.[19] In many cases, the hours inscribed in calendars are based on a standard system which sets the length of daylight in December at 9 hours, and increases it by one hour each month until June, when it is 15 hours long. This is an artificial scheme, but widely adopted throughout the Mediterranean and Byzantine world.[20]

But which *mensura* (length) was Isidore referring to: the length of the gnomon shadow or the length of the day in equal hours? It is not certain,

18 Bernard Teyssèdre, 'Les Illustrations du *De natura Rerum* d'Isidore', *Gazette des Beaux-Arts* 56 (1960), 24–26 and figs 1–2. Obrist, 'Le Diagramme isidorien', 100–01, offers a different hypothesis, namely that the human figures are in some instances echoes of ancient monumental versions of the diagrams; see below, p. 190.

19 *Calendar & Cloister*, fol. 16r (calendar for month of January in Oxford, St John's College 17); see also Peter S. Baker and Michael Lapidge (eds.), *Byrhtferth's Enchiridion*, Early English Text Society SS 15 (Oxford: Oxford University Press, 1995), 392.

20 For a medieval attempt to adjust the hours for more northerly latitudes, see McCulloh, '*Martyrologium excarpsatum*', in King and Stevens, *Saints, Scholars and Heroes*, 2:187, 197–237 and *passim*.

but at least one medieval tradition opted for the gnomon shadow, because this chapter was sometimes excerpted to accompany a diagram found in computistical manuscripts. The diagram is based on the principle of unequal hours – that is, that the day, however long or short it is, is deemed to have twelve hours, only these hours will be of shorter duration in the winter and longer in the summer. The diagram specifies the length of the measurements of these shadows: for example, in January and December, the shadow at the first and eleventh hours is 29 pedes, that of the second and tenth hours is 19 pedes, and so forth.[21]

Finally, this chapter gives Isidore an opportunity to indulge in some pleasant rhetorical variation by displaying every synonym he could muster for 'is equivalent': *concordat* (January–December), *spatium aequale consummat* (February–November), *consentit* (March–October), *aequat* (April–September), *respondit* (May–August) and *conpar est* (June–July).[22]

CHAPTER 6 THE YEARS

This chapter is structured in two parts. In the first part (sections 1–2) Isidore explores the defining feature of the generic 'year' – its completeness. The second part (sections 3–7) distinguishes the diverse periods which can be designated as a 'year'.

Isidore starts by defining a year as the sun's revolution over twelve months. From a purely astronomical or calendrical point of view, this 'year' is identical to the 'solstitial year' in section 4, which is defined as the circuit of the sun through the signs of the zodiac, and which is synonymous with the solar or civil year of 365 days. In *Etym.* 5.36.1, Isidore will combine all this information, specifying the sun's annual period as 365 days, and defining 'return' as a return to the same position vis-à-vis the backdrop of the fixed stars. But in *DNR* he chooses to begin with a pared-down definition of a generic year – a definition which stresses the completeness of the closed circle. This allows him to expatiate on the allegorical use of 'year' as a symbol of entirety, be it of human life, or of world history. The Latin word *annus* is etymologically related to *anus* ('ring'), and because a

---

21 Barbara Obrist, 'The Astronomical Sundial in Saint Willibrord's Calendar and its Early Medieval Context', *Archives d'histoire doctrinale et littéraire du moyen âge* 67 (2000): 71–118.

22 On such synonymy as a feature of late Antique rhetoric, see Fontaine, *Isidore et la culture*, 296–97 and 336–37.

ring has neither beginning nor end the year can begin wherever one wishes. Different peoples have arbitrarily chosen different points at which to begin the year. Moreover, the unqualified word 'year' can refer to a natural year, the 'great year', or the civil year. The rather odd definition of a civil year as the 'revolution of a single star', derived from Censorinus, will be dropped from the account of years in *Etym.* 5.36: its meaning is very obscure, but probably refers to the apparent annual rotation from west to east of the sphere of the fixed stars or constellations. It is particularly curious, because Isidore equates the solar and civil years in section 4. But what attracts his attention in section 2 is the idea of the 'revolution'. The same is true of the 'natural year', which is unexpectedly defined as a lunar eclipse. This may point to one possible motivation behind the book's composition, namely to present eclipses as unambiguously *natural* and hence deflate anxiety concerning them. This is reinforced by the reference to Thales of Miletus, who was famous for his ability to predict lunar and solar eclipses using astronomical calculations (as Isidore's probable source, Augustine, observed). Isidore may have intended to gesture here towards the idea that eclipses recur in cycles – in other words, there is an 'eclipse year'. This information was also dropped from the *Etymologies* – a fact which reinforces the hypothesis that unusual cosmic phenomena played a particularly significant role in *DNR*, but not in the later work.

The figure of 19 years for the Great Year is also peculiar, as it refers to the Metonic luni-solar cycle, when the phases of the moon will occur on the same solar calendar dates. It was also the period on which the cycle of Easters favoured by Isidore, though by no means by everyone in western Christendom, was based (*Etym.* 6.17.5). Isidore's support of the 19-year cycle as the basis for the Paschal *computus* may have persuaded him to dignify it with the title 'Great Year', even though this normally referred to a period when all the planets, and not only the sun and moon, would return to their initial positions with respect to the fixed stars. Was Isidore simply careless or confused when he claimed that 'all the heavenly bodies' come back to their initial places after 19 years? Or was he deliberately trying to substitute a Christian *annus magnus* for the unacceptable astral determinism of the pagan Great Year, after which events would repeat again in the same order? In *Etym.* 5.36.3, Isidore is more faithful to classical tradition, but less precise: the Great Year is when all the planets 'return to the same place', but he does not specify its length, only that it may occur 'after many solstitial years'. The difference between *DNR* and the Etymologies on this point seems to indicate Isidore's intention in

DNR to attack astral determinism in some form, perhaps pagan, perhaps Priscillianist. The classical Great Year doctrine was strenuously opposed in works by the Christian Fathers, including some which were well known to Isidore, such as Clement of Alexandria's *Stromata* and Augustine's *The City of God*.[23]

In the remaining sections of this chapter, Isidore will list types of years which vary according to their duration. The generic year which began the chapter now reappears as the solstitial, solar, or civil year, defined astronomically by the path of the sun through the signs of the zodiac, and calendrically by 365 days. Lunar years are 'common' (i.e., 12 months of 29½ days, or 354 days) or 'embolismic' (13 months, or 384 days). In every 'common' year, the new lunar year begins 11 days before the end of the solar year. After about 3 years, these 11 days have cumulated into a whole extra lunar month, the 'embolismic' month. When Isidore claims that Easter is 'delayed' during an embolismic year, what does he mean? In the Alexandrian–Dionysian system of Paschal *computus* he favoured, Easter in embolismic years always falls in April and never in March, but the verb 'delayed' (*protenditur*) strongly suggests that (as in *Etym.* 6.17.22) Isidore is merging the embolismic year with the postponement of Passover to the second month mandated for those who are ritually defiled or on a journey (Numbers 9:9–11).

The leap year signals a shift from different types of years to time-periods formed of groups of years: the Jubilee, the Olympiad, the lustrum, the indictional cycle, and the era.

CHAPTER 7 THE SEASONS

Isidore's hemerology plays consistently on the ambiguous nature of units of time: are they human artifices of measurement? Or are they entities in nature? This chapter begins by considering the seasons as the physical products of the sun's course and the motion of the heavenly bodies. Almost immediately, however, we are vaulted into the less certain realms of human convention. The Old Testament prophecy of Daniel underscores the ambiguity of the term tempus, which can mean 'season' or 'time'; the

---

23 Godefroid de Callataÿ, *Annus Platonicus: A Study of World Cycles in Greek, Latin and Arabic Sources* (Louvain-la-Neuve: Université catholique de Louvain, Institut orientaliste, 1996), esp. pp. 88–97.

Patristically sanctioned reading of Daniel 12:7 – *tempus et tempora et dimidium tempus* – was that *tempus* referred specifically to a year, and the time until 'the fulfilment of these wonders' would be three and a half years. For 'the Latins', it is the reverse: a season is a part of a year, and there are four of them: winter, spring, summer, and autumn. Following Ambrose, Isidore defines the seasons in terms of events in nature, correlating the sun's annual swing to the north and south with the variable length of day and night, and the concomitant weather and growing conditions.

The circular diagram of the four seasons in relation to the four primal qualities of hot, cold, wet, and dry has a distinctive form which seems to bespeak an established tradition. It is the first true diagram in *DNR* – that is, an abstract (non-mimetic) schema that includes words, and is intended to illustrate a proposition. This diagram invokes an intelligible reality using a geometrical sign (the circle), a mathematical symbol (the number four), and a pattern of spatial arrangement (opposing seasons, linked qualities).[24]

We have 'translated' Fontaine's rendering of the image in Munich 14300, which shows a circle with four arcs inscribed equidistantly around the inside of the rim. There is space between the arcs, in which are inscribed the names of the four seasons and their corresponding cardinal directions – something Isidore does not mention in his text – beginning with spring/east at the top. From the mid-point of the arcs, a second set of four arcs is inscribed, bisecting the first set of arcs, and forming a diagonal cross with a central space. The arcs of this second set slightly overlap, so that the arms of the cross look like leaves. The form conveys the impression of contrastive forces held in balance.[25]

The bisected arcs are inscribed with the four qualities, each quality being written twice, once in each half of the arc. Thus *Ver/oriens*, 'spring/east', is flanked on the left by *humidus*, 'moist', and on the right by *calidus*, 'hot', and the continuity to summer is represented by the second *calidus* lying to the left of *Aestas*, 'summer'. In the central space is the word *Annus*, 'year'. There is some perfunctory foliage decoration in the centre, as well as foliage and interlace around the rim. The schematic foliage in the Munich diagram might seem to echo the description of the four seasons

24 Wallis, 'What a Medieval Diagram Shows?', 3; Barbara Obrist, *La Cosmologie médiévale: textes and images*, vol. 1, *Les fondements antiques*, Micrologus' Library 11 (Florence: SISMEL/Edizioni del Galluzzo, 2004), 20–24.

25 Bianca Kühnel, 'Carolingian Diagrams, Images of the Invisible', in Giselle de Nie, Karl F. Morrison, and Marco Mostert (eds.), *Seeing the Invisible in Late Antiquity and the Early Middle Ages* (Turnhout: Brepols, 2005), 363–64.

which Isidore took from Ambrose's *Hexaemeron*, where the changes in the seasons are marked by the growth patterns of plants. But the presence of the cardinal directions argues that Isidore did not invent this diagram, but took it from an earlier source – indeed, a source that might have been available to Ambrose himself.

The *rota* of the seasons and year illustrates the genesis and passage of the seasons as a progress through combinations of qualities. In this sense it is quite different from conventional Roman iconography of the seasons, which linked the year to the zodiac and/or *Aion*. Barbara Obrist expresses surprise that no ancient schema of the qualities has survived, despite the strong interest in this subject. She explains this as due to the fact that such an image would only be included in a school manual, not in monumental art.[26] Fontaine identifies it as a '[f]igure scolaire destinée à illustrer un rudiment de physique aristotélicienne',[27] and proposes that its source was a commentary on the *Physics* mediated through a late antique Latin compendium. However, the resemblance to the *rota* of the macrocosm and microcosm in chapter 11 also raises the possibility that Isidore's source was a medical work.

A manuscript source seems all the more probable given that this *rota* unlike others in Munich 14300 does not have a human head in the hub, or any personification of the seasons. Obrist regards these representational elements as clues to the influence of monumental rather than manuscript art,[28] and there are numerous surviving representations of the personified seasons grouped in ways which evoke Isidore's *rota*, particularly in mosaics.

Isidore's dates for the beginning of the seasons do not conform to any known classical source. Varro set the beginning of each season on the 23rd day after the sun's entry into Aquarius, Taurus, Leo, and Scorpio, i.e., 7 February, 9 May, 11 August, and 10 November.[29] Pliny shifted the dates slightly, and based them on different criteria: spring begins on the 25th day of the sun's course through Aquarius (8 February); summer on the heliacal rising of the Pleiades (10 May); their setting on 11 November marks the beginning of winter. Autumn begins at the autumn equinox.[30] Bede broadly follows Pliny, but again changes the rationale: the seasons commence at the

26 Obrist, 'Le Diagramme isidorien', 100–01.

27 *Traité*, 16.

28 Obrist, 'Le Diagramme isidorien', 102.

29 Varro, *De re rustica* 1.28; see Lehoux, *Astronomy, Weather, and Calendars in the Ancient World*, 51.

30 *NH* 2.47.122–25.

midpoints between solstice and equinox or equinox and solstice.[31] Isidore does not tell us why he chose to begin the seasons on 22 February, 24 May, 23 August, and 25 November, though it is noteworthy that these dates fall one month before the solstice or equinox. Coincidentally, modern meteorological seasons fall about 20 days before the solstice or equinox, but these are moveable dates based on annual temperature cycles.

Finally, Isidore signals the end of the hemerology with a 'recapitulation' of the divisions of time. These parallel chapters 1–7, save for the mention of the hour and its subdivisions at the end, for Isidore in fact discusses no division of time smaller than the day.

CHAPTER 8 THE SOLSTICE AND THE EQUINOX

The solstices and equinoxes serve as a logical bridge between the hemerology and the cosmology, because they have precise dates, but are not divisions of time. Isidore's dates for the solstices and equinoxes are those of the Julian calendar adopted in Rome in 45 BC, and they were accurate at the time the calendar was established. However, it was known even in Antiquity that the mean tropical year is slightly shorter than 365.25 days (it is now about 365.2422 days), which means that the equinoxes and solstices arrive a bit earlier with respect to the calendar every year. By the early fourth century, the vernal equinox was falling on 21 March, and this date, confirmed by the astronomer Ptolemy, was adopted for the purposes of Paschal *computus* in the Greek-speaking world. In the Latin-speaking part of the Christian world, however, the Julian 25 March date persisted until the middle of the fourth century; on the westernmost fringes, in Ireland and Britain, it lasted even longer.[32] Hence it is not surprising that Isidore records the traditional dates, even though they were no longer astronomically accurate, nor did they apply to Paschal reckoning. As Bede will later observe in *The Reckoning of Time* 30, the weight of secular authority, and many of the Church Fathers, would back Isidore up. So does the symbolism of Christmas at the old winter solstice and Midsummer Day (24 June) at the old summer solstice as the celebrations of the birth of Christ and John the Baptist respectively, for the waxing and waning of daylight after these turning points exemplify John's prophecy that 'He (Jesus) must increase,

---

31 See Bede, *DT* 8, Kendall/Wallis, 111 and n. 35.
32 Wallis, *Reckoning*, xxxix.

but I must decrease' (John 3:30). It is a telling illustration of how little religious allegory Isidore actually intrudes into *The Nature of Things* that he does not mention this parallel.

Isidore's explanation of the solstices and equinoxes is exceedingly compressed. He does not explain what it means for the sun to 'stand still' (i.e., for its daily path from sunrise to sunset to halt in its annual advance to the north and south, and apparently 'mark time' for several days before reversing direction) or how this produces the lengthening and shortening of days. By not explaining the celestial mechanics of the solstices and equinoxes, Isidore misses his opportunity really to knit the theme of time reckoning into his cosmology, which begins in the next chapter.

Isidore remarks that at the equinox, the day and night are of equal length, and so their conventional twelve hours each are 'hours of equal duration'. In Antiquity and in the Middle Ages, the periods of daylight and night were for some purposes of time-telling divided into twelve hours respectively. Except at the equinoxes, these hours were 'unequal', since in summer the twelve hours would be distributed across a longer span of time than in the winter. The hours of day and night were equal in length only at the equinoxes. Ancient and medieval people also divided the day into twenty-four equal hours (see Commentary on ch. 5); equal hours were sometimes called 'equinoctial hours', because they matched the natural length of the hours at the equinoxes. This terminology is used, and explained, by Bede, *The Reckoning of Time* 3.

Isidore's statement that the summer solstice is called the 'lamp' (*lambada*) is puzzling, as it does not come from any identified source. And while the sun may produce greater heat after the solstice, it cannot be said to become brighter, as the days are actually shortening. This information is omitted in the parallel passage in *Etymologies* (5.34.1).

CHAPTER 9 THE WORLD

This chapter restarts the narrative of *DNR* from a fresh point of view, namely that of the material universe, its structure and contents. Isidore begins with the *universitas* itself, the sum of everything you get when you add together heaven and earth. Isidore's term for this ensemble is *mundus* – a word which, somewhat confusingly, he will later apply to planet Earth. Here, however, he focuses on explaining how *mundus* connotes both 'entirety' and 'shape' or 'form'. Both elements appear is his quotation from St Paul, 'for the fashion (*figura*) of this *mundus* passes away'. The *mundus* here is evidently the whole of creation, but what passes away is its *figura*. The primary connotation in classical Latin of *figura* is 'shape' or '(what is cast in a) form'. It glosses the Greek *schema* and when applied to images usually denotes representations that are defined by an armature of lines, e.g., a geometrical figure or an architectural plan. For example, in Manilius' *Astronomica* (first century AD), a constellation, a group of stars that can be represented by a natural form when bounded or joined by lines is a *figura*. Like a rhetorical 'figure' (already a commonplace expression by Quintilian's day), a *figura* could be a (visual) metaphor or allegory. By the Middle Ages, however, the word *figura* had acquired a complex burden of theological and exegetical significance. It came to denote a way of reading scripture that interpreted Old Testament people and events as phenomenal prophecies or prefigurations that were fulfilled in the New Testament.[33] Thus the *figura* of this world is passing into the fulfilment of the world to come.

The idea of shape, however, resonates with Isidore's second paragraph, where he equates *mundus* with the Greek *cosmos*, a term which denotes 'array' (though Isidore does not hint that he knows its meaning). Here he introduces the theme of the microcosm, one of his signature ideas. When section 3 inaugurates a discussion of the universe's *formatio* or design, Isidore actually turns the analogy inside out. The cosmos is a human being writ large. It is 'upright', with the north celestial pole at the 'top'; it has a head and a face, and parts that, like human limbs, are four in number. These are the cardinal directions. Isidore has captured here the ancient connection between the concept of direction and the form of the human body. When one 'faces' the sun, and extends one's arms, the axes of the body define the directions.

It is worth noting the curious inversion of authorities in this chapter. The scientific definition of the cosmos is conveyed in a passage from the Bible, while the macrocosm–microcosm analogy, based entirely on classical physics and physiology, becomes the 'mystical sense'. Isidore's characteristic blend of secular and religious knowledge reaches the point of fusion here.[34]

---

33 See Erich Auerbach, 'Figura', in Erich Auerbach, *Scenes from the Drama of European Literature*, trans. Ralph Manheim (Minneapolis: University of Minnesota Press, 1984), 11–78, esp. 13–23.

34 Fontaine, *Isidore et la culture*, 546.

CHAPTER 10 THE FIVE CIRCLES OF THE WORLD

The shape or form of the universe, its fundamental architecture, was one of the two stock topics about science that were proposed to students of rhetoric in classical schools (the other being the size of the sun), and it was ubiquitous in the scholia and commentaries on Vergil.[35] Notwithstanding, this chapter and its accompanying diagram bristle with difficulties. Isidore begins by stating that the ancient philosophers divide the cosmos into five parallel regions. Since he uses the terms 'circles' and 'zones' (*zona* = 'belt'), the assumption is that he is speaking of the Hellenistic spherical universe, and the quotation from Vergil's *Georgics* indicates that he is referring particularly to the celestial sphere. He explicitly states that he is talking about the heavens in the parallel passage in *Etym.* 3.43.1.

To visualize these 'belts', he invites his reader to use a flat surface, namely his hand. Each of the five fingers represents one of the zones. At the 'top' is the thumb, standing for the Arctic Circle.[36] However, Isidore has silently shifted from thinking of the celestial sphere to thinking of Earth, and instead of notional *circles* projected onto the heavens he is visualizing climatic *zones* on our planet.[37] The Arctic Circle is 'uninhabitable because of the cold', but of course it is the whole region of the planet north of the Arctic Circle which is intended. Similarly, the index finger represents the 'summer circle', but what is meant is not the Tropic of Cancer, where the sun reaches its maximum elevation in summer. Rather, it is the band of lands somewhere between the torrid equatorial zone and the Arctic Circle that are 'temperate' between extreme heat and extreme cold, and therefore habitable. The middle finger is the uninhabitable equatorial 'circle', i.e., zone, though how far it extends on either side of the equator is not specified. The ring finger is the winter 'circle' corresponding to the summer 'circle', and likewise habitable – though Isidore does not say whether it is in fact inhabited. The little finger is the Antarctic Circle. Then Isidore abruptly shifts back to the celestial sphere in order to define the circles as imaginary lines on the inner surface of the spherical universe. The second circle is now the 'solstitial circle', that is, the line traced by the sun when it is at the northernmost point of its elevation. We would call this line the Tropic of

---

35 Fontaine, *Isidore et la culture*, 470–71.

36 Hermann, 'Die astronomischen Metaphern', 445–46. The image was probably derived from a commentary on *Georgics*, as there is a rough parallel in Probus; see Fontaine, *Isidore et la culture*, 487.

37 Fontaine, *Isidore et la culture*, 489.

Cancer, because the constellation which is the backdrop of the sun at this point is Cancer.

The reader's uncertainty as to whether he is dealing with the universe/ *mundus* or the Earth/*mundus* is amplified by Isidore's diagram illustrating the five circles. Given his analogy between the five fingers and the five zones, we might expect this *figura* to look like the representation of the spherical earth with its five zones found in Macrobius' *Commentary on the Dream of Scipio* 2.5.13 *sqq.* and 2.6.2 *sqq.* If one imagines one's hand held open with the five fingers together, the fingers could serve as a reasonable mnemonic for the climates. Notice that the relative length of the fingers mimics the proportions of the lines on the globe: the thumb and little finger are the shortest, and stand for the Arctic and Antarctic Circles; the longest circle, the equator, is represented by the longer middle finger, and the more or less equal index and ring fingers stand for the equal and intermediate tropics.[38]

But this is not what Isidore chooses to represent. Instead there is a framing circle within which are evenly distributed five roundels, containing the texts on the five zonal circles of the world, beginning at 10 o'clock with the Arctic Circle, and moving clockwise. Jacques Fontaine finds this schema 'absurd' but declines to ascribe it to Isidore, suggesting instead that it might stem from a school introduction to Aratus.[39] It is hard to imagine how it could have been devised. It is unlikely that someone simply misunderstood 'circles' to mean the detached and distributed circles depicted here, because the inscribed text retains the designation of the equitorial circle as *medius* or 'middle'. Did someone (Isidore's source? Isidore himself?) take hold of the hand image, but imagine the hand with its fingers spread open rather than closed, and then transform the fingertips of this open hand into five circles arranged around a central 'palm'?[40] This hypothesis finds some support in a variant form of the schema found in

---

38 The Macrobian zonal map is based on the one in Willis's edition of the *Commentarium in Somnium Scipionis*, p. 165. The Macrobian type of world map that represents the whole globe of the earth, divided into latitudinal zones or *climata*, enjoyed wide diffusion in the Middle Ages; see David Woodward, 'Medieval *Mappaemundi*', in J.B. Harley and David Woodward (eds.), *The History of Cartography*, vol. 1, *Cartography in Prehistoric, Ancient and Medieval Europe and the Mediterranean* (Chicago: University of Chicago Press, 1987), 295–97.

39 *Traité* p. 16; *Isidore et la culture*, 488.

40 Stahl's claim in *Roman Science*, 221, that '[Isidore's] juxtaposition of arctic and antarctic circles and his accompanying diagram indicate that he was conceiving here of five circular areas on the earth's flat surface', finds no support in the evidence.

the Carolingian 'imperial computus encyclopaedia', e.g., Vienna NB 387, fol. 133, where the zone-circles are arranged in a semi-circle around a square *mundus* like fingertips around a rectangular palm. On the basis of this Carolingian diagram, Obrist posits that the centre of the Isidorian schema represents *terra*,[41] though she does not attempt to relate the human head found in the Munich manuscript, and a number of others, to the personification of Earth. In any event, Isidore seems to have intended this image, and none other, as an illustration of chapter 10; his claim (10.3) that 'the northern and southern [polar] climates are associated with each other (*sibi coniuncti*, literally 'joined to one another') makes literal sense as a description of this schema, where the Arctic and Antarctic zones are side by side. Indeed, perhaps he selected it precisely because it provided visual confirmation of Hyginus' statement that the two polar zones are alike in being uninhabitable due to extreme cold.

In books 3 and 13 of the *Etymologies* Isidore designates the 'antarctic' circle, which he explicitly states is opposite to the Arctic Circle, as the *fourth* circle, and the 'winter tropic' as the *fifth*. Since he repeats this 'mistake' in both passages in the *Etymologies*, we should entertain the hypothesis that it was perhaps deliberate, i.e., that he counted down from the north pole to the equator, and then up from the south pole to the equator. He would thus approach the middle circle from both extremes. A Macrobian-type zonal map would simplify this visualization, but, as we have seen, Isidore chooses not to represent the zones in this way.

Isidore's attempt in section 4 to associate what is in effect the north temperate zone with the east and the south temperate zone with the west is cryptic, and is not repeated in *Etymologies*. In his previous references to the 'summer' circle he employed the Greek term *therinos*. The apparent internal contradiction in the first sentence of section 4 ('The solstitial [summer, second] circle that is located in the east between the northern [first] and summer [*aestiuum*, second] circle and the one that is placed in the west between the summer [*aestiuum*, second] and southern [fifth] circle …') can be resolved if we assume that Isidore was thinking of the equatorial circle when he wrote *aestiuum*. Reading the diagram in this way, circle 2 [summer solstitial] in the east (top) is between circle 1 [Arctic] and circle 3 [equatorial]. Circle 4 [winter solstitial] in the west (bottom right) is between circle 3 [equatorial] and circle 5 [Antarctic]. The assumption that 'summer

---

41 Obrist, 'Le Diagramme isidorien', 148, esp. n. 202. The Vienna diagram appears as her fig. 29.

circle' in this section means the equatorial circle also makes sense of the final sentence: the Ethiopians are burnt black because they are nearest to the equatorial zone (which is uninhabitable). It is not irrelevant to note that *oekumene* world maps, like the T-O map that accompanies chapter 48(47), are oriented with east at the top.

CHAPTER 11 THE PARTS OF THE WORLD

The notion that the material world is composed of the four elements, fire, air, water, and earth, was undisputed scientific dogma by Isidore's day. However, there were two theories of how the elements join together to constitute compound entities that can cohere, but also can dissolve and reassemble – in other words, two ways to explain both stability and change. Isidore will discuss both views in this chapter.

In book 2 of *On Generation and Corruption*, Aristotle argued that each of the primal elements was composed of two qualities, one from the binary hot–cold and another from the binary wet–dry. Isidore will present this model in the second part of this chapter.

First, however, Isidore addresses the rival explanation first propounded by Porphyry in his (lost) commentary on Plato's *Timaeus* 31B–32B, and then taken up by Proclus and Calcidius.[42] There are three qualitative binaries arranged on the axes of acuity (sharp–blunt), density (thin–thick), and capacity for motion (mobile–immobile). Each element represents a different combination of these qualities, and they form a graduated hierarchy: earth is blunt/thick/immobile, water is blunt/thick/mobile, air is blunt/thin/mobile, and fire is sharp/thin/mobile. The advantage of this schema is that it replicates the natural physical location and behaviour of the elements; fire tends to rise and earth to fall, while water sits on top of earth, and air on top of water, but beneath the fires of the heavenly bodies.

But what will keep the elements together, particularly since fire and earth share no common qualities? In the *Timaeus*, Plato describes how fire and earth are joined not by means of shared qualities but by mathematical proportions. The first numbers to emerge from the Monad, 2 and 3, are

---

42 Obrist, *La Cosmologie médiévale*, 270–301. The relevant passage in Proclus is *Commentaire sur le Timée* 3.39.22–28, ed. A.J. Festugière, 5 vols. (Paris: J. Vrin, 1966–1968), 3:67–68; in Calcidius, *Timaeus a Calcidio translatus* 22, ed. Waszink, 72–73; see also the translation accompanying the new edition of Calcidius, *Commentaire au Timée de Platon*, trans. Béatrice Bakhouche (Paris: J. Vrin, 2011), 229.

transmuted into the three dimensions of the material world through cubing: 2 cubed is the number of earth (8), and 3 cubed the number of fire (27). The two intermediates are rectangular solids formed by different combinations of these primary numbers: $2 \times 2 \times 3 = 12$ (water) and $2 \times 3 \times 3 = 18$ (air).

Isidore, however, says nothing about numbers. Instead, he explains the ability of elements to combine in quasi-Aristotelian terms as the linking of corresponding qualities of acuity, density, and mobility. Earth and fire cannot share any qualities, but Isidore implies that they can combine through the mediation of one of the intermediate elements, though he does not explain how. Instead, he appends a diagram, 'lest these bewildering matters not be understood'.

This *figura* (Diagram 4) is challenging to read, even in early and reliable manuscripts. The version in the Munich codex from which Fontaine derived his rendering (p. 212 *bis*), and which we have translated, shows two partially overlapping rectangles, one higher up and to the right, the other lower down and to the left, positioned so that their lower-left and upper-right corners respectively are at the centre point of the other rectangle. The effect is of a transparent three-dimensional rectangular box with open sides and top; it is also an optical illusion, since either the lower or the upper rectangle could be the 'front' of the box. Diagonals from the bottom corners of the lower left-hand rectangle to the lower right corner of the upper right-hand rectangle outline the 'floor'. The diagonal running from the lower left corner of the lower left box is continued through to the upper right corner of the upper right box, and the upper left and lower right corners of the upper rectangle are also joined by a diagonal, making a St Andrew's cross on the upper rectangle. The overlap between the two rectangles thus forms a smaller rectangle, divided into two triangles by a diagonal running from lower left to upper right. 'Water: thick, blunt, and mobile' is in the upper left triangle, 'Earth: thick, blunt, and immobile' is in the lower right. The top and right-hand triangles formed by the St Andrew's cross in the upper right rectangle are occupied by 'Fire: thin, sharp, and mobile' and 'Air: mobile, sharp, and thick' respectively. The lower left rectangle has no additional diagonal to form a St Andrew's cross. In the space remaining in its upper left triangle after the 'Water' triangle is deducted is the inscription, 'This figure is solid according to geometrical proportion'. This leaves a number of blank spaces: three smaller triangles, and a square directly beneath the Water and Earth triangles. In the Munich manuscript, two of the triangles are inhabited by dragons, while the square contains a rayed disk; a similar rayed disk is positioned in the L-shaped space outside the diagram at the upper left. These look like representations of the sun, and in many Carolingian and later renderings of the diagram they are explicitly labelled as sun and moon.

Jacques Fontaine notes that this is the only diagram in *DNR* that Isidore introduces with a first-person statement. Fontaine initially concluded that the diagram was Isidore's invention,[43] but later modified his view, arguing that it was equally likely that that it stems from a commentary tradition on Plato's *Timaeus* 31B–32B.[44] *The phrases 'figura solida' and 'secundum geometricam rationem'* echo chapter 22 of Calcidius' commentary: mathematical science demonstrates how *solida corpora* are knit together from the qualities of acuity, density and mobility 'according to geometrical proportion' (*geometrica iuxta rationem*).[45] However, there is no diagram in Calcidius which matches Isidore's. Obrist points out that Isidore's diagram bears some resemblance to the parallelepipeds which Calcidius deploys to illustrate how four cubes can be arranged into a continuous solid figure in such a way that two of them do not actually touch one another (like the incompatible elements earth and fire) but are linked through two intermediate cubes, as in Figure 1.[46]

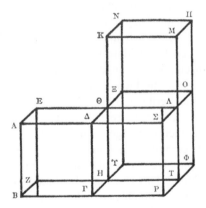

Figure 1: Calcidius' Parallelepiped

43 See Introduction, n. 9; this is also the view of Eastwood, 'Diagram of the Four Elements'.

44 See Fontaine, *Isidore et la culture*, 657–59 and Smyth, *Understanding the Universe*, 58 and 71 (who regards this derivation as possible, but doubtful).

45 *Timaeus a Calcidio translatus*, ed. Waszink, 73; *Commentaire au Timée de Platon*, trans. Bakhouche, 228.32; Eastwood, 'Diagram of the Four Elements', 551–52.

46 Calcidius, *Timaeus a Calcidio translatus* 18–20, ed. Waszink, 68–71; the diagram reproduced here is on p. 69; *Commentaire au Timée de Platon*, trans. Bakhouche, 223–27.

The Isidorian schema distorts Calcidius' form by introducing diagonals, but also displaces it from a mathematical model to a physical representation. The elements are assigned to the triangles within the 'solid'.[47] Obrist argues that the *figura* was originally an attempt to harmonize what Plato says about the *physical* bonds holding the four elements together in three-dimensional bodies and what he says about the *geometric* form of the elements (all being composed of triangles, fire = tetrahedron, air = octohedron, water = icosahedron, earth = cube). As Calcidius points out (*Timaeus a Calcidio translatus*, chs. 20–23), the problem was how to form a cycle from these elements, because fire/pyramid does not share an angle with earth/cube. He invokes the notion of the three qualities (mobility, density, acuity) as a way of joining the elements. The Isidorian *figura solida* was the remnant of an ancient commentator's efforts to resolve this problem.[48] Isidore draws heavily on Cassiodorus, and the *Excerptum de quattuor elementis* appended by a later editor to book 2 of the *Institutions* of Cassiodorus contains a discussion of the geometry of the four elements and their three primal qualities, but the surviving diagrams do not resemble the *figura*.[49]

The proponents of the 'copy' hypothesis – Fontaine, Obrist, and Gorman ('Diagrams', 533) – conclude that the form of diagram which appears in early manuscripts like the Munich codex closely represents the design that Isidore borrowed and that reconstructions or prototypes are speculative. An ingenious attempt at such a reconstruction is proposed by John E. Murdoch, *Album of Science: Antiquity and the Middle Ages* (New York: Scribner, 1984), 280–81, no. 247. The idea is that each of the six faces of a cube represents one of the six qualities. These are aligned so that their convergence at four of the eight corners of the cube will produce the combinations found in the elements. The other four corners represent impossible combinations, e.g., immobile – thin – blunt, or mobile – thin – blunt. This is a logical solution, but, as has been observed,[50] it has no textual or graphic cognate in the ancient or medieval record.

Bruce Eastwood on the other hand argues that Isidore did not adapt a diagram from Calcidius or any other source but created the *figura* himself based on analogy with the second diagram illustrating this chapter of *DNR*,

---

47 Obrist, *La Cosmologie médiévale.*

48 Obrist, 'Le Diagramme isidorien', 155–57.

49 Cassiodorus, *Institutiones*, ed. Mynors, 167–68; for discussion, see Obrist, *La Cosmologie médiévale*, 284–89.

50 Obrist, *La Cosmologie médiévale*, 276; Gorman, 'Diagrams', 531–34.

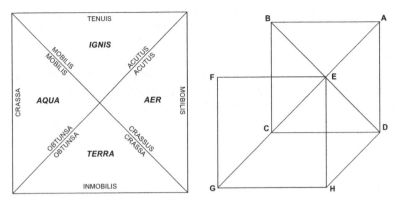

Figure 2: Eastwood's Reconstruction

the wheel of the elements (see Figure 2). Eastwood's reconstruction puts the square divided diagonally into four triangles representing the four elements at the beginning rather than the end of the process of creating the diagram. It is, he argues, a graphic argument demonstrating how the six qualities both link and distinguish the elements. The lines shared between the triangles represent the qualities shared by the elements, while their unique qualities (e.g., earth's immobility) form the outer boundary lines. The resulting square is then turned into a three-dimensional cube by superimposing the second square and drawing in the 'floor'.[51]

Eastwood's reconstruction is attractive; his argument that Isidore invented the diagram explains both the problem of the absence of any traceable model and the difficulty that later scribes and illustrators encountered in reading and replicating this image. The weakness of the hypothesis lies in the fact that if Isidore designed this diagram it was the only one he designed. Hence, Obrist's position that he copied the illustration from some untraced commentary on the *Timaeus* remains at least plausible.

The lack of any explanatory context or established iconographic tradition for this confusing diagram led to further loss of integrity in the *figura* over time. Isidore himself may have quietly abandoned the *figura* as

---

51 Eastwood, 'Diagram of the Four Elements', fig. 1.2; his argument is laid out on pp. 548–54.

a way of representing the elements later in his career.[52] As time passed, its internal structure broke down even further, until the diagram was reduced to two slightly overlapping squares.[53] Isidore's medieval readers and copyists also tried to reinterpret the diagram as a *rota*, thereby aligning it with the second, more comprehensible illustration of the elements presented in this chapter. In Exeter 3507, for example, the cube metamorphoses into a circle enclosing a Z; a marginal rubric even describes it as a *circulus*. In the late-eleventh-century MS Bodleian Auct. F.2.20 (possibly also from Exeter), the elements are presented as a stacked hierarchy, with earth at the bottom and fire at the top, enclosed in a circular frame.

In sections 2–3 of this chapter, Isidore presents the Aristotelian model of the elements, illustrated by a very different kind of diagram, with a very different *fortuna*. The text, taken from Ambrose's *Hexaemeron*, not only explains how the four qualities are shared amongst the elements, and thus links them into a whole, but it also evokes ideas of love and harmony. The elements join hands through their shared qualities, and their circle becomes a dance.[54]

Formally, this diagram (Diagram 5) is very close to the *rota* of the seasons, year and qualities in chapter 7, with two overlapping sets of four arcs forming spaces between the qualities, and a central hub. The spaces between the qualities contain (reading from the rim towards the centre) the four elements (with *ignis* at the top), the corresponding seasons, and the corresponding humours. In the central hub is *Mundus – Annus – Homo*.

Like Ambrose's explanation, itself derived through St Basil of Caesarea from the traditions of the ancient philosophical schools, this schema summarizes, but also elaborates, Aristotle's concept, adumbrated in *On Generation and Corruption*, of the how oppositions and combinations of

---

52 In the Spanish family of manuscripts of the *Etymologies*, there is an image of the elemental solid as a rectangle divided into four by diagonals, with the names of the elements in the quarters: El Escorial, Real Monasterio de San Lorenzo, Biblioteca MS &.I.3, fol. 59r, reproduced in Obrist, *La Cosmologie médiévale*, fig. 107; this version supplanted the original *figura* in some manuscripts of *DNR*, e.g., Vatican City Reg. 255, reproduced ibid., fig. 109.

53 These transformations are documented in the plates of Eastwood, 'Diagram of the Four Elements', and Gorman, 'Diagrams', plates III–VIII.

54 This is visually evoked by the variant version of Isidore's diagram found in a collection of excerpts from Bede's *DNR* and *DTR* in Cava de' Tirreni, Archivio e Biblioteca della Badia della SS. Trinità 3, fol. 199r. The elements are personified as four bearded figures in classical garb (philosophers?) linking hands in a circle: Mario Rotili, *La miniatura nella Badia di Cava*, 2 vols. (Naples: Di Mauro, 1976–1978), vol. 1, no. 1; cf. Obrist, 'Le Diagramme isidorien', 103.

elemental qualities maintain stability through perpetual cyclical change in the physical world, including the seasons and the human body. The basic form of the schema was probably devised to illustrate an ancient commentary on Aristotle's text,[55] though the inclusion of the four humours of the human body – something not mentioned in Isidore's text – suggests some cross-fertilization from a medical source.[56]

This diagram was destined to achieve unparalleled diffusion in the medieval period. While its core arrangement was stable, it could be varied, elaborated, and decorated in myriad ways.[57]

CHAPTER 12 HEAVEN AND ITS NAME

This chapter marks a second threshold in *DNR* – the first being the transition from hemerology to cosmology in chapter 8. From this point onwards, Isidore's text will follow the vertical order of the cosmos from top to bottom, beginning with the heavens and ending in the middle of the earth.

Isidore begins by establishing the identity of 'heaven' and 'the heavens' by calling on biblically based symbolism: 'heaven' (singular) is the Church, and 'the heavens' is its citizens. He then turns to Ambrose for an evocative etymology: heaven (*caelum*) is 'engraved' (*caelatum*) with the stars 'like signs', like a plate of silver engraved with signs in relief. The image evoked here is of a silver bowl or dish, with *repoussé* and chased decoration, like the great dish of the Mildenhall Treasure (fourth century). Indeed, in *Etym.* 3.31 (Gasparotti and Guillaumin, *Etym.* 3.30), Isidore explicitly likens the heavens to an engraved bowl (*uas caelatum*). The comparison

55 P. Vossen, 'Über die Elementen-Syzygien', in Bernhard Bischoff and Suso Brechter (eds.), *Liber Floridus: mittellateinische Studien Paul Lehmann gewidmet* (St. Ottilien: Eos Verlag der Erzabtei, 1950), 33–46.

56 Obrist, *La Cosmologie médiévale*, 300–04; Fontaine thinks the schema originated with Isidore (*Traité*, 297), but also speculates concerning a medical intermediary. Isidore does not actually mention the microcosmic dimension of the diagram in his text, and indeed there are versions of this schema which omit the four humours. He does, however, discuss humours in a manner which essentially conveys the content of this diagram in *Etym.* 4.5.3 (Fontaine, *Isidore et la culture*, 666 and n. 2).

57 Obrist, 'Le Diagramme isidorien', 95–164. Obrist compares versions of this diagram in Laon 423, fol. 12 (fig. 3); Munich 14300, fol. 8 (Fontaine's model); and Einsiedeln 167, p. 92 (where *mundus–annus–homo* is framed in a square: Obrist regards the square as a symbol of the three-dimensional universe). On the Laon version, see also Kühnel, 'Carolingian Diagrams', 364.

of the heavens to a work of virtuoso craftsmanship evokes both pagan and Christian ideas of the divine Artificer,[58] but it also calls to mind by anticipation Einhard's description of Charlemagne's silver table which depicted the Ptolemaic system of the universe in three concentric circles.[59] Such prestigious representations of the heavens in engraved form, whether working astrolabes or purely decorative pieces, must have struck ancient writers with a deep feeling of awe.

The image of the heavens as an inverted bowl of silver slides imperceptibly into Isidore's description of heaven. It is of a *subtilem naturam*, 'exquisitely fine', and it is ambiguously both diaphanous and solid, 'firm like smoke', in Isidore's oxymoronic formulation based on Isaiah 51.6. In chapter 13.2, Isidore uses the biblical term 'firmament', *firmamentum*, to stress its solid support of the supercelestial waters. But, by separating the terms *caelum* and *firmamentum*, perhaps Isidore is evading the question of whether the firmament is really solid.[60]

In 12.3, Isidore describes the 'parts' of what is really a model of the heavens. First there is the vault (*cohus*) – a poetic term for the visible dome that seems to sit like an inverted bowl over the earth. The 'axis' runs from the north to the south celestial 'poles' – the ever-visible Boreus and its inferred southern cognate, Austronotius. The 'pivots' are the ends of the axis.[61] The climes (as Isidore explains in *Etym.* 3.41) are not the 'climates' of the latitudes but the four cardinal directions.[62] Less easy to decode are the 'domes' (*convexa*), which Isidore characterizes rather vaguely as 'the extremities of heaven'.

In ch. 12.4, Isidore sets this machinery in motion. The *sapientes* hold that the whole sphere turns on its axis once a day, from east to west. This diurnal motion (although this is not explained by Isidore) carries the sphere of the fixed stars, as well as all the planets, around the earth once a day. Isidore is more concerned to establish, using the words of Ambrose and

---

58 Fontaine, *Isidore et la culture*, 473; Hermann, 'Die astronomischen Metaphern', 444–45.

59 Charlemagne's Testament, which Einhard purports to be quoting, refers to the representation as a 'description of the whole world' (*totius mundi descriptionem*). That it was a representation of the Ptolemaic system is known from other sources. See Einhard, *Vita Karoli Magni: The Life of Charlemagne*, ed. Evelyn Scherabon Firchow and Edwin H. Zeydel (Dudweiler: AQ-Verlag, 1985), 116 and n. 7 (p. 138).

60 See Smyth, *Understanding the Universe*, 109–13.

61 On Isidore's sources for and use of these terms, see Fontaine, *Isidore et la culture*, 480–84.

62 On the double meaning of the term, see Fontaine, *Isidore et la culture*, 485–86.

the Pseudo-Clementine *Recognitions*, that the universe is well and truly spherical, and that earth lies at its centre. Indeed, the sphericity of the universe proves the necessity of the biblical 'waters above the firmament', which act as a sort of lubricating fluid, preventing the build-up of excess heat under the revolving orb – a Stoic concept that, with some interpretive effort, could be made to harmonize with Genesis.[63] Isidore will return to the waters above the firmament in chapter 14.

The ultimate reason for the emphasis on the sphere, however, becomes evident in chapter 12.5: the spherical shape of the universe is proof of the rational mind of its Creator – and even the pagans, like Plato, acknowledge this.[64] Isidore adduces three arguments to support this assertion. First, the zodiac 'consists of one line' although 'drawn from the angles formed from five lines'. This is an extremely puzzling claim. The Latin is itself opaque: *ex linearum quinque angulis zodiacus ductus* – literally, 'the zodiac, extended from the angles of the five lines' – or, possibly, 'the zodiac, extended from the five angles of the lines'. This second construction is probably incorrect: if Isidore is imagining the zodiac intersecting the equator, and touching the two tropics, eight angles would be formed, and not five, as in Figure 3.

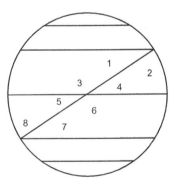

Figure 3: Zone Diagram
Showing 8 Angles

It seems that *quinque* (five) must apply to the lines. The 'angles', then, refer to the 'corners' from which the oblique zodiacal band is 'extended' in such a manner as to connect five lines.

63 Fontaine, *Isidore et la culture*, 478–79.
64 On the importance of this theme for Isidore, and his dependence on commentaries on the *Timaeus* for the arguments which follow, see Fontaine, *Isidore et la culture*, 476–78, and our Introduction, p. 15.

Exactly the same phrase is found in *Etym*. 3.45. Fontaine translates this sentence by supplying a relative clause: the zodiac is drawn from the angles *which it forms* with five lines ('tracé partir des angles *qu'il forme* avec cinq lignes').[65] He justifies this by claiming that Isidore is describing a two-dimensional projection of the zodiacal band intersecting the five circles of latitude of chapter 10 (the equator, the tropics of Cancer and Capricorn, the Arctic and Antarctic circles) on the celestial sphere, as in the Figure 3. The five circles would be represented by five parallel straight lines through which the zodiac would cut at an oblique angle. The line connecting the five lines (i.e., the zodiac) would be straight.[66]

The problem with this explanation is that the zodiac does not touch the Arctic and Antarctic circles: it cuts through the equator, and touches the tropics of Cancer and Capricorn. Moreover, as we have shown, it is not five but eight angles that result.

Gasparotto and Guillaumin, in their edition of book 3 of the *Etymologies*, accept Fontaine's emendation, but not his explanation, precisely because the 'five lines' would have to include the Arctic and Antarctic circles. Since the zodiac joins the two tropics and crosses the equator, it touches only three circles, not five. They suggest that Isidore may have had a defective diagram in front of him – one which showed the zodiac (more precisely, the ecliptic) as in Figure 4.[67]

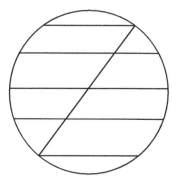

Figure 4: Zone Diagram
Showing 5 Lines

65 Fontaine, *Traité*, 220, lines 43–44.
66 Fontaine, *Isidore et la culture*, 490–91.
67 Isidore, *Etym*. 3, ed. Gasparotto and Guillaumin, 108 n. 264.

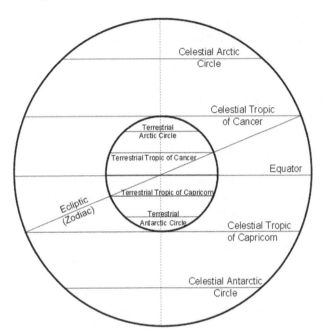

Figure 5: Macrobian Spheres

The problem with this explanation is that no such defective diagram has ever shown up in the manuscript record. Moreover, the fact that Isidore repeated this information verbatim in the *Etymologies* indicates that he never saw any reason to modify or correct what he said in *DNR*. Isidore never says that the zodiac or ecliptic is in fact bounded by the tropics; hence he may, indeed, have imagined that it ran at an angle from the Arctic to the Antarctic circles. There is, however, another possibility: Isidore may have been visualizing a diagram such as the one associated with Macrobius' *Commentary on the Dream of Scipio* 2.7.4 *sqq*. This diagram shows the terrestrial sphere with its equator, tropics, and Arctic and Antarctic circles, nested inside the celestial sphere, with its equator, tropics, and Arctic and Antarctic circles. In principle, the circles on the celestial and terrestrial spheres are projections of each other, but in order to depict the two unequal-sized spheres in two dimensions, the circles on the smaller, terrestrial sphere will not overlap with those on the larger, celestial one, except for the equator. The ecliptic or zodiac in the Macrobius diagram is depicted joining the two *celestial* tropics, crossing the two *terrestrial* tropics and the (overlapping) celestial/terrestrial equator. In consequence, the ecliptic or zodiac appears to touch or intersect with five lines: the celestial tropic of Cancer, the terrestrial tropic of Cancer, the celestial/terrestrial equator, the terrestrial tropic of Capricorn, and the celestial tropic of Capricorn, as in Figure 5.[68]

In the end, what Isidore meant by the 'five lines' remains elusive. The force of his argument lies in another direction entirely. He wants to prove the *rationality* of the celestial architecture by asserting its *sphericity* using three arguments, of which this is the first: that the zodiac is one single circular line, although it is represented as a straight one, 'extended from the angles (or corners) of the five lines'. Despite the fact that a diagram on the page of a book will show the zodiac as a straight line intersecting with other straight lines, it is in fact a circle wrapped around a sphere. This is totally consonant with Isidore's insistence on the sphericity of the *mundus*, but also reflects a certain awkwardness when it comes to describing, and perhaps even conceptualizing, three-dimensional objects.[69]

Isidore's second and third arguments also aim to identify sphericity with the rationality of the Artificer. The second argument concerns the infinite character of a sphere, which has neither beginning nor end; the third is somewhat more opaque, namely that the sphere 'is constructed beginning from a point'. Moreover, it is self-moving, and finally it incorporates the heavenly bodies. This seems to be an echo of *Timaeus* 34a, where Plato describes the Demiurge's plan for the sphere of the universe. This sphere is a uniform outer 'body' of the world, and at the same time the projection outwards of the world-soul from the centre of the universe to its extremities; it is self-moving and complete.[70]

For Isidore, the universe's circular motion is rationality and intelligence itself, in contrast to the linear motions of forward, back, right and left, up and down. Plato expresses the same idea, namely that circular motion, 'belongs to reason and intelligence'.[71] But Isidore's summation is difficult

68 Diagram adapted by Kendall Wallis from the illustration in James Willis's edition of Macrobius' *Commentarii in Somnium Scipionis* (Leipzig: Teubner, 1963), 166.

69 This is noted by Fontaine, *Isidore et la culture*, 403–04, in his discussion of the geometry section of the *Etymologies*.

70 Plato, *Timaeus* 34A, trans. Francis M. Cornford, *Plato's Cosmology: The Timaeus of Plato translated with a running commentary* (London: K. Paul, Trench, Trubner & Co./New York: Harcourt, Brace, 1937), 58.

71 *Timaeus* 34A, trans. Cornford, *Plato's Cosmology*, 55; the six linear motions are actually enumerated in 43B.

to understand: 'it necessarily follows that this line cannot be drawn beyond the circle'. To which 'line' is Isidore referring? It could be any of the 'lines' inscribed on the diagram of the universe – the ecliptic, equator, etc. – none of which projects beyond the sphere of the world, but the demonstrative *haec* points to the *linea* discussed at the beginning of this paragraph, namely the zodiac or ecliptic. It cannot be extended beyond the sphere of the world because it is actually a circle notionally inscribed on the sphere of the world, though Isidore does not offer this explanation.

Isidore likewise does not make it plain that the 'two axes on which heaven revolves' are in fact a single axis with two poles. He does, however, allude to the apparent dual motion of the universe. The entire sphere of the heavens appears to rotate once each day from east to west, carrying the constellations of the fixed stars and the planets as well. However, the planets also seem gradually to move in an eastwards motion along the line of the zodiac, so that each diurnal rotation of the outermost sphere brings the planet back to a position a little bit behind its starting point of the previous day. Isidore refers to these planets as *astra* and his quotation from Lucan calls them *sidera*, but they are not the constellations fixed in the dome of heaven, which is turned by the axis running from pole to pole. Rather, they are the 'wandering stars' or planets, and Isidore imagines their slow west to east progress as a break on the extreme velocity of the rotation of the universe. These form the subject of Isidore's next chapter.

CHAPTER 13 THE SEVEN PLANETS OF HEAVEN AND THEIR REVOLUTIONS

Chapters 13–21 form an exceptionally coherent and tightly structured block, each chapter following closely on the previous one, with the motifs and arguments referring back and forth. One is left with the impression that Isidore is building up to his explanation of the eclipses by reinforcing the themes of divine rationality and providential care, exhibited in the order in the heavens. The first paragraph of chapter 13 doubles back to the discussion at the beginning of chapter 12, where Isidore invoked the biblical meanings of (singular) 'heaven' and (plural) 'heavens'. In this chapter, he turns to Ambrose to explicate the natural distinction between singular and plural: there is one single heaven, but the seven spheres of the planets contained within it are also 'heavens'. Ambrose's statement that the spheres of the planets are 'connected with (*innexos*) and as it were nested inside (*insertos*) each other' calls to mind the concentric orbits of the diagram of the planets in *DNR* 23.

Hilary is then summoned to echo the claim of chapter 12 that the heavens exhibit the rational plan of a divine Creator. The rationality of his design, however, is exhibited in the 'fire-proofing' built into the form of the universe. Hilary distinguishes between the superior heaven, the perfectly spherical and uniform home of the angels,[72] and the lower heavens of astronomy, which exhibit diverse types of motion. Between the two lie the 'waters above the firmament' of Genesis 1.7–8; indeed, the outer edge of the lower heavens *is* the firmament, so-called because it holds up these superior waters, which temper the heat of the upper heaven and prevent it from unleashing a cosmic conflagration.

CHAPTER 14 THE HEAVENLY WATERS

The 'waters above the firmament' of Genesis 1:7 constituted a line in the sand for many of the Church Fathers – a point on which Scripture and pagan science could not be reconciled, and on which the superior authority of Scripture would have to prevail. Origen attempted to resolve the problem of how water could rest on top of the fiery lower heavens, in defiance of the physics of the elements, by interpreting the supercelestial waters as an allegory. This gambit provoked a strenuous response from Basil of Caesarea in his *Hexaemeron* – the model and source of Ambrose's work of the same name. Basil saw it as a craven surrender of the authority of revelation: the waters had to be literal waters, and no scientific explanation of their existence was required. Augustine was less of a rigorist, endorsing an allegorical approach in *On Genesis against the Manicheans* (*De Genesi contra Manichaeos*) while offering a thoughtful and cautious defence of the reality of the supercelestial waters in *The Literal Interpretation of Genesis*. Though Isidore knew both these works, he chose here to follow in Ambrose's footsteps.[73] Ambrose offers no positive explanation of the

---

72 Though Isidore does not dilate on this point, it is noteworthy that Hilary depicts the heaven where the angels dwell as part of the material universe. Fontaine, *Isidore et la culture*, 547 ascribes this somewhat shocking inability to distinguish between spiritual and material as an index of the cultural weight of pagan cosmology, and of the persistent influence of the idea that the blessed departed went to live amongst the stars.

73 Thomas O'Loughlin, '*Aquae super caelos* (Gen 1:6–7): The First Faith–Science Debate?', *Milltown Studies* 29 (1992): 92–114; Fontaine, *Isidore et la culture*, 548.

waters. Instead, he dismisses the objections of 'worldly philosophers' who complain that the liquid waters would run down the sides of the sphere of the universe: God who created the universe *ex nihilo* could surely make the heavenly waters stable. At this point, Isidore inserted two words not found in Ambrose: *glaciali soliditate* 'with the solidity of ice'. This phrase comes from Augustine's *The Literal Interpretation of Genesis* 2.5. Augustine is laying out the position of a hypothetical Christian who tries to refute the pagan view that the heavens, being the realm of the element fire, could not sustain water. Augustine's imaginary Christian replies that pagans actually believe the highest reaches of the cosmos to be cold, since they claim that Saturn, the planet closest to the outermost zone of heaven, is by nature cold. How could Saturn be cold unless there is water in its vicinity, congealed into ice (*glaciali soliditate*)? The irony is that Augustine thinks this a rather foolish argument; Isidore, however, has read it as celestial physics – the waters above the firmament are solid, like ice, or because they *are* ice. This almost accidental hypothesis about the nature of the supercelestial waters was destined to have a long career in Christian treatments of cosmology. It is also an interesting example of how Isidore mined his science out of Patristic commentary.[74]

As was mentioned above in connection with chapter 12, the Stoic notion of waters bathing the celestial sphere and thereby moderating its heat was available to Isidore *via* Ambrose and other Christian sources as a means of rationalizing the supercelestial waters in scientific terms, without selling out on the literal truth of Genesis. Isidore recycles the idea here to confound pagan critics of the Bible with their own scientific theory. He even alludes to the ancient controversy over whether the outermost sphere of the universe rested on water, or was bathed in water.[75] However, he drops Ambrose's idea that the celestial waters were the reservoir from which the biblical Deluge poured down. He seems to want to emphasize the benign and protective function of these waters, not their potential for destruction;

74 Thomas O'Loughlin, 'The Waters above the Heavens: Isidore and the Latin Tradition', *Milltown Studies* 36 (1995): 104–17. While O'Loughlin's analysis is very persuasive, we are not willing to concede that Isidore has simply assumed that the supercelestial waters are true, and effaced even the memory of the ancient pagan–Christian controversies over the supercelestial waters. Were this the case, he would not have found it necessary to allude to the fact that some people did not accept their existence. Indeed, Isidore's book may actually have kept alive the idea that the waters above the firmament are not undisputed fact.

75 Noted in the Pseudo-Clementine *Recognitions*, one of Isidore's sources for this chapter; see Fontaine, *Isidore et la culture*, 478.

this may also reflect a Stoic providentialist reading of the design of the universe. Bede, in his own *DNR*, will mention the Deluge reservoir theory, but only to dismiss it in favour of the 'more correct' explanation of the waters as cosmic temperature control.[76]

CHAPTER 15 THE NATURE OF THE SUN

Ambrose again takes the 'philosophers' to task in this chapter, but not, this time, for opposing biblical truth; rather, he accuses them of being unscientific. These philosophers (i.e., the Peripatetics) claim that the sun is not hot *by nature* because (1) it is white, not red and (2) something naturally hot requires liquid fuel, which it will inexorably exhaust unless it is replaced, and the sun does not consume moisture. The buried image here is the oil lamp, with its liquid fuel, wick, and flame. The flame will drink up the oil through the wick until the oil is exhausted, and then the flame itself will die. The oil lamp was adopted as a model for explaining why patients with fevers and wasting diseases die; the heat of the fever intensifies their normal body heat, and prematurely exhausts the body's moisture. As Ambrose says, heat which consumes moisture need not be 'natural'; it can also be a pathological accident.[77] And in any event, the sun can replenish any moisture it needs by vaporizing water, a phenomenon we can observe on any hot and misty day. Isidore pauses to reinforce this claim of how the sun can be 'fed' by a contrary element, water; though he does not make this explicit, it strengthens his case that there is no argument in the physics of the elements against the 'waters above the firmament'. Then he returns to Ambrose: on purely common-sense grounds, it stands to reason that something which produces both light and heat must be made of fire. Thus do Ambrose and Isidore (probably with additional support from scholia on Lucan's *Pharsalia*) uphold the Stoic theory of the fiery nature of the sun.[78]

Isidore closes this chapter with some observations on the biblical

---

76 Kendall/Wallis, 77 and n. 40.

77 This idea is discussed in Galen's *Ad Glauconem, Methodus medendi* and, above all, *De marasmo* ('On wasting'), where it is extended into an explanation for why old people eventually die, even without being ill. The medieval fortunes of this idea were important; see Michael McVaugh, 'The *humidum radicale* in Thirteenth-Century Medicine', *Traditio* 30 (1974): 259–83.

78 Fontaine, *Isidore et la culture*, 491–93.

and spiritual symbolism of the sun as an emblem of Christ. He quotes Malachi 4:2, but avoids using Jerome's commentary on this passage, with its eschatological interpretation[79] – another indication, albeit from silence, of Isidore's discretion in associating natural phenomena with apocalyptic speculation. Instead, he picks up Jerome's theme of fevers: the sun warms the healthy in temperate weather, but causes fevers during the dog days, just as Christ strengthens the faithful and torments unbelievers.

CHAPTER 16 THE SIZE OF THE SUN AND THE MOON

This chapter seems to open with a digression. Ambrose is intent on persuading his readers that the sun and moon are each at a uniform distance from the earth; in other words, the tropical regions are not hotter because the sun is closer to them. All inhabitants of earth are equidistant from the sun, and likewise from the moon.

Ambrose then proceeds to claim that the sun 'is seen when it rises at the same moment [*eodem momento*]' by both the inhabitants of India in the east and Britain in the west. This conveys the impression that neither Ambrose nor Isidore (see also *Etym.* 3.47) appears to have grasped the concept of the sun's rising and setting progressively later as one travels from east to west. Bede explicitly refutes this misconception in *DNR* 23, quoting Pliny (see Kendall/Wallis, 88). However, such a claim may not have been what Ambrose intended. He is focused on the issue of the *equidistance of the sun from all parts of the earth*, as is evident from the fact that he follows this curious statement about sunrise by asserting that the setting sun is no smaller and hence no 'further off' to Orientals than to Occidentals. He may have intended to say that the sunrise looks the same in India and in Britain, at whatever time it rises; nonetheless, the passage is confusing.

Finally, Ambrose claims that the sun is slightly (*aliquot partibus*) larger than earth, and that the moon is smaller than the sun. In *Etym.* 3.46, Isidore is less precise: the sun is simply bigger than the earth (*Magnitudo solis fortior terrae est ...*). The question of the size of the sun was vigorously debated in Antiquity: Isidore follows the Stoic line that the sun is definitely larger than the earth, but is unwilling to commit to Ptolemy's precise figure of five times as large.[80] In *Etym.* 3.47, Isidore reiterates that the moon is smaller than the sun, using the same arguments as in *DNR*. However, he adds information omitted from *DNR*, namely that the earth is larger than the moon.[81]

CHAPTER 17 THE COURSE OF THE SUN

In this chapter, Isidore explains that the sun is a planet, and hence has a motion of its own. Even though it is swept around from east to west once a day by the revolution of the heavenly sphere, it also moves in the opposite direction around the zodiac, making a complete circuit once each year. Unlike the constellations of the fixed stars, which always rise and set at the same point on the horizon at any given latitude, the sun's rising and setting points shift from south to north and back again over the course of a year, producing days and nights of different lengths at any given latitude. The sun's daily revolution around the earth (from a geocentric perspective) brings it back to the east every morning, but in winter that sunrise point will be in the south-east, and the sun's daily path will 'nearer to the earth' (i.e., closer to the horizon, for inhabitants of the northern hemisphere). In summer it rises in the north-east and is 'raised up high', that is, its daily path will be higher up in the sky. It is these swings of the sun's course that produce the four seasons (*tempora*), God's providential way of moderating the temperature of the air (*aeris temperies*).[82] Isidore thus gestures back to ch. 7 of *DNR*, where the seasons are linked to the annual journey of the sun from north to south, and the affinity of *tempora* with *temperamentum* (tempering, moderation) is illustrated by the *rota*. Isidore also slips in an echo to chapter 1 on how the day and its hours are produced by the sun, and recalls the double symbolism of the sun as both a sign of God's loving care and a token of punishment in chapter 15. At the same time, he announces in advance his treatment of pestilence in chapter 39. This remarkable series of allusions and cross-references speaks eloquently to Isidore's care in the closely textured construction of his work.

---

80 Fontaine, *Isidore et la culture*, 493–94.

81 In this case, Isidore is drawing on Cassiodorus, *Institutions* 2.7.2; see *Etymologies* 3, ed. Gasparotto and Guillaumin, 111 and n. 273. While Isidore's claim about the size of the moon is in fact correct, it contradicted some ancient authorities such as Pliny, *NH* 2.8.49; see Bede, *DNR* 19 (Kendall/Wallis, 85 and n. 108).

82 On Isidore's providentialism in this chapter, see Fontaine, *Isidore et la culture*, 494–96.

But this chapter is all about the sun's annual course, and the final section draws out the religious allegory of this phenomenon.

CHAPTER 18 THE LIGHT OF THE MOON

Given Isidore's compressed style of exposition, it is remarkable that he presents (through a passage from Augustine) two views concerning the source of the moon's light with two concomitant explanations of lunar phases, when he intends to support only one position – namely that the moon's light is a reflection of the sun's. This was the consensus view of ancient authorities.[83] The point of this apparent digression becomes evident at the end of section 4: if the moon shone with its own light, it would never be eclipsed. Isidore is driving the reader carefully but inexorably towards the crucial chapters on the lunar and solar eclipses.

Furthermore, the borrowed light of the moon presents Isidore with an opportunity for religious allegory, as does its waxing and waning – potent and visible symbols of the reversal of life and death represented by fall and redemption, mortality and immortality. Ancient writers were also in agreement that the moon generates and regulates moisture on earth, just as the sun produces heat and dryness. Isidore will cite a brief passage from Ambrose in chapter 19 about how the moist bodies of marine animals swell in sympathy with the waxing moon – a mere sliver of Ambrose's very detailed discussion about the power of the moon over all things connected to water such as the production of dew, and the growth of trees, and the sprouting of seed. At this point, however, Isidore is content with a general allusion to the moon and water, because he is unrolling a series of allegories comparing the moon to the Church, likening the moon's authority over water to the Church's unique control of baptism. The seven phases of lunar visibility are rather vaguely aligned to the seven 'graces of merits'. Isidore does not actually say what these graces or gifts of the Spirit are (presumably those set out in Isaiah 11:1–2), but he does enumerate the visible phases: waxing crescent, waxing half-moon, full moon, waning half-moon, and waning crescent. The half-moons occur halfway through each 15-day period separating conjunction from full moon, that is: seven and a half days into the lunation, and again on day twenty-one and a half. As for the crescent

---

83 e.g., Pliny, *NH* 26.45; see M.R. Wright, *Cosmology in Antiquity* (London and New York: Routledge, 1995), 43–44, who traces this insight to the pre-Socratic thinker Parmenides.

moons, they are imprecisely sited at 'proportional' points. While a diagram of the phases might position the crescents halfway between the new moon and the half-moons, a crescent moon can be seen at any time from the first day of the lunation up to the sixth. Isidore's phrasing suggests that he did indeed have a diagram in mind, and in a number of manuscripts of *DNR* his list of visible lunar phases is actually accompanied by drawings. We refer to them collectively as Diagram 5A (see Introduction, p. 29).[84] Their absence in some manuscripts might be attributed to the desire of scribes not to waste space on what perhaps seemed obvious. In fact, St Gall 238 drops this section of the text completely.

CHAPTER 19 THE COURSE OF THE MOON

In this chapter, Isidore edges closer to the issue of the eclipse by comparing the course of the moon to that of the sun. When he says that both planets 'move', he (or more precisely, his source Hyginus) is referring to their proper motion from west to east around the band of the zodiac (as distinct from their 'enforced' daily motion from east to west, propelled by the outermost sphere of the universe). The moon's proper motion is much easier to perceive than that of the sun, for it takes only a month to circle through the signs, while the sun takes a year. However, Isidore is not correct about the length of this circuit, as Bede points out (without naming him) in *DTR* 16.[85] The moon requires about 27 days and 8 hours to complete a tour of the zodiac (a sidereal month), while Isidore's figure of 30 days (actually about 29.5 days) is the length of the lunation, that is, the period from one conjunction with the sun to the next (a synodic month). Although Isidore distinguishes the synodic from the sidereal month in ch. 4.1, there is no indication here or in the *Etymologies* of what he thought their precise durations to be.

Twelve synodic lunar months will fall short of a solar year of 365 days, and hence the Jewish lunar calendar intercalates an extra month every two or three years. Isidore is vague about the details of both the shortfall and the intercalation. Twelve months of 30 days would leave only a five-day gap, as in the Egyptian calendar described in chapter 4; in fact, the shortfall is about 11 days ($29.5 \times 12 = 354$). The embolismic or intercalary lunar month

---

84 For an illustration from Paris 6413, fol. 13v, and discussion, see Teyssèdre, 'Les Illustrations du *De natura rerum* d'Isidore', 27 (fig. 12): this version also shows the moon as Diana, with a crescent 'crown'.

85 See Wallis, *Reckoning*, 58 and n. 179.

is inserted when this shortfall accumulates into a full lunar month after a year or two. Undoubtedly, Isidore was at the mercy of his source, Ambrose, here; but if he observed the problem, it was not his immediate purpose to correct it. His principal point is to underscore the parallel between the behaviour of the sun and that of the moon. This prepares the reader for the following chapters, where the eclipses of both planets are described using a similar logic.

CHAPTER 20 THE ECLIPSE OF THE SUN
AND CHAPTER 21 THE ECLIPSE OF THE MOON

Isidore's explanation of how a solar eclipse takes place is clear and accurate, but lacks detail: he does not, for example, explain why an eclipse does not occur at every conjunction, though he hints at it when he says in chapter 20.2 that the moon must not only be in the same *place* as the sun (or, as he phrases it earlier, in the same zodiac sign) but on the same *path* (*lineam*). The orbit of the moon is tilted at 5 degrees to that of the ecliptic, so a solar eclipse can only take place when the moon crosses the ecliptic at conjunction. A full lunar eclipse can likewise only occur when the full moon is positioned on the ecliptic; a partial lunar eclipse occurs when the full moon is slightly above or below the ecliptic.

This explanation of why an eclipse does not occur every month was well known in Antiquity (cf. Pliny, *NH* 2.7.48)[86] and appears in Sisebut's poem, so it is puzzling that Isidore does not address this question in this chapter. The issue of solar and lunar eclipses was evidently significant in *DNR* – one has only to compare the relatively circumstantial explanation in this book with the exceedingly jejune description in *Etym.* 3.57–58 – and yet Isidore does not explicitly refute 'superstitious' explanations of the phenomenon, as Sisebut does. Instead, and rather surprisingly, he offers alternative explanations which might even seem to encourage them. In the case of the solar eclipse, he says that 'philosophers and wise men of this world' argue that the hole (*foramen*) in the atmosphere through which the sun's light passes can be narrowed or blocked by 'some exhalation' (*aliquo spiritu*).[87] The 'exhalation' almost hints at the atmospheric corruption that

86 So was the cyclical character of eclipses; cf. Pliny, *NH* 2.10.50.
87 Aetius or Pseudo-Plutarch's *Placita philosophorum* ('Opinions of the Philosophers') 2.24 notes the theory (not ascribed to any individual) that a solar eclipse can be caused by an 'invisible concourse of condensed clouds which cover the orb of the sun'. The translation

is 'pestilence' (see ch. 39), and even suggests some arbitrary or malevolent cause. If one compares this chapter with chapter 21 on the eclipse of the moon, moreover, the same pattern emerges: an explanation based on celestial mechanics is succeeded by a doctrine, ascribed to the Stoics, that suggests that a lunar eclipse is a shadow cast by mountains on the periphery of the earth. Not only has the source of this information never been traced, but it is not really supported by the quotation from Lucan.[88] And, again, why is Isidore offering an alternative explanation at all?

The answer may lie in the spiritual interpretations that close both chapters 20 and 21.[89] In chapter 21, Isidore compares Christ's death and resurrection to the eclipse and re-emergence of the sun. However, built into this allegory is an explanation of the eclipse that took place at the Crucifixion. Isidore explains that it was 'unnatural' and out of 'order', though he does not dilate on the reason why. The Crucifixion eclipse was a miracle, because Christ was executed at the Jewish Passover, when the moon would have been full, not in conjunction. The 'darkness' that covered the earth was the 'sacrilegious conspiracy' of the Jews against Jesus – a sort of spiritual corruption or pestilence which infected the elements. Likewise, in chapter 21, lunar eclipses are likened to persecutions inflicted the Church, whose earthly enemies might be imagined as hedging her around like mountains. This is very speculative, of course; but the parallel patterns of these two chapters suggest that less predictable and mechanistic explanations for eclipses were invoked by Isidore in order to do some particular exegetical work in relation to his allegories. Sisebut, on the other

---

is by John Dowel, in *The Complete Works of Plutarch: Essays and Miscellanies*, vol. 3 (Boston: Little, Brown and Co., 1911), p. 144. Though he is cited as a parallel by Fontaine, Aetius mentions no *formamen*. Hence, even were the *Placita* translated or abbreviated into a Latin compendium, it probably would not qualify as Isidore's source for this curious theory. In *Isidore et la culture*, 497, however, Fontaine suggests that the theory is a misreading of a model propounded by Anaximander, and recorded in the Christian doxographical compendium of Hippolytus, *Philosophumena*; it might have arrived in Isidore's library through a similar source. Anaximander's view was that the sun was obscured by thick clouds, from which its light emerged by a narrow orifice.

88 Fontaine, *Isidore et la culture*, 500 argues that the mountain theory must derive from a scholium on Lucan, but there is no evidence that this is the case. Curiously, the mountain model bears some resemblance to Cosmas Indicopleustes' explanation for night, namely that the sun disappears behind a huge mountain in the centre of the (flat) earth; Anderson, 'Description of the Miniatures and Commentary', 42–44.

89 On the originality and finesse of these allegorical passages, see Fontaine, *Isidore et la culture*, 556–57.

hand, may not have entirely approved of this potentially dangerous equivocation; hence his more robustly rationalist poem.

Isidore did not include diagrams to illustrate lunar and solar eclipses, though a model was probably available to him in chapter 89 of Calcidius' commentary on the *Timaeus*. However, Isidore never uses diagrams to simulate how the heavens actually work through a graphic reduction, as Calcidius does. Instead, his diagrams are visual 'translations' of his arguments (e.g., the *figura solida* or *a fortiori* the diagram of the five zones) or else they are renderings of actual instruments, like the wind-rose, or reference tables, like the *rota* of the months.[90]

CHAPTER 22 THE COURSE OF THE STARS

In *DNR*, Isidore uses a number of terms for stars, planets, constellations, and heavenly bodies collectively: *sidera*, *astra*, *stellae*, and *signa*. In principle these words have distinct meanings for him, but in practice he seems to have changed his mind over the course of his career about what the criteria of distinction were. In his early *Differentiae* 1.2, Isidore writes: '*Sidera* are what sailors take into account (*considerant*) when they plot their course [cf. *DNR* 4.1]. *Astra* are large stars, like Orion; but *stellae* are smaller … like the Hyades and the Pleiades. A *signum* is in the shape of a living creature, like Taurus, Scorpio, and the like'.[91] Thus *astra*, *stellae*, and *signa* all refer only to constellations, but constellations of different sizes and forms. *Sidera* connote any heavenly body, but with respect to a specific human application or use. In his last great work, the *Etymologies*, Isidore adopts a slightly different mode of distinguishing these words. *Stella* refers to a single star, *sidera* to a group of stars or constellation. *Astra* (he says) are large *stellae* – but the examples he gives (Orion, Bootes) are constellations, not individual stars. Authors (says Isidore) confuse these terms and use *astra* in place of *stellae* and *stellae* in place of *sidera*.[92]

In this chapter, Isidore silently grants himself this literary privilege, using *stellae* and *sidera* to denote either fixed stars or planets or both, and

---

90 Wallis, 'What a Medieval Diagram Shows', 10.

91 *Inter* **sidera** *et* **astra,** **stellas** *et* **signa**. *Sidera illa dicuntur quibus nauigantes considerant quod ad cursum dirigant consilium; astra autem stellae grandes, ut Orion; stellae autem minores uel multiuages* [sic]*, ut Iadas, Pliades; signum uero quo animantis imago formata est, ut Taurus, Scorpio et eiusmodi* (*Diferencias Libro I*, ed. Codoñer, 86).

92 *Etymologies* 3.60 (Gasparotto and Guillaumin, 3.59).

may also have employed *sidera* to mean 'constellations'. This ambiguous terminology produces some confusion, particularly because Isidore shifts from discussing stars to discussing planets without warning.[93]

Isidore begins by distinguishing the *stellae* (in this case, both the fixed stars and the planets), which are borne from east to west once each day by the revolution of the outermost heavenly sphere. As Hyginus says, it is actually the sphere which turns; the stars and planets are carried along with it. The *stellae* are of two kinds: the planets ('wandering') and the fixed or non-wandering (*aplanes*) stars. The fixed stars never vary their position or direction. Isidore will discuss some forms of planetary 'wandering' in chapter 22.3, and also in the following chapter; but he nowhere explains the principal form of planetary motion, which is a gradual west-to-east creep along the line of the ecliptic – in other words, through the signs of the zodiac.

In section 2, Isidore deals with anomalies in the motion of *sidera*. Some have paths that are 'higher' (*superius*) and others paths that are 'lower' (*inferius*). Here *sidera* seems to mean 'constellations'. At any latitude on earth, some constellations in their nightly journey across the sky will pass relatively high overhead, while others will rise and set closer to the northern or southern horizon. Some which circle closely around the pole will never rise or set at all, but, depending on the observer's latitude, even these circumpolar constellations will be higher overhead, or closer to the horizon. However, Isidore derails the reader by introducing a sentence from Hyginus that applies to planets – indeed, he has already invoked it in explaining the difference between the size of the sun and that of the moon in chapter 16.4: *sidera* closer to the horizon appear larger because they are closer to earth. Isidore, in short, has made a lateral move from 'nearness to the horizon' (of constellations) to 'nearness to the earth' (of the planetary orbits). When Isidore turns to the 'different orbits' which cause some *sidera* to return 'more swiftly' and others 'more slowly, to the beginning of their course', however, he is referring exclusively to the planets. The fixed stars all return to the same starting position together every day, but the planets in their west-to-east trajectory of the zodiac come back to any given sign at varying speeds. It takes the moon a little over 27 days to pass through the whole zodiac; Saturn requires 30 years.

Terminological confusion extends into the closing section of this chapter. The subject here is the constellations: Hyginus explicitly calls

93 Fontaine, *Isidore et la culture*, 507–09.

them *signa*, but Isidore substitutes the less precise *sidera*. Then Isidore changes the topic to the planets, the 'wandering' stars, some of whom seem on occasion to come to a halt or reverse direction.

Fontaine chastises Isidore for his apparent ineptitude in this chapter: given what *DNR* had already said about the architecture of the universe, this confusion seems inexcusable. However, Isidore did not offer any revisions in the *Etymologies* and the quotation from Lucan which appears in both works may offer a clue as to why. In section 2 of this chapter, Isidore states that the *sidera* move 'within themselves' (*inter se*). In *Etymologies* 3, he claims that the 'stars … are carried about by the different orbs of the heavenly planets'.[94] It seems quite possible that Isidore imagines the stars not as lights fixed in or to the outmost sphere, but as layered between the planets, in order to ensure the equilibrium of the universe by balancing the effect of their contrary motions, rather as the supercelestial waters regulate the temperature of the cosmos. Hence the attraction of Lucan's image of the sun putting the brakes on the planets.[95]

CHAPTER 23 THE POSITION OF
THE SEVEN WANDERING STARS

In contrast to the preceding chapter, this chapter is explicit that the 'stars' under discussion are the 'wandering' ones, that is, the planets, including the sun and the moon. The discussion of planets in general is sandwiched between two blocks of text of a structural nature, laying out the order of the planets, and detailing the length of their courses.

The order of the planets is the straightforward 'Chaldean' series, which places the sun in the middle position, with three planets below and three above.[96] Only the locations of the sun and moon are provided with a providentialist explanation: the moon is closest to the earth so that earth

---

94 'Stellae pro eo quod per diuersos orbes caelestium planetarum feruntur …', *Etym.* 3.64 (Gasparotto and Guillaumin, 3.63, p. 125). On p. 124 n. 299, Gasparotto and Guillaumin claim that *stellae* refers to the planets, citing Fontaine, *Isidore et la culture*, 508; however, Fontaine argues that Isidore meant 'stars', and that it is the confusion of stars and planets that is the issue here.

95 Fontaine, *Isidore et la culture*, 508–09.

96 The distinction between the 'Chaldean' order of the planets and the 'Egyptian' one, which positioned the sun below Mercury and Venus is explained by Macrobius, *Commentary* 19.2.

may benefit from her light, while the sun's central location speaks to divine reason (*ratione* ... *diuina*), which assigned the most honourable position to the most noble planet.

The description of each of the other planets, however, poses particular problems. What does Isidore mean when he says that Mercury possesses 'a kind of contrary force'? The phrase seems to be lifted from Cicero's translation of the *Timaeus*, but what Isidore thought it meant is not evident: retrograde motion? or some astrological influence?[97] The unusual term used for Venus's orbit – *circumuectio* – seems also to come from Cicero's *Timaeus*;[98] but Isidore's information that Venus casts a shadow is only recorded in Pliny, and Isidore may not have had access to Pliny at this time. Where Isidore acquired his unusual alternative name for Mars – *Vesper* (normally applied to Venus, and already remarked on above in connection to ch. 4) – is unknown. Indeed, the names of all the planets vary considerably between *DNR* and the parallel section of *Etym.* 3.71 (Gasparotto and Guillaumin, 3.70) (see Table 2).

TABLE 2 THE NAMES OF THE PLANETS IN ISIDORE'S *DE NATURA RERUM* AND *ETYMOLOGIAE*

	De natura rerum	*Etymologiae*
Moon	Luna	Luna, Lucina
Mercury	Mercurius	Stilbon (cf. Cicero, *De natura deorum* 2.20.53)
Venus	Lucifer (cf. Cicero, *De natura deorum* 2.20.53)	Hesperus (cf. Cicero, *De natura deorum* 2.20.53)
Mars	Vesper	Pyrion (cf. Cicero, *De natura deorum* 2.20.53)
Jupiter	Phaeton (cf. Cicero, *De natura deorum* 2.20/52	Phaenon (cf. Cicero, *De natura deorum* 2.20.53, where it applies to Saturn)
Saturn	Saturnus	Phaethon

---

97 Fontaine, *Isidore et la culture*, 512–13.
98 Cf. *OLD*, s.v. *cirumuectio*.

The names used in *DNR* may have been borrowed from the source of the diagrammatic *rota* of the planets (Diagram 6). Similar schemata are found in Aristotle's *De mundo* as well as Macrobius and Calcidius.[99] The diagram displays seven concentric circles, with *'terra'* at the hub (superimposed on a human head). Each circle contains the name of the relevant planet, and the number of years required for it to complete its course, i.e., to return to the same sign and degree whence it began. The figures are very peculiar: the moon is said to take 8 years, Mercury 20 years, Venus 9 years, the sun 19 years, and Mars 15 years. Only Jupiter and Saturn are correctly stated to have periods of 12 years and 30 years, respectively.

It is difficult, or impossible, to understand Isidore's figures, which he will reiterate in *Etym.* 3.66(65).[100] These clearly are intended to represent the circuit of the planet through the zodiac through 'the same signs and regions'. The figures for Jupiter and Saturn match those found in conventional sources accessible to Isidore, such as Macrobius, *Commentary on the Dream of Scipio* 1.19.3.[101] But Mars makes its journey in about two years, not 15, while Venus and Mercury hover close to the sun and hence require about a year (cf. Macrobius 1.19.4). The sun, of course, makes the circuit in a year – indeed, its circuit defines the year, as Isidore himself points out in chapter 6. The moon passes through the zodiac in 28 days according to Macrobius (1.19.5), in 30 days, according to Isidore elsewhere (see Commentary on ch. 19), but not in eight years.

The nineteen-year period for the sun and the eight-year period for the moon might be confused echoes of the luni-solar periods used for reckoning Easter. Luni-solar cycles were important for *computus* because the date of Easter depended on both a solar phenomenon (the vernal equinox) and a lunar one (the full moon). Hence it was necessary to plot full moons across the Roman solar calendar in order to locate the first full moon after the

99 Fontaine, *Isidore: Génèse*, 299. See also Fontaine, *Traité*, 17; Macrobius, *Commentarii*, ed. Willis, 164; Calcidius, *Timaeus a Calcidio translatus*, ed. Wazsink, 149. However, as Eastwood remarks, Isidore redirects the schema away from representing the planets in spatial relationship and towards depicting their periodicity: Bruce S. Eastwood, 'Celestial Reason: The Development of Latin Planetary Astronomy to the Twelfth Century', in Susan J. Ridyard and Robert G. Benson (eds.), *Man and Nature in the Middle Ages* (Sewanee, TN: University of the South Press, 1995), 164.

100 Ed. Gasparotto and Guillaumin, 126–27, who comment (n. 303) on the curious figures, and the lack of scholarly attention they have elicited. Fontaine does not remark on these figures, either in his notes in *Traité* or in *Isidore et la culture*, pt. 4, ch. 3 (pp. 503–39).

101 The parallels cited by Fontaine (*Traité*, 260), namely Vitruvius 9.1.6 and Cicero, *De natura deorum* 2.20.51–52, agree with Macrobius.

equinox. This is difficult, because the true astronomical lunation and the mean solar year are incommensurable. It is possible, however, to achieve an approximation. Two such formulae were commonly in use among Christians: the *octaëteris* or eight-year cycle, whereby three embolismic lunar months were inserted over eight solar years, and the more accurate 19-year or Metonic cycle, which inserted seven embolismic lunar months across nineteen solar years. Isidore was certainly familiar with the latter, as it is the basis of the Alexandrian Paschal cycle described in *Etym.* 6.17. The 19-year cycle combined the *octaëteris* with another cycle, which spread four embolisms over 11 years, thereby offsetting the error accumulated by the eight-year cycle. The notional subdivisions of the 19-year cycle into *ogdoad* (8-year portion) and *hendecad* (11-year portion) were duly recorded on Paschal tables and in the *computus* literature (cf. Bede, *Reckoning of Time*, ch. 46). Since in *Etym.* 6.17 Isidore describes the 19-year cycle as a *circulus,* his importation of this figure into the discussion of the astronomical *circulus* of the sun in this chapter of *DNR* may derive from a confusion of the two *circuli.*[102]

The figures for Mercury, Venus, and Mars remain unexplained. Perhaps the source of Isidore's diagram of the planets already contained the mistake, and Isidore transferred it from the diagram into his text.

CHAPTER 24 THE LIGHT OF THE STARS
AND CHAPTER 25 THE FALL OF THE STARS

The subject of chapter 24 is declared to be the source of starlight, but in fact that topic is dismissed in the opening words of the first sentence: the stars (like the moon) shine with reflected light from the sun. Isidore does not allude to any alternative theory, though this question was the subject of debate among ancient philosophers.[103] The second half of the sentence exposes the real concern which underlies this discussion, namely that the stars might depart (*abscedere*) from the heavens, or that they could fall (*cadunt*). Isidore's message is that the reflected light of the stars is drowned out by the radiance of daylight, and that they merely appear to vanish.

---

102 The influence of the Metonic cycle is likewise suggested by Gasparotto and Guillaumin, *Etym.* 6, p. 126 n. 304, citing André Le Boeuffle, *Astronomie, astrologie: lexique Latin* (Paris: Picard, 1987), no. 781.

103 Fontaine, *Isidore et la culture*, 516.

The stars are nonetheless always in their wonted place, as is made evident during a solar eclipse, when they are temporarily visible.

The disjunction between the stated subject of this chapter and its actual content becomes evident when we pass to chapter 25. Falling stars are optical illusions – just fragments of celestial fire that fall from the ether and are blown about by the winds. Isidore's citations from Lucan and Vergil point to the origin of this rationalization in the literary scholia on these poets.[104] Isidore's dependence on these sources is ironic in that he disdains the poets' testimony as mere pandering to vulgar opinion; the philosophers 'whose business it is to examine the system (*rationem*) of the world' know better.

Isidore's allusion to the ether is one of two in *DNR* (for the other, see ch. 32.2). He does not define what he means by 'ether', but in *Etym.* 13.5.1 he states that ether is 'the place where the stars are', and as 'that fire which is separated high above from the entire world'.[105] In short, the ether is both a place – a 'part of the heavens' – and a kind of fire, though Isidore does not go so far are to claim that it is a fifth element to be distinguished from the fire of the four elements of earth, water, air, and fire.[106] On the other hand, Isidore proposes a distinction between *caelum* and *aether* in *Differentiae* 1.1 which indicates that he thought *aether* could also refer to a layer of the atmosphere. This would accord with the view expressed in this chapter, that so-called falling stars are essentially meteorological phenomena, since they are driven by wind.[107]

104 Fontaine, *Isidore et la culture*, 517.

105 Trans. Barney, et al., *Etymologies*, 272.

106 This is Aristotle's 'ether', as defined in *On the Heavens* 1.2–3 and *Meteorology* 2.2; see also Smyth, *Understanding the Universe*, 97. Aristotle did, however, postulate that the motion of the celestial ether produced heat in the lower world; see Taub, *Ancient Meteorology*, 85.

107 *Diferencias Libro I*, ed. Codoñer, p. 86: '*Inter caelum et aethera ita distinguitur: quod non tantum ille astriferus locus, sed etiam iste aer caelum uocatur; aether autem sublimior caeli pars est in quo sidera constituta sunt. Sane aether aer igneus est superior, aethra uero lux et splendor est aetheris*'. (*Caelum* refers not only to the heavens populated by stars, but also to the atmosphere. The ether is the higher zone of the heavens, in which the stars are situated. In reality, *aether* is the higher fiery air, and *aethra* is the light and radiance of the *aether*.)

CHAPTER 26 THE NAMES OF THE STARS

Isidore frames his discussion of the names of the constellations with considerable caution. The book of Job indeed refers to the constellations by name, but the mythological explanation of these names – the 'catasterisms' – are just conventional labels, like the planetary names of the weekdays. The Bible is merely adapting itself to this convention for the sake of intelligibility. Once again, the 'nonsense of the poets' is demoted beneath the authority of 'wise men'.

But Isidore in fact made no such distinction when discussing the planetary weekdays in chapter 3.5. The fact that he does so here is a silent reflection of his famous distinction between astronomy and astrology in *Etym.* 3.27(26).1–2: *astronomia* is the science of 'the revolution of the heavens, the rising and setting and motion of heavenly bodies, as well as the reason behind their names', and is without qualification a legitimate science; *astrologia* is a matter of fact (*naturalis*) when it measures the course of the sun and moon and planets, but *superstitiosa* when practised by astrologers (*mathematici*) who divine which zodiac sign rules which part of the human body or mind, or presume to predict the 'nativities' and characters of men from the course of the stars.[108] Here it suffices to note that Isidore includes catasterisms within the impeccable science of astronomy. In fact, Isidore's definition of astronomy takes Cicero's definition in *De oratore* 1.42.187 – '*in astrologia caeli conuersio, ortus, obitus motusque siderum*' – and expands it precisely by adding catasterisms.[109]

While the accounts of the origins of the names of the constellations in the transformation of mythological figures is part of late antique astronomical tradition (e.g., Aratus, *Phaenomena*; Hyginus), and also reflects Isidore's dependence on literary scholia for much of his astronomical information, it is nonetheless worth remarking that he did not have to include this information at all. To be sure, most of the star names in this chapter are the ones found in the passage from Job 38:31–32 with which Isidore opens, which gives the chapter the air of an exegetical aid. Indeed, much of his material on the Job constellations is derived

108 See above, p. 23, n. 31; see also Max Lejbowicz, 'Les Antécédents de las distinction isidorienne *astrologia/astronomia*,' in Bernard Ribémont (ed.), *Observer, lire, écrire le ciel au moyen âge: actes du colloque d'Orléans 22–23 avril 1989* (Paris: Klincksieck, 1990), 175–212.

109 Fontaine, *Isidore et la culture*, 467; Hermann, 'Zwischen heidnischer und christlicher Kosmologie', 324–27.

from Gregory and Ambrose.[110] By contrast, Isidore pointedly omits any discussion of the signs of the zodiac – the target of his blistering assault on pagan 'folly' in *Etym.* 3.71(70).31–32.

But if this catalogue of star names was supposed to aid the reader of Scripture, it exceeded its mandate. Boötes and Sirius are not in the Bible. Boötes slips into this chapter because he is the Ploughman who drives the Seven Ploughing Oxen (*Septentrio*), and because he is mentioned in the passage from Gregory that Isidore is exploiting for this section of his exposition. The discussion of comets (ch. 26.15), on the other hand, seems to come out of nowhere. Comets in turn 'cue' Sirius, another star which 'burns' and brings on disease. Neither comets nor Sirius are given allegorical meanings, and Isidore's motivation for including them is obscure.[111] Sirius, it can be argued, fits into Isidore's strategy of replacing mythological origins with explanations focused on practical navigation and astro-meteorology:[112] as the Pleiades and Orion delimit the summer and winter seasons, so Sirius marks the height of summer heat. But Sirius shares with the comet an implied malevolence, a hint of active intention. Between the two of them, comets and Sirius nudge Isidore's reader towards the problem of whether stars are living beings.

CHAPTER 27 WHETHER THE STARS HAVE A SOUL

Given Isidore's clear and sometimes strenuous rejection of both pagan astral worship and astrology, it is disconcerting to find him entertaining the possibility that the celestial bodies are alive at all, let alone entertaining it in such an inconclusive way. Yet his agnostic stance is precisely that of Augustine, who in the passage from *The Literal Interpretation of Genesis* excerpted at the beginning of this chapter states that this difficult question admits of no easy solution, and that he will not propose one. Augustine's reserve reflects a vigorous Patristic controversy over the status of the stars. Origen, drawing on Platonist tradition, endorsed the

110 The only one left out is the Hyades, though this omission is rectified in the cognate passage in *Etym.* 3.71.12 (Gasparotto and Guillaumin, 3.70.12); see Fontaine, *Isidore et la culture*, 519 and 531.

111 Fontaine, *Isidore et la culture*, 524–25 suggests that the material derives from a scholium on book 10 of the *Aeneid*, but offers no hypothesis to explain why it was included at all.

112 Fontaine, *Isidore et la culture*, 526–27.

idea that the stars were animated beings, and Jerome in his commentary on Ecclesiastes seems to back him up; Basil, John Damascene, and others argued that the stars were not themselves animated, but were instead corporeal bodies propelled by separate intelligent spiritual beings. Isidore lays out both these possibilities, but also proposes a third scenario, namely that the stars are purely material entities. This was the position adopted by Augustine in his attack on the Priscillianists, but Isidore strangely enough seems not to know about this text; instead, he leans on *The Literal Interpretation of Genesis*.[113] No one, it seems, was willing to argue for this materialist option, and Isidore on his own provides the explanation why: the stars move autonomously and in a strikingly regular manner, and this points to a living entity endowed with the power of self-movement and a rational soul. But if that is the case, what will become of the stars at the resurrection? Isidore does not answer this question, or even explain its implications. For if the stars are mere material bodies moved by spiritual intelligences or angels, nothing will change at the resurrection because the movers are immortal. If the stars are animated, rational, but embodied, then they are equivalent to human beings, who alone among mortal creatures are embodied rational souls. Was Isidore in contact with Jewish circles in Spain who held that the stars were angels whose judgement was reserved to the end of time?[114] Or is his question a distant echo of the controversy surrounding Origen's speculations that our bodies after the resurrection will be 'heavenly bodies' and thus similar in some way to the bodies of the star? Origen's astral body had a long pedigree in Platonist and Neoplatonist thinking, as well as in Gnosticism, though he explicitly rejected the idea that humans will *become* stars after the resurrection. We are destined for higher things. But neither did he explain what the final destiny of the stars themselves would be.[115]

CHAPTER 28 NIGHT

Isidore devoted two widely separated chapters (chs. 2 and 28) to 'night'. What at first glance seems an unnecessary and awkward duplication proves on reflection to have a structural justification. In chapter 2, Isidore

---

113 Fontaine, 'Isidore de Séville et l'astrologie', 284–85.
114 Fontaine, 'Isidore de Séville et l'astrologie', 295.
115 Alan Scott, *Origen and the Life of the Stars: A History of an Idea* (Oxford: Clarendon Press, 1991), ch. 9.

considers night in its relation to day, considered as a *unit of time*, whereas in chapter 28, he deals with night as the product of a *physical phenomenon*, namely the shadow cast by the earth as it intercepts the rays of the sun. Viewed from this perspective, the way in which earth itself produces night makes night a terrestrial phenomenon. Thus, this chapter forms a bridge between the chapters of *DNR* which deal with celestial cosmography, and those concerning earthly phenomena of 'meteorology'.[116]

This chapter illustrates some of the problems posed by Isidore's concern to compress his source material. In section 2, he paraphrases a passage from Hyginus explaining the shadow cast by the earth as the sun sets below the horizon. Hyginus is clear that the shadow is that of the sphere of the earth itself:

> even though some have said that this happens because, by the course of the sun, and when the sun arrives at the place where it sets, the magnitude of the mountains there [i.e., *on the western horizon*] blocks the light from us, and thus it seems like night. If this were so, we would obviously call it an eclipse of the sun rather than night, but one may understand it to be otherwise from the sphere itself.[117]

Isidore actually makes Hyginus say the opposite: 'as the day wanes, and the sun reaches the place where it is said to set, it is separated from us by the bulk of the mountains there'. This leaves the reader with the impression that Isidore understood 'the shadow of the earth' which makes night to be a shadow cast *onto* the earth by a mountain range, not *by* the earth. The perplexing statement that the shadow falls to the north does nothing to clarify the explanation.

---

116 When Bede divided Isidore's book into two treatises, *On Times* and *On the Nature of Things*, he omitted this second chapter on night. This was part of Bede's general reorganization of this section of Isidore's treatise, in the course of which he moved Isidore's chs. 22–26 up to their 'natural' place above the planets (Bede's ch. 11), and pulled material from chs. 24 and 25 together to create his own ch. 24 on comets. It is comets, and not night, which for Bede form the link between astronomy and meteorology, but Bede is operating with the same logic as Isidore: comets are liminal phenomena, star-like but generated in the atmosphere.

117 '*Cum enim traditum sit nobis prius noctem quam diem fieri. noctem dicemus umbram terrae esse eamque obstare lumini solis, etsi nonnulli dixerunt id solis cursu evenire et, cum pervenerit ad eum locum ubi occidere dicatur, ibi montium magnitudine a nobis lumen averti solis, et ita noctem videri. Quod si ita sit, nimirum eclipsin solis verius quam noctem dixerimus, sed aliter esse ex ipsa sphaera intellegere licebit*'. Hyginus, *De astronomia* 4.9 (Viré, 136).

CHAPTER 29 THUNDER
AND CHAPTER 30 LIGHTNING

The remainder of *DNR* is devoted to meteorology in the ancient sense of the term: weather conditions of all kinds, but also geological and seismological issues. Aristotle defined meteorology as the study of 'everything which happens naturally, but with a regularity less than that of the primary element of material things, and which takes place in the region which borders most nearly on the movement of the stars'. Its subject also includes 'all phenomena that may be regarded as common to air and water, and the various kinds and parts of the earth and their characteristics'.[118] Meteorology is thus defined by the 'irregular regularity' of its phenomena, their location, and the elements involved – air, water, and earth. It thus takes into consideration comets; atmospheric phenomena like rainbows, winds, clouds, thunder and lightning; precipitation of all kinds; rivers and springs; the sea; and seismic and volcanic activity. Modern scientists would not class these things together, but Aristotle did, because in his view they were all the products of some form of 'exhalation'; moreover, they all involved a change from one element into another.[119] Isidore's meteorology hews closely to this model, though he certainly did not read Aristotle's *Meteorology*, and seems not to have made use of Seneca's *Natural Questions* – virtually the only works dedicated exclusively and extensively to meteorology composed in Antiquity. Instead, he seems to have absorbed his meteorology largely through scholia on the poets and the writings of the Church Fathers.

Isidore's focus on the stars as indicators of seasonal weather conditions prepares the reader for this transition, and bridges the time-reckoning of the opening chapters to the terrestrial phenomena with which the book closes. The exposition follows the natural hierarchy of air (chs. 29–39), water (chs. 40–44), and earth (chs. 45–48), but within the airy zone, Isidore chooses to start with phenomena which have traditionally provoked scientific curiosity and debate, but also fear and 'superstition': thunder, lightning, and the rainbow.

Explaining thunder and lightning was one of the core projects of early Greek natural philosophy, from the Presocratics to Aristotle. It was a point where questions about the presence or absence of the gods, the elemental constitution of the world, the methodological problems of inference from

---

118 *Meteorologica* 1.1.338b, trans. Lee, p. 5.
119 Taub, *Ancient Meteorology*, 76–78.

observation, and the debates over how change took place converged. Thunder and lightning also figured in literature in more mythological and religious guise. Isidore's account of thunder and lightning reflects this rich and ambiguous backdrop. On the one hand, his explanations are strictly physical: thunder is the sound of agitated and pent-up winds exploding from the confinement of a cloud;[120] lighting is a spark produced by clouds colliding like flints or rubbing together like dry sticks of wood. 'In another sense' – or perhaps 'in other circumstances' (*alias*) – thunder is a divine rebuke. Fontaine argues that Isidore here compresses his allegory to the point where the natural phenomenon and the spiritual reality to which it points merge into one: thunder *is* the divine voice.[121] This interpretation seems rather strained, because thunder also *is* the preaching of the saints, and Isidore certainly did not mean his reader to take this literally. Note that lightning in the following chapter *signifies* 'the miracles of the saints'. To the contrary: Isidore here is at pains to demonstrate that even the most alarming of natural phenomena are *symbols* but not *omens*.[122]

Few other sections of *DNR* illustrate so clearly Isidore's dense interweaving of scientific, literary, and Christian materials. Ambrose is the (unacknowledged) source for the scientific explanation of thunder, and Jerome for lightning. Lucretius silently furnishes the account of the time-lapse between seeing lightning and hearing the thunder-clap; but, in his own words, he is the authority for the idea that lightning is composed of atoms. Lucretius does not, however, use the word *atomi* in the passage on which Isidore is drawing; rather he refers to lighting as a 'subtle fire composed' of small *figurae* ('*dicere enim possis caelestem fulminis ignem / subtilem magis e parvis constare figuris*': *De rerum natura* 2.384–85). Not only did Isidore intensify *parvis* into *minutis* but he substituted *semina* for *figurae*. *Semina* means 'seeds', but it is used as a technical Epicurean term for 'atoms' by Lucretius in several other passages in *De rerum natura*, and also by Ovid, Vergil, and Manilius' *Astronomica* (1.487). This would seem

---

120 An explanation which goes back to the Presocratic Anaximander; see Taub, *Ancient Meteorology*, 774.

121 Fontaine, *Isidore et la culture*, 544.

122 Ancient Roman augurs and haruspices interpreted thunder-claps, lightning-bolts, and other meteorological events as well as the cries and flights of birds, as Cicero observes in *De natura deorum* 2.5; even the Christian John Lydus (490–560) in his *De ostentis* discusses thunder prognostics; see Taub, *Ancient Meteorology*, 66–68; Michael Maas, *John Lydus and the Roman Past* (London: Routledge, 1992), ch. 8. Isidore does not, however, include thunder and lighting prognostication in his discussion of pagan augery in *Etymologies* 8.9.20.

to suggest that Isidore understood *semina* as atoms in the Epicurean sense, particularly when he adds that atoms of lighting can pass through objects. Isidore gives a neutral account of classical atomic theory in *Etym*. 13.2, but he also associates atomic theory with the despicable philosophical sect of the Epicureans in *Etym*. 8.6.15–16.[123] In both cases, however, he uses the term *atomi* not *semina*. Indeed, Lucretius does not himself use the term *atomi*. Hence Isidore may have thought he was dealing with something less specific than atomic theory here, and hence was conveying a scientific fact untainted by Epicurean folly. This does not, however, solve the problem of why he included this fact at all. A clue can be found in the allegory in section 3: the miracles of the saints penetrate to the inmost parts of the human heart. Given their potential as a natural symbol, Lucretius' lightning-atoms were simply too attractive to overlook.

CHAPTER 31 THE RAINBOW

Isidore culled his explanation of how rainbows are produced from the pseudo-Clementine *Recognitions* (whose protagonist is allegedly the martyred Pope Clement I), behind which stands the theory laid out by Aristotle in *Meteorologica* 3.4–5, 373a32–377a28. The account of the rainbow in *Etym*. 13.19 is somewhat different, in that Isidore slips in the detail that the cloud is hollow (*caua*), a point made in Seneca's *Natural Questions* 1.3.11 and also in Pliny's *NH* 2.60.150.[124] Aristotle discusses the colours of the rainbow in *Meteorologica* 374b–375b, but claims that there are only three: red, yellow, and green. Isidore, on the other hand, names four, and relates them to the four elements: 'fiery colour' (red or yellow) from the heavens, purple (water), white (air), and black (earth). Isidore also refers to the rainbow as 'four-coloured' (*quadricolor*) in his *Liber numerorum* 5.24, but without specifying the colours; however, this statement is immediately preceded by an enumeration of the four categories of living beings, arranged according to their native element: heavenly creatures, feathered creatures, water animals, and land animals.[125] Hence the connection of the rainbow's colours to the elements seems to be an

---

123 Fontaine argues on the basis of *De haeresibus* that most of Isidore's knowledge of Epicurean tenets came from Christian heresiological literature: *Isidore et la culture*, 722–24. However, the authenticity of this treatise has been contested: cf. Martín in *Te.Tra*. 2:411–17.

124 Taub, *Ancient Meteorology*, 154.

125 *Liber numerorum*, ed. Guillaumin, 31.

established fact with him, though whether he understood this in a literal and physical sense, or as a kind of natural analogy (like the linkage of the humours, elements, and seasons in ch. 11) is not evident. The source of Isidore's colours, and his correlation of the colours with the elements, is not known.[126] In the *Etymologies* 13.10.1, Isidore says that the rainbow exhibits 'various colours', but does not explain how many, or what they are.

CHAPTER 32 CLOUDS

Following in the broad tradition of Aristotelian meteorology (see above), Isidore's treatment of clouds stresses the metamorphosis of the elemental air into opaque clouds that shed water. This theme guides the allegory as well: unbelievers whose minds are as vacuous as air are solidified in faith when they convert. The juxtaposition of Job and Vergil is a striking example of his predilection for aligning biblical and classical authorities on an equal footing, but the barrage of etymologies at the close of this chapter is unusual in *DNR*.

CHAPTER 33 RAINS

The concept of the rain cycle was well understood in Antiquity, being invoked by Aristotle in *On Sleep, Posterior Analytics,* and *The Parts of Animals* as well as in *Meteorology.* To explain the process, Isidore borrows a striking medical analogy from Jerome. Cupping glasses (*cucurbitae* – literally 'gourds') were small dome-shaped vessels of glass or metal which were heated and applied to the skin. As they cooled, the cups contracted, creating suction that pulled blood to the surface. This was regarded as a less violent and invasive form of bloodletting, but its aim was the same:

---

126 Fontaine's note (*Traité*, 285) referring to a scholium on Lucan's *Bellum civile* 4.80, appears to convey a hypothesis rather than an identification. No such scholium has been located the *Scholia Vetera* edited by Hermann Genth (Berlin, 1868), in Hermann Usener's edition of the Bern *Scholia in Lucani Bellum Civile* (Berlin, 1869) or in J. Endt's edition of the *Adnotationes super Lucanum* (Stuttgart, 1969). The passage in Lucan refers to a rainbow in pale sunshine whose colours are indistinct (*vix ulla variatus luce colorem*); it mentions neither the specific colours nor the elements. Aristotle (*Meteorologica* 3.2, 372a) says that there are three colours – red, blue, and green, with sometimes yellow. Bede in *On the Nature of Things* 31 reproduces Isidore's colours and explanation, but substitutes the more intuitive blue for (airy) white and green for (earthy) black; Kendall/Wallis, 92.

to evacuate the blood, and with it the corrupted humours causing disease from the body. Isidore adds the detail that seawater, when evaporated and condensed into clouds, becomes sweet.

CHAPTER 34 SNOW AND CHAPTER 35 HAIL

The theory that snow was the precipitate of a frozen cloud reaches back to Aristotle.[127] Just as the cooling of water evaporated by the heat of the sun produces the clouds from which rain falls, so the freezing of these clouds generates snow. When it comes to hail, though, the scientific sources are much less circumstantial than Isidore. Aristotle regards snow and hail as products of the same process, except that hail is formed higher up in the atmosphere.[128] Isidore's explanation goes beyond how hail is formed to discuss its fall, and its spherical shape. No single source can account for all of his information, and much of it finds no analogue in any other text. For example, Isidore states that the round shape of hail is due to the warmth of the sun, plus the slow descent of the hail through the atmosphere. This is somewhat reminiscent of Seneca's model (*Natural Questions* 4B.3), which also invokes a prolonged, tumbling fall, but does not mention the sun: Seneca thinks that hail is round because water droplets are globular. Fontaine has observed that Isidore outpaces Ambrose (his principal source in this section of the text) both in distinguishing the different properties of snow and hail and in crafting nuanced allegories based on those specific differences.[129]

CHAPTER 36 THE NATURE OF THE WINDS

The winds, their names and nature, played an exceptionally important role in ancient cosmology and meteorology. Aristotle set the tone in *Meteorologica* 2.4–5. Here, as in Isidore's book, the discussion of the winds follows on that of precipitation, because the winds were understood to accompany rain or snow from different points of the compass. More importantly, the winds link earth to the celestial sphere. For Aristotle, they

127 *Meterologica* 1.11.347b12–28.
128 *Meterologica* 1.11.347b28.
129 Fontaine, *Isidore et la culture*, 552.

are set in motion by the rotation of the heavens, but also by the special influences of the sun's heat. The blowing of certain persistent winds was linked to the annual appearance of constellations like Orion and Sirius. The winds, in sum, were indices of the seasons. Finally, the directions of the winds mapped onto the zones of the earth demarcated by the poles and the tropics. Hence the winds were a primary mode of orientation.[130]

Isidore's approach is considerably less coherent and holistic: wind is simply air in motion, though, thanks to Augustine, he captures Aristotle's notion that the heavenly bodies transmit their motion to the atmosphere as wind. The Aristotelian concept of interlocking cycles of water and air (*Meteorologica* 2.4, 361b) is briefly invoked in section 2, but dismissed in favour of the Pseudo-Clement's idea that wind is air compressed by mountain ranges and subsequently expelled.

CHAPTER 37 THE NAMES OF THE WINDS

Chapter 37 and Diagram 7 were independently excerpted from *DNR* more frequently than any other chapter or diagram. Moreover, of all the diagrams in *DNR*, this one has the clearest scientific pedigree, and seems to be the most closely integrated into the text. A circular diagram of twelve winds pinned to the cardinal directions and to intervening points corresponding to the tropics and the Arctic and Antarctic circles appears in Aristotle's *Meteorologica* 2.6; Fontaine argues persuasively that it was the ultimate model for Isidore's wind-rose.[131] Not only do Isidore's Greek wind-names broadly match those on Aristotle's diagram (but not exactly or consistently, as we shall see), but his text clearly visualizes such a wind-rose, referring to the collateral winds as 'on the left' or 'on the right' of the cardinal winds. These directions only make sense from the perspective of the diagram: for example, the wind 'to the right' of the east wind Subsolanus is the ENE wind Vulturnus/Caecias. If one were *facing* the east, Vulturnus would be on one's left; but if one imagines oneself standing at the Subsolanus point of the wind-rose and looking inwards to the centre of the diagram, Vulturnus is indeed on one's right. Oddly, the diagram in the Munich manuscript

130 Vegetius, the author of a late fourth-century handbook, *Epitoma rei militaris*, on the art of warfare, compiled a chapter (4.38) on the subject because of its importance for sailors going into battle.

131 Fontaine, *Traité*, 17; see also Lee's edition of Aristotle's *Meteorologica*, p. 187; Taub, *Ancient Meteorology*, 104–06.

reproduced in Fontaine's edition (p. 296 *bis*) includes a human head in the middle, facing Subsolanus – which would reverse all the directions.

And yet, surprisingly, there are good reasons to believe that Isidore did not compose chapter 37 with a wind diagram in mind. There are several aspects of Diagram 7 that are unusual. It is the only diagram that is found in the second part of *DNR*, which is devoted to 'meteorology' and terrestrial phenomena. Unlike the other diagrams, it is not introduced by a textual formula. If the diagram were omitted, a reader or a scribe copying a manuscript would have no reason to suspect that it was missing. In fact, Diagram 7 is rarely omitted, but scribal uncertainty is evident. It is sometimes, as in Fontaine's text, inserted without an introduction after chapter 37.4, and sometimes after chapter 37.5. It is also sometimes inserted in either of these positions with one or another of several different introductory formulas.

The form of this diagram is also unstable. The Munich version shows the *rota* divided into twelve equal 'petals' inscribed with Isidore's double names for the winds. The rose is oriented so that the line between Subsolanus and Eurus is at the top. This is interesting, because it suggests that the designer did not try to align the cardinal directions orthogonally. Note that Isidore's text reads these winds counter-clockwise, beginning with Septentrio (north).[132]

But other quite different forms of the wind-rose appear in some very early manuscripts. This suggests that Isidore's text did not originally contain an illustration, but seemed to cry out for one, and that scribes were casting about for suitable candidates. Sometimes the centre of the diagram is filled with a Latin/Greek inscription MVNDVS/KOCMOC arranged to form a cross, with or without a human head. A wind-rose of this type appears *c*.600 in a manuscript of excerpts from Vegetius' *Epitoma Rei Militaris*.[133] Isidore, along with Vegetius, adopted Aristotle's 12-wind scheme (an eight-wind schema was preferred by Pliny, Aulus Gellius, and Vitruvius), but while Isidore's wind-rose was a circular table of names Vegetius' grouped the winds around the cardinal directions, and centred them on the KOCMOC–MVNDVS. The Vegetian KOCMOC–MVNDVS version with or without cardinal directions 'colonizes' chapter 37 of

132 Fontaine has rather confusingly inscribed the names of the winds with north at the top around his drawing (cf. *Traité*, 150).

133 The MS is Vatican City, BAV, Reg. lat. 2077 (fol. 99r). See Obrist, 'Wind Diagrams', 43–45 and figs. 3 and 5.

DNR at an early date. It was popular in the Carolingian miscellanies, but disappeared thereafter.[134]

Secondly, the wind-names were supplemented with personifications. In Laon 422, fol. 5v, the primary winds at the cardinal points are represented by larger, frontally oriented winged figures with their feet planted on a cube labelled KOCMOS–MVNDVS, while the secondary winds are represented as smaller winged figures floating to either side of the primary winds, and turning towards them.[135] This elaboration at least makes Isidore's left–right statements more comprehensible. Bianca Kühnel goes so far as to argue that the 'Vegetius' KOCMOC–MVNDVS version is a visual gloss on Isidore's text. 'Inhabited' wind-roses like the one in Laon 422 make visible the identity between winds and angelic spirits (*DNR* 37.1–4; *Etym.* 13.11.3–13).[136] Nonetheless, it seems more likely that Isidore did not write chapter 37 with the intention of including a diagram. Not only did he fail to provide an introduction to the image, as was his usual custom, but there are discrepancies – contradictions even – between what he says in the chapter and the names of the winds embedded in the diagram that became attached to it.

The wind-rose described in *DNR* 37 and shown in *rota* 7 both name twelve winds. These are the winds blowing from the four cardinal points, north (N), east (E), south (S), and west (W), and from two intermediate points between each pair of adjacent cardinals, which we may for convenience's sake label as north-north-east (NNE), east-north-east (ENE), east-south-east (ESE), south-south-east (SSE), south-south-west (SSW), west-south-west (WSW), west-north-west (WNW), and north-north-west (NNW).[137]

However, the text of chapter 37 starts with the north wind, whereas *rota* 7 (the manuscripts are quite consistent in this) puts the east wind at the top.[138] A comparison of the names of the winds as given in *DNR* 37 and *rota* 7 turns up several more discrepancies (see Table 3).

134 Obrist, 'Wind-Diagrams', 47–48; for examples of variant versions, see her figs. 4, 6–8, and 25.

135 Obrist, 'Le Diagramme isidorien', fig. 8. A comparable figure is found in Biblioteca Apostolica Vaticana, Reg. lat. 1263, fol. 78 (fig. 6), the Micy *computus* anthology.

136 Kühnel, 'Carolingian Diagrams', 363.

137 The names of these intermediate points ('north-northeast', etc.) are terms drawn from the modern 16-point wind-rose. They are only approximately correct for a 12-point compass.

138 Obrist, 'Wind Diagrams', 45 n. 68 observes that this 'discrepancy between diagram and text has been corrected' in Strasbourg 326 (s. x), where the diagram has been rotated to put north at the top.

TABLE 3 THE NAMES OF THE WINDS IN ISIDORE'S
DE NATURA RERUM 37 AND DIAGRAM 7

Wind	*De natura rerum* 37	Diagram 7
N	Septentrio [Aparctias]	Septentrio Aparctias (Gr)
NNE	Aquilo Boreas	Aquilo Boreas (Gr)
ENE	Vulturnus Caecias	Vulturnus Caecias (Gr)
E	Subsolanus Apeliotes	Subsolanus Apeliotes (Gr)
ESE	Eurus –	Eurus Eurus (Gr)
SSE	Euroauster –	Euroauster Euronotus (Gr)
S	Auster Notus	Auster Notus (Gr)
SSW	Euronotus –	Austroafricus Libonotus (Gr)
WSW	Africus Lips	Africus Lips (Gr)
W	Zephyrus Favonius	Favonius Zephyrus (Gr)
WNW	Corus Argestes	Corus Argestes (Gr)
NNW	Circius Thrascias	Circius Thrascias (Gr)

In the Graeco-Roman tradition from which *DNR* 37 and *rota* 7 derive, the names of the twelve winds are usually given in both Greek and Latin. These are faithfully reproduced in *rota* 7, Latin above, Greek beneath. Isidore's text gives the Latin first (*Zephyrus* being an exception), followed by the Greek, but he omits (at least initially) the Greek for N (*Aparctias*) and he gives only the Latin name for SSE (*Euroauster*) and for SSW (*Euronotus*). In fact, *Euronotus* is the Greek term not for the SSW but for the SSE wind, as in the diagram. Combining as it does the terms for ESE (*Eurus*) and S (*Auster*), it is transparently incorrect for SSW, but totally consistent in the text tradition. Since *Euronotus* is the only name given for the SSW wind in *DNR* 37, Isidore was apparently unaware of either the conventional Latin (*Austroafricus*) or the Greek (*Libonotus*) names. And yet *rota* 7 faithfully preserves the authentic terms.

We may reasonably conclude that Isidore did not have this particular wind diagram before him when he wrote chapter 37, and that he probably did not foresee including a diagram with the chapter. Nonetheless, some form of Diagram 7 is almost universally present in the manuscript tradition. An omission was perceived, and scribes acted to fill it, and very early on; but there was no wind-rose in the manuscript that originally left Seville for Toledo. This supports the model of phased construction of the text outlined in the Introduction.

CHAPTER 38 SIGNS OF STORMS OR FAIR WEATHER

This chapter and the next follow on logically from Isidore's treatment of the winds and their directions. The connection between weather systems, prevailing winds, and geographic directions, as well as with the rising and setting of prominent constellations, and the extrapolation of weather prediction from these data, lies in the deepest substrate of ancient observation about the physical world, and forms the foundation of classical meteorology.[139] To these were added the appearance of the sky itself, the sun and moon, as well as the behaviour of animals. Though Isidore draws from his usual eclectic range of sources in this chapter, he touches on all the major themes of ancient weather prognostication.

Isidore's remarks about how sea-creatures presage coming storms draw on a minor, but fairly consistent thread of classical natural history. As part of a lengthy catalogue of animals as weather-indicators, Pliny, *NH* 18.87.361, mentions dolphins indicating the direction of an oncoming storm, and slapping the water with their tails, as well as cuttlefish leaping from the water. Isidore's claim that dolphins leap out of the waves of an oncoming storm to resist being driven onto the shore has not been traced, but could be a distant echo of Aristotle's observation (*History of Animals*

---

139 Taub, *Ancient Meteorology*, ch. 1; Lehoux, *Astronomy, Weather, and Calendars in the Ancient World*.

4.8) that that dolphin-hunters make a loud noise to drive the animals onto the beach.[140]

The fact that Isidore, unlike Pliny, confines himself to marine animals points to the importance of weather-signs for navigation, and to the dominant influence of his source, Varro. The significance of the direction of thunder and lightning comes from Varro's now fragmentary treatment of seafaring. When it comes to the aspect and colour of the sun and moon, Isidore claims as his authority Publius Nigidius Figulus (d. 45 BC) whom Jerome in his *Chronicon* called 'a Pythagorean and a *magus*' because of his interest in divination and the occult. However, Isidore is careful to disinfect his source by explaining that changes to these planets are the consequence of thickening air, and hence an accidental result of the changing atmospheric conditions.

Probably the most obscure passage in this chapter concerns weather prognostication from the angle of the horns of the moon, i.e., the tips of the crescent moon. Isidore names Aratus of Soli's *Phaenomena* as the source of his information. This enormously popular verse astronomical manual was translated into Latin by Cicero, Germanicus Julius Caesar (15 BC–AD 19), and Rufius Festus Avienus (fourth century); the last is the only extant one to include the entire section on weather signs. Another translation, the so-called *Aratus latinus* (in an original and a revised version), was composed around 750 in northern France, and may, indeed, have drawn upon Isidore himself.[141] Latin versions of Aratus were also equipped from a very early stage with scholia and commentaries; indeed, the *Aratus latinus* was in some measure assembled from such glosses.[142]

According to Isidore, Aratus states that if the northern horn of the moon is straighter (*correctius*), the north-north-east wind Aquilo will

140 Charles Speroni, 'The Folklore of Dante's Dolphins', *Italica* 25 (1948): 1–5 traces the history of the belief that leaping dolphins forecast approaching storms. See also Ashley Montagu and John C. Lilly, *The Dolphin in History. Papers Presented ... at a Symposium at the Clark Library, 13 October 1962* (Los Angeles: William Andrews Clark Memorial Library, 1983).

141 Aratus, *Phaenomena*, ed. Kidd, introduction; H. Le Bourdellès, *L'Aratus Latinus. Étude sur la culture et la langue dans le nord de la France au VIIIe siècle* (Lille: Université de Lille, 1985); Taub, *Ancient Meteorology*, 47–53; Fontaine, *Isidore et la culture*, 577–78; Elly Decker, *Illustrating the Phaenomena: Celestial Cartography in Antiquity and the Middle Ages* (Oxford: Oxford University Press, 2013), 4.

142 This passage from *DNR*, including the quotation from Vergil, is found *quasi verbatim* in the 'Basel Scholia' on Germanicus's *Aratea*, ed. Alfred Breysig, *Germanici Caesaris Aratea cum scholiis* (Berlin, 1867), 202.

blow, and if the southern horn is more upright (*erectius*), the south wind Notus will blow. The Greek text of Aratus, in fact, says nothing about a 'northern' or 'southern' horn of the crescent moon. There, the passage rendered by Isidore corresponds to the italicized portion of the following larger discussion:

> But if, when she brings the third day, the moon does not lean forward from the line of the two horn-tips, or shine inclining backwards, but instead the curve of the two horns is upright, westerly winds will blow after that night. But if she brings in the fourth day also similarly upright, she will certainly give warning of a gathering storm; *if the upper one of the horns should lean well forward* [*epineuē*], *expect a northerly; when it* [*i.e., the upper horn*] *inclines backward* [*huptiaēsi*], *a southerly.*[143]

Aratus' text contrasts three types of waxing crescent moon: one which is upright like a backwards C, with the upper horn literally 'nodding forwards', one where the upper horn 'inclines backwards', tilting the crescent to a slightly supine position, and a fully supine moon, with 'the two horns ... upright'. Latin translations of Aratus, however, rendered this italicized passage (as Isidore did) to imply that Aratus was referring to the two horns of the moon, not the same upper horn, and that these two horns were pointing respectively, north and south.[144]

The notion that the angle of the crescent moon's horns was an index of weather is an ancient one, though there were varying views as to what they signified and how. According to one view, the horns of the moon signified weather by pointing to the quarter of the horizon whence winds would arise.[145] This seems to be the idea Isidore is trying to convey here: the northern horn of the moon points to the NNE wind, and its southern horn to the south wind. However, the horns of the crescent moon never point to the north; the mistranslation of Aratus has produced an astronomical impossibility.

As the moon pulls away from conjunction with the sun, it is visible as a crescent moon above the west for a few hours after sunset. Just before conjunction, it will again appear as a crescent shortly before sunrise on the eastern horizon. The horns of the moon thus always face away from

---

143 Aratus, *Phaenomena*, trans. Kidd, 131.

144 e.g., Rufi Festi Avieni Aratea, lines 1475–83, ed. Breysig, 60: '*istius* in borean *quod se sustollit acumen ... nam subrigat* auster *acumen inferiore plaga*'.

145 Neugebauer, *History of Ancient Mathematical Astronomy* 1.141; see Taub, *Ancient Meteorology*, 34 and n. 87.

the sun, which illuminates its bow. Whether the crescent moon (waxing or waning) looks like a C or more like a U depends on the time of year.[146] However, whether waxing or waning, the horns of the moon at northern latitudes will always point south; to put it another way, the bow of the crescent will be on the viewer's right in the west at sunset (waxing) and on his left in the east at sunrise (waning), but in both cases the horns point south, except at the vernal equinox.[147]

This is, of course, what Aratus knew, and what the Greek text implies. Isidore, influenced by a Latin translation or scholium, has made two horns out of Aratus' single, upper horn, and made them point to the north and south respectively. His *correctius* refers to the upper horn, and competently renders 'rather forward', but *erectius* – 'rather more upright' – departs from Aratus' original meaning.

Taub comments that weather prognostication, somewhat like medical prognosis, was treated as an inference from natural signs, and not as a form of divination. Cicero, in fact, contrasts the track record of diviners unfavourably to that of experienced pilots and physicians. Divine or supernatural agency was not part of the picture, and this probably contributed to its adoption by Christians. Isidore's casual juxtaposition of Varro and Nigidius with Christ's words concerning weather signs in the

---

146 See helpful explanation and diagrams in Norman Davidson, *Astronomy and the Imagination: A New Approach to Man's Experience of the Stars* (London: Routledge & Kegan Paul, 1985), 3 and 73.

147 The celestial equator intersects our horizon at due east and due west; the angle of that intersection varies with latitude, but it is always constant, and in northern latitudes, always tilts to the south. The ecliptic that carries the sun and moon is set at a 23½-degree angle to the celestial equator; the two circles intersect in the constellations of Aries and Libra. The angle that the ecliptic forms with the celestial equator will vary throughout the day as different parts of the ecliptic lying to the north and south of the equator come into view. Over a period of 24 hours, the ecliptic will be (a) parallel to the equator and lying to the north of it; (b) crossing the equator at due west, but slanting towards the south; (c) parallel to the equator and lying to the south of it; and (d) crossing the equator at due west, but slanting to the north. At the vernal equinox in Aries when the ecliptic at sunset will be crossing the celestial equator precisely at the horizon, inclined at its maximum angle to the north of the equator, the crescent moon will be almost directly above the setting sun. It will thus 'lie on its back' pointing both its horns upwards from the horizon, towards the east. It will look like the letter U. At other times of year, however, the crescent moon will appear to stand more or less erect on its lower horn. See helpful explanation and diagrams in Davidson, *Astronomy and the Imagination*, 3 and 73 and Kim Long, *The Moon Book: Fascinating Facts about the Magnificent, Mysterious Moon* (Boulder, CO: Johnson Books, 1988), 26.

Gospel speaks to this hospitable attitude.[148] Nonetheless, Bede pointedly omits this chapter of *DNR* when he creates his own book of the same title, and sharply attacks the notion that the angle of the horns of the moon can presage weather in chapter 25 of *The Reckoning of Time*.[149]

CHAPTER 39 PESTILENCE

The first paragraph of this chapter captures all the complexity of ancient thinking about epidemic disease to which Isidore's culture, in both its classical and Christian aspects, was heir.

For the ancient Greeks, pestilence (*loimos*) is not just disease (*nosos*), but disease that kills suddenly and massively. As Isidore puts it, pestilence 'spreads widely', affecting 'whatever it touches'. There is no interval between presenting symptoms and final outcome; victims seem to drop dead. Above all, pestilence is something sent from the gods. Homer's *Iliad* opens with a *loimos* that decimates the Greek army besieging Troy. This is the vengeance of Apollo for the capture by the Greek leader, Agamemnon, of the daughter of his priest Chryses. In Hesiod's *Works and Days*, loimos (plague), along with *limos* (famine), is the punishment sent by Zeus upon a whole city, because of the misdeeds of one man:

> Often a whole city is paid punishment for one bad man who commits crimes and plans reckless action. On this man's people the son of Kronos out of the sky inflicts great suffering, famine [*limos*] and plague [*loimos*] together, and the people die and diminish.[150]

The crime of one 'contaminates' all by a sort of moral 'contagion' or contact. It is guilt by association.

In Sophocles' play *Oedipus the King*, a *loimos* devastates the city of Thebes, and not just its people: it is a total environmental catastrophe, blighting crops and causing sterility in animals and humans, as well as epidemic sickness. The king of Thebes, Oedipus, sends Creon to the oracle

---

148 Fontaine, *Isidore: Génèse*, 338–39 singles out this chapter as an exceptional example of Isidore's *compilatio* method and its consequent alignment of pagan and Christian sources on the same plane. Perhaps Isidore intended to give the Saviour the last word, but it is noteworthy that his authority confirms that of the pagan sources.

149 See Commentary in Wallis, *Reckoning*, 301–04.

150 Hesiod, *Works and Days*, trans. Richard Lattimore (Ann Arbor: University of Michigan Press, 1973), 47.

of Apollo to find out why. Creon returns with Apollo's answer: the city – even the land – has suffered ritual pollution (*miasma*) because of the sin of one person. That person, as it turns out, is Oedipus himself, who has caused *miasma* by inadvertently killing his own father and marrying his own mother – parricide and incest. To cure the disease, Thebes will have to get rid of the cause.

Isidore's reading of the Bible would have reinforced this model of pestilence. 'Plague' is sudden, often gruesome death sent by God as punishment upon a whole population, frequently for the sin of a single individual, whether upon Egypt because of Pharaoh's stubbornness (Exodus 11), or Israel because of David's proposed census (2 Samuel 24).

The medical texts ascribed to Hippocrates offer a significantly different view of disease as a whole. One of the hallmarks of Hippocratic medicine was its resolutely naturalistic approach to disease of all kinds. The human body was made of the same stuff as the material world, and behaves in the same way. Disruptions in nature occur because one factor – heat, for example, or wetness – gets out of balance, and the same is true with the body. But humans also are embedded in their environment, so when there is an abnormal disequilibrium in the environment (for example, an unusually wet winter), this can taint the air we all breath, and cause an entire population to fall ill. For this kind of disease, the Hippocratic texts coin a new word, 'epidemic'.

> ... there are two kinds of fevers: one is epidemic, called pestilence [*loimos*], the other is sporadic, attacking those who follow a bad regimen. Both of these fevers, however, are caused by air. Now epidemic fever has this characteristic because all men inhale the same wind; when a similar wind has mingled with all bodies in a similar way, the fevers too prove similar ... So whenever the air has been infected with such pollutions [*miasmasin*] as are hostile to the human race, then men fall sick.[151]

'Epidemic' ('around/upon' the 'demos' or population) is still a disease which affects a whole community, but its source is now unusual atmospheric disequilibrium. The first book of the Hippocratic *Epidemics*

---

151 Hippocrates, *Breaths* 6, trans. W.H.S. Jones, in Hippocrates, *Prognostic; Regimen in Acute Diseases; The Sacred Disease; The Art; Breaths; Law; Decorum; Physician (Ch. 1); Dentition* (Cambridge, MA: Harvard University Press/London: Heinemann, 1923), 233–35. *Nature of Man* 9 expresses a similar idea, but adds that bad air is the consequence of some exhalation or excretion (*apókrisis*), a notion that suggests Aristotle's 'exhalations' model of meteorology.

is a chronicle, year by year, of the weather in Thassos, and the consequent outbreaks of disease. At the same time, every location has its own 'endemic' diseases – the ones which regularly occur there because of local climatic and geographical conditions. But the Hippocratic writer is a man embedded in his culture, even though he is also quite self-consciously innovating. He has to explain the word 'epidemic', and he does so by glossing it as 'pestilence' [*loimos*]. Secondly, he has to borrow a term from religious language to convey the notion of an atmospheric taint that transmits itself like a poison from the atmosphere to the body, and from body to body, and the term he chooses is *miasma* – 'pollution'. He wants to give *miasma* a novel, secular, and natural meaning, but, of course, it was still (and most commonly) used in its ritual sense. This is significant, because the Hippocratic author has embedded the idea of 'pestilence' into the concept of 'epidemic'.

Isidore does this as well. Pestilence 'strikes' – it is a *plaga* or blow (hence our English word 'plague') – and this always carries the valency of punishment; at the same time, it is 'from some [natural] cause', namely the 'corruption' of the atmosphere through an excess of one of the four elemental qualities of heat, cold, moisture, and dryness. The air envelops the whole earth, the corruption is born on the winds, and everything that comes into contact with it is 'infected' or tainted. It is important to grasp that, for Isidore, 'contagion' means contact with the air, and 'infection' is the tainted quality of that air that can be passed to whatever is in contact with it. *Pestilentia* was primarily a problem with the air itself, and not the humans who were made sick by it.[152]

Isidore records in section 2 the view of 'others' that 'plague-bearing seeds (*semina*)' borne on the wind are the cause of pestilence. The word *semina*, as was noted above, is a favoured Latin synonym for the 'atoms' in the Epicurean sense. Indeed, this entire section, including the phrase '*semina rerum multa*' and other *verbatim* passages, summarizes Lucretius' explanation of plague in *De rerum natura* 6.1090–132. Lucretius argues that every climate and region has its endemic diseases. If we travel to another land, and are unaccustomed to the local air, we expose ourselves to unfamiliar ailments. Conversely, when the air in some distant place is

---

152 Isidore, *Differentiae* 1.143 (ed. Codoñer, 158): '*Inter pestem et pestilentiam. Pestis ipsum nomen est morbi, pestilentia uero id quod ex se efficit. Pestilentiae autem tres modi sunt: aut ex terra, aut ex aqua, aut ex aere*'. ('*Pestis* is the name of a disease, but *pestilentia* is what produces it from itself. There are three modes of *pestilentia*: from earth, from water, and from air.')

set in motion, and makes its way to our part of the world, it infiltrates our air, water, and soil, and taints them, making living creatures sick. Lucretius then goes on to paraphrase Thucydides' account of the Plague of Athens; Isidore's remark that 'enfeebled by illness the body [of the plague victim] expires either from foul sores or from a sudden stroke' is a verbal mosaic from this section of the poem.

For Lucretius, the seeds were definitely Epicurus' atoms; in *De rerum natura* 6.769–830, he describes how atoms harmful by their nature and structure penetrate bodily orifices and cause pain, illness, and death. Whether Isidore grasped or accepted this atomic theory is doubtful. The agricultural writings of Varro, as well the medical treatises of Galen or the Methodists, made the notion of sub-visible morbific particles in the air relatively common currency, without invoking atomic theory explicitly.[153]

CHAPTER 40 THE OCEAN

In Aristotle's view, 'exhalations' both 'dry' and 'wet' constituted the driving dynamic of all change in the sublunary world, and thus were the foundation of meteorology.[154] This did not, however, include the tides, of which he seems to have had little knowledge; indeed, tides in the Mediterranean are fairly inconspicuous. Nonetheless, Aristotle's principle was applied to the tides by later thinkers, and their views are conveyed in this chapter. Isidore concentrates on explaining the cause of the tides, and does not, like Pliny, for example (*NH* 2.99.215), remark on their periodicity or variable height. All the explanations he collects are oriented towards a pneumatic or respiratory model of tides; in this respect, *DNR* contrasts sharply with *Etym.* 13.28, where tides are treated as special cases of currents, such as are found in estuaries (an echo of Aristotle's *Meteorologica* 2.1, 354a). From Solinus he takes the notion that tides are the effect of winds which travel through passages under the ocean floor and erupt through vents, 'the nostrils of the world, as it were'. The flux and reflux of the winds, like the rhythm of a breathing animal, produces the ebb and flow of the tides. Ambrose credits the moon, but imagines its influence as 'respiration', and the pushing and pulling of the waters as inhalation and exhalation. Even

---

153 Vivian Nutton, 'The Seeds of Disease: An Explanation of Contagion and Infection from the Greeks to the Renaissance', *Medical History* 27 (1983): 1–34, on Lucretius, see pp. 9–11; on Isidore, see pp. 20–21.

154 See above, p. 230.

the third alternative – that tides ebb because of solar evaporation – invokes Aristotle's exhalations.

Just as the origin of the tides lies beyond human understanding, so the dimensions of the ocean are beyond our ken. There are whole worlds (*mundi*) out there, but we cannot access them. Moreover, their very nature is ambiguous: there is no land there, and yet there *is* land there, but shrouded in impenetrable fog. Whether Isidore is imagining the western ocean, or the southern one – the dubious 'antipodes' – is not clear. His main point seems to be to convey the idea of the Ocean as a natural symbol of the mystery of the divine mind.

CHAPTER 41 WHY THE SEA DOES NOT GROW IN SIZE AND CHAPTER 42 WHY THE SEA HAS BITTER WATERS

These two chapters form an interconnected unit, as the volume of the sea and the saltiness of its waters were closely related topics for Aristotle. In *Meteorologica* 2.3 he discusses the problem of the sea's saltiness at length. The constant volume of the sea would support the theory that the sea is inherently salty, because the amount of water that enters the sea equals the amount evaporated by the sun: this 'evaporation hypothesis' is represented by Isidore's quotations from Pseudo-Clement in chapter 41, and Ambrose in chapter 42. The alternative view is the 'admixture hypothesis', namely that earthy material is swept into the sea by the rivers, making it salty. Aristotle favours his own 'admixture' theory, namely that 'dry exhalations' of the earth are condensed into clouds, and fall as rain into the sea. That the volume of the sea is constant does not nullify this theory, because evaporation and rainfall balance one another. This hypothesis is not discussed by Isidore.[155]

In chapter 41.2, however, Isidore presents another model to explain the constant volume of the sea: the waters that the rivers pour into the sea are carried back to their sources again through subterranean passages. This view is adduced by a number of Patristic sources, notably Ambrose in his *Hexaemeron* and Augustine in *De Genesi ad litteram*, the latter

---

155 A useful synopsis of ancient and medieval speculations about the relationship of salt and sweet water is presented by Joëlle Ducos, 'Eau douce et eau salée', in Danièle James-Raoul and Claude Thomasset (eds.), *Dans l'eau, sous l'eau: Le monde aquatique au moyen âge* (Paris: Presses de l'Université de Paris-Sorbonne, 2002), 121–38. She does not, however, mention the link between the volume of the sea and its salinity.

to explain how the waters of the Flood could be said to spring 'from the abyss'. But Isidore seems to have drawn his material from more secular sources. Seneca's terminology in *Natural Questions* finds some distinctive echoes in Isidore: 'and for this reason the oceans do not grow larger (*maria non crescere*) because they do not assimilate the water that flows into them but immediately return it to the land. For the water enters the land by hidden (*occulto*) routes'.[156] Furthermore, this phenomenon, in Seneca's view, explains why the rivers that flow into the sea are fresh: the salt has been filtered out. The pairing of the two problems that Isidore weaves together in this chapter may point to Seneca as Isidore's source.[157] However, Isidore's phrase 'in their own rivers' (*per suos amnes*) echoes *De rerum natura* 6.631–38. Here Lucretius offers the same paired explanation as does Seneca, stating that the percolated seawater flows back to the stream-heads (*et ad caput amnibus omnis confluit*: 6.636–37).

The contrast between the smoothness of the surface of the sea and its uneven bed is repeated in *Etym.* 13.14.2, but without the curious claim than the depths of the sea exceed the highest points on earth. Pliny (*NH* 2.105.224) records the estimate of one savant, Fabianus, of two miles (3.2 km). Whether this was Isidore's inspiration or not, this evocation of inestimable and extraordinary depth makes an interesting thematic antiphon to the indeterminate extent of the ocean discussed in the closing lines of chapter 40.

CHAPTER 43 THE NILE

The Nile was the subject of curiosity and debate for many ancient philosophers and scholars, because its annual flood occurs in high summer, when other rivers flowing into the Mediterranean are at their lowest level. Inquiry into this anomaly led to a second question: where were the sources of the Nile? Isidore, however, is focused here on meteorology rather than geography. Hence his account reproduces some, but not all, of the explanation found in Lucretius' *De rerum natura* 6:712–37. Lucretius opines that the flooding of the Nile could be caused by the Etesian winds, which blow in a direction

---

156 Seneca, *Natural Questions* 3.4.5, trans. T.H. Corcoran (Cambridge, MA: Harvard University Press/London: Heinemann, 1971), 216–17.

157 Pliny also invokes the underground channels (*NH* 2.66.166) but does not mention desalinization. On the afterlife of this material in early medieval Irish scholarship, see Smyth, *Understanding the Universe*, 239–40.

opposite to the current of the Nile and push its waters back (Lucretius' Etesian winds blow from the north, as do Aristotle's [*Met.* 2.5, 362a], not from the west as Isidore states). Lucretius also mentions the sand dune that allegedly blocks the mouth of the Nile when the Etesians blow. Isidore does not follow Lucretius into his further speculations that the Etesians may drop a greater amount of rain in the south, resulting in the Nile's anomalous summer flood, or that the heat of summer melts the snows on the Ethiopian mountains. Theories such as these would open up the question of the sources of the Nile, and Isidore seems unwilling to discuss this matter.

Some details of Isidore's account, however, suggest an acquaintance with Seneca's famous essay on this subject, notably the statement that the Nile flood serves the Egyptians in lieu of rain (cf. *Natural Questions* 4A 2.2). But Seneca rejects the Etesian winds theory, and does not mention the sand dunes (4A2.23). The detail that the Etesian winds blow from the sixth to the tenth hour (i.e., noon to 4 p.m.) has not been traced: Pliny, *NH* 2.47.127 says that they pick up at the third hour of the day.

CHAPTER 44(–) THE NAMES OF THE SEA AND THE RIVERS

This chapter, found only in the long recension of *DNR*, seems to abandon the meteorological agenda of the previous chapters in favour of etymology; indeed, it is broadly replicated in book 13 of the *Etymologies*. Only chapters 2 ('Night') and 4 ('The Months') offer similar extended chains of terms and their origins.

CHAPTER 45(44) THE POSITION OF THE EARTH

With this chapter, Isidore continues in the vein of meteorological explanation. The problem expounded here is that the spherical earth is enveloped by the atmosphere, so its heavy mass implausibly rests on a foundation of air which is much lighter. How does it remain steady? Ambrose's answer is that the Creator 'suspended the earth in the void' (*in nihilo*). This would suggest that the earth is not sitting on anything at all, but hanging in space. Nonetheless, Isidore prefers to align Ambrose (and by association, the Bible) with 'the philosophers' (e.g. Pliny, *NH* 2.4.10) who argue that the earth actually floats on a mass of thick air. In the end, Isidore takes the agnostic position that both the floating and the hanging models are equally reasonable, and equally unprovable. This is the second time in *DNR* that Isidore invokes the mystery of God's creative plan to bring closure to an argument, the other being the discussion of the tides in chapter 40.

Isidore's imagery of the earth floating like a sponge has not been traced. It is an interesting one, because in the next chapter he will again compare the earth to a sponge in order to explain not its stability, but its instability.

CHAPTER 46(45) EARTHQUAKE

After the fashion of ancient writers on natural phenomena, Isidore provides a number of alternative explanations for earthquakes. These match closely the ones promoted by Lucretius (*De rerum natura* 6.557–95), namely (1) gusting winds in subterranean caverns, (2) collapse of underground caves, and (3) buried rivers.[158] Like Lucretius (and, before him, Aristotle), Isidore gives pride of place to the wind theory; his account of how winds that are trapped in underground caves build up turbulence and shake or split open the earth to make their escape matches in sense, if not wording, Lucretius 6.577–84. Here again, the core idea of 'exhalation' provided a secure conceptual berth for earthquakes within the capacious science of meteorology.[159]

Isidore's allegorical remarks are carefully worded to avoid any implication that earthquakes are omens: an earthquake is not a sign of the Last Judgement, but alludes or refers to it [*pertinet*]. Moreover, Isidore assigns a more positive meaning to earthquakes, namely that they symbolize the upheaval of conversion – though his source, Gregory's *Homilies in Ezekiel*, ascribes the impulse to conversion to fear of the Last Judgement.

CHAPTER 47(46) MOUNT ETNA

The fires of Mount Etna were the stuff of poetic allusion and scientific speculation in the classical world. The most famous work on the subject is the anonymous Latin poem *Aetna*, possibly composed by a contemporary

---

158 The hydraulic model resurfaces in *Etym.* 14.1.3, where Isidore names Lucretius as his source.

159 Taub, *Ancient Meteorology*, 75, 88, 90, and 165.

of Seneca. Like the treatment of Etna by Lucretius (*De natura rerum* 6.639–702), the poet's intention in *Aetna* was to replace superstitious fear with rational explanation. While Isidore turned to a work of history for his information – Justinus' *Epitome* of Pompeius Trogus – his account accords with the model prevalent in other scientific and didactic accounts. As with earthquakes, winds are the principal cause; thus even volcanoes are within the purview of meteorology. However, while earthquakes were treated generically, volcanoes were particular and local phenomena, not cases of 'vulcanology'.[160]

That being said, Etna's eruptions can be understood using the logic of earthquake. Isidore may have favoured Justinus' account precisely because its details echoed his chapter on earthquake. The crumbly texture of the soil in Sicily recalls Isidore's otherwise untraced claim that sandy soils are more prone to earthquakes than heavy, compacted ones. But what was at least as attractive was Justinus' fixation on the persistence of the fires of Etna, which (like those of the sun) are 'fed by the nutriment of moisture'. This reinforces his allegory of Etna as a natural symbol of the real and eternal fires of hell.

CHAPTER 48(47) THE PARTS OF THE EARTH

De natura rerum closes with a verbal evocation of two images of planet Earth. The first (ch. 48.1) pictures Earth positioned in the centre of the cosmos. The second (ch. 48.2–3) depicts the inhabited lands known to ancient and medieval people. Between the two, the final sentences of chapter 48.1 evoke the ocean as both covering the surface of the globe and ringing the perimeter (*circumductio*) of the lands.

Isidore likewise visualizes the *oekumene* in two different ways. From Hyginus he derives a mental map of three continents separated by conspicuous bodies of water; from Augustine, he borrows the more abstract concept of the continents occupying sectors defined by the cardinal directions. Finally, he cites Ambrose's geometrical formula: Asia occupies half the world, Europe and Africa the other half.

Manuscripts of *DNR* are the oldest witnesses to a form of stylized map that encompasses all three of these approaches: the so-called T-O map of the inhabited lands of Earth. This 'map' is in fact a diagram or schema.

160 Taub, *Aetna and the Moon*, 45–55, esp. p. 49.

It represents the land mass of the known world as a disk (the O) divided horizontally, and with its lower half bisected vertically (the T).[161] Though Ambrose seems to be describing such a map, this does not prove that the map pre-dated Isidore. Indeed, whether the map was devised on the basis of the idea of the tripartite world, or vice versa, and if the former is the case, what role *DNR* may have played in its genesis, has excited scholarly controversy. While the oldest manuscript of *DNR* (Escorial R.II.18) has two maps, Patrick Gautier Dalché thinks that they were both later additions.[162] Woodward, however, thinks that Isidore himself conceived the minimalist form of the T-O map (such as one found in the St Gall manuscript, which we have adapted for our illustration), from which more elaborate forms were devised.[163] The T-O map migrated into manuscripts of the *Etymologies*; conversely, information from the *Etymologies* on the origins of the names of the continents (*Etym.* 14.3.1, 14.4.1), as well as the dispersal of the three sons of Noah and the number of nations descended from them (*Etym.* 9.2.1–10), was inserted into the map in *DNR*.[164]

Isidore is not generally interested in quantitative information, but he closes his treatise by citing Ambrose's claim that the circumference of Earth measures 180,000 stades. This would be roughly the equivalent of 20,685 miles (33,290 km) (based on one modern estimate that a Roman stade = 606¾ feet (185 m)). The modern figure for the earth's circumference is about 24,900 miles (40,075 km). Ambrose's figure derives ultimately from Posidonius' later revision of his own original estimate of 240,000 stades (27,580 miles). For Eratosthenes' earlier estimate of 252,000 stades (28,959 miles (44,385 km)) and the survival of these competing figures in the handbooks of antiquity, see Stahl, *Roman Science*, 38–41 and 49. The fact that Isidore does not report Eratosthenes' estimate for the

---

161 It bears repeating that this schema represents the inhabited lands of Earth, and not the planet itself. In other words, Isidore, and medieval people generally, understood Earth to be a sphere. That the land mass of the known world could be schematically represented as a circle, the *orbis terrarum*, in no way implies that they thought Earth was a flat disk. See Rudolf Simek, 'The Shape of the Earth in the Middle Ages and Medieval Mappaemundi', in P.D.A. Harvey (ed.), *The Hereford World Map: Medieval World Maps and their Context* (London: British Library, 2006), 293–303; Stevens, 'The Figure of the Earth in Isidore's *De natura rerum*', *Isis* 71 (1980): 268–77.

162 Patrick Gautier Dalché, 'De la glose à la contemplation. Place et fonction de la carte dans les manuscrits du haut moyen âge', *Testo e immagine nell'alto medioevo*, Settimane di studi del Centro italiano di studi sull'alto medioevo 41 (Spoleto, 1994), 707.

163 Woodward, 'Medieval Mappaemundi', 301.

164 See Appendix 5.

earth's circumference, which Pliny gives in book 2 of his *Natural History* (2.112.247), offers additional support to Fontaine's conclusion that Isidore only came into the possession of Pliny after the completion of *DNR*, and that it was Pliny's chapter heading in the table of contents (book 1) – *Terrae universae mensura* – not the text of book 2, that inspired him to add what is now chapter 48 to his work (Fontaine, *Traité*, 42).

Appendices

1 THE VERSE EPISTLE OF KING SISEBUT

Here begins
The Letter of Sisebut, King of the Goths,
Sent to Isidore, author of
The Book of Wheels

Perhaps in the woods you are without hurry composing idle songs and amid warbling waters and musical breezes you are infusing the serene mind with Pierian nectar.

But a turbulent swarm of affairs clouds our sight **(5)** and the cares of our thousands of ironclad soldiers weigh us down. The noisy prattle of lawyers is deafening, the marketplaces bark, the trumpets of war cast everything into confusion, and then we are borne all the way over the Ocean,[1] since the snow-covered Basque holds fast and the frightful Cantabrian shows no mercy.[2]

Lo, by these [songs] you may proclaim that a garland of leaves girds the hair of Phoebus **(10)** or, more grandly,[3] that ivy overshadows his locks; lo, you may bid them fly through the flaming upper air.[4] But it is more likely that the sluggish strength of elephants will outrun swift eagles, and the irritated tortoise the fleet hound, than that we may track the dewy moon in song. /**331**/

---

1 = the Bay of Biscay?

2 Sisebut and his generals were involved in, possibly, two poorly understood campaigns in Asturia and Cantabria against the *Ruccones* and a duke named 'Francio'. Unfortunately, the campaigns cannot be dated with any precision. For details, see Collins, *Visigothic Spain*, 75 and 77.

3 Fontaine, unnecessarily in our view, emends the reading of the MSS from *augustius* 'more grandly' to *angustius* 'more simply'. His reason (*Traité*, 362 n. 4) depends on the assumption that *frondea* (line 10) refers to a *laurel* garland, sacred to Apollo. But by itself it simply means a leafy garland (cf. Pliny, *NH* 16.5).

4 i.e., it would be easy for you (Isidore) to make up a fanciful poetic explanation of an eclipse of the sun (Phoebus) by the moon (Phoebe).

(15) Nevertheless, bent down by these affairs, struggling on account of earthly burdens, I will explain why the circle of the sick orb becomes dark and why the brilliant light of its snowy countenance vanishes. It is not, as people believe, that a frightful shrieking woman in a black cave **(20)** takes it down from her high-wandering mirror under the infernal shadows, nor that, conquered by an incantation or by the moisture of the Styx or the vapour of an herb, it seeks the noisy earth and the triumphant clangour.[5]

For in fact the moon moves inviolate though the upper air, where its nearest boundary separates disorderly things from pure. But when the earth with its vast body, **(25)** which occupies, far below, the middle of the axis,[6] interrupts the apogee of the brother[7] with the cone of its shadow, then the deprived moon grows pale, like the round shadow of a wheel, until in its swift course it transits the axis[8] of that piled-up mass[9] and with its rose-coloured mirror, moving unimpeded through the sky, recovers the fraternal flames. /333/

(30) But since you wonder why, when the enormous mass of the sun is reputed to be 18 times greater than the earth's disk, it does not engulf the terrestrial cone with light, consider this well-established line of argument: notice how Phoebus moves aloft through the golden vault of the universe **(35)** and how it illumines the earth below with its high chariot.

Whatever the great size of the former may be, whether it scatters its fires down vertically or shines its beams on an oblique axis, its rays are scattered onto the earth; the remaining light of the sun, in as much as its greater [body] escapes with its shining shafts **(40)** unobstructed by the globe [of the earth], extends through the vast emptiness, until the vanquished shadow [of earth] ends in the tip of the pyramid.

When moist Phoebe at her lowest point drives her icy team through this shadow,[10] sometimes, deprived of colour by the neighbouring shadow [= *penumbra*], she is cut off from her brother, and disappears, bloodless, with an empty countenance.

---

5 i.e., the victorious clangour 'of brazen implements in an eclipse of the moon': Lewis and Short, s.v. *vincibilis*.

6 The axis of the universe: i.e., the earth is at the very centre (the lowest point) of the spherical universe.

7 Phoebus, the sun, is the brother of Phoebe, the moon.

8 Reading *axem* with the MSS instead of Fontaine's conjecture *imum*.

9 The cone of shadow cast by the earth in a lunar eclipse.

10 *Quam* refers to *uicta umbra*, 'the vanquished shadow', or cone of total darkness (see Fontaine's diagram, p. 335 *bis*).

(45) Moreover, why only the moon is deprived of light, is nothing to be wondered at: for, lacking light herself, /**335**/ another's light warms her. When the nearest part of the cone begrudges it, sickly grey, she awaits the rays of her brother. But the rest of the chorus of stars is not touched by the shadow, **(50)** and they all have their own light and they do not glow red from the sun, but the pure firmament with its starry rays is forcibly borne far aloft beyond the sun with the pole of the heavens.

Now as to why the disk of the full moon does not always lose its colour, its bent courses on an oblique path are responsible. **(55)** For then, when the moon, wandering by a calculated miscalculation off its path, choses tortuous circuits, the distant sun passes beyond the moon's axial point, and wraps up the mantle of the night and illumines his sister.

It is for the same reason that the golden brilliance of the majestic sun is broken by sudden shadows, **(60)** when, her disk deprived of light, the nurturing body of the moon moves between the earth and the sun, blocking her brother by her direct interposition.

COMMENTARY

Sisebut was in his late 40s when he was elevated to the Visigothic throne after the death of King Gundemar in early 612.[11] He probably spent much of his life as an adult on military campaigns.[12] The *Epistle* alludes to battles that he personally directed, perhaps in his first year as king, against enemies in Asturia and Cantabria (lines 4–8).

Sisebut's authorship of the *Epistle* has never been disputed. The scion of a powerful Visigothic noble family, he rose to royal power on the strength of his military prowess. But, like others of his class, evidence of whose libraries and literary activity survive, his education was of a high order. It included thorough instruction in classical metrics, together with deep immersion in classical authors and technical treatises on astronomy, sufficient to create this remarkable poem.[13]

---

11 Isidore, *Historia Gothorum Wandalorum Sueborum* 60, ed. Mommsen, MGH: AA 11.2:291; García Moreno, *Historia de España Visigoda*, 147. Sisebut was born around 565.

12 As Collins, *Visigothic Spain*, 75 points out, Sisebut 'must have been sufficiently competent militarily to have been chosen by the nobility who had supported his predecessor'. Fontaine, 'Sisebut's *Vita Desiderii*', 98 n. 2 speaks of 'the king's brilliant military career'.

13 See Fontaine, *Traité*, 151–52; Riché, *Education and Culture*, 258–59; Roger Collins, 'Literacy in Early Medieval Spain', in Rosamond McKitterick (ed.), *The Uses of Literacy in*

Sisebut's contemporaries were full of admiration for his intellectual talents. According to Isidore himself, Sisebut 'was refined in speech, learned in judgement, and imbued with the knowledge of letters to a large extent'.[14] The Frankish chronicler called Fredegar and the anonymous Spanish author of the Chronicle of 754 concurred, and their perceptions have been confirmed by recent scholarship.[15]

Early Medieval Europe (Cambridge: Cambridge University Press, 1990), esp. p. 115. Sisebut's letters have been edited by Juan Gil, *Miscellanea Wisigothica* (Seville: Publicaciones de la Universidad de Sevilla, 1972), who also surveys other works attributed to the king (pp. ix–xx, 3–28, 53–68). On his life of Desiderius of Vienne, see Fontaine, 'King Sisebut's *Vita Desiderii* and the Political Function of Visigothic Hagiography', in James, *Visigothic Spain: New Approaches*, 93–129.

14 Isidore, *Historia Gothorum* 60, MGH: AA 11:291; trans. Donini and Ford, *History of the Kings*, 28.

15 Hen, 'Visigothic King', 89–99 (on the *Epistle*, see pp. 94–99); Hen, *Roman Barbarians*, ch. 5.

2 INTRODUCTORY FORMULAS FOR
THE DIAGRAM OF THE WINDS (DIAGRAM 7)
IN CHAPTER 37

(a) *De qualitate*. Some MSS add this formula of introduction (with minor variations) for Diagram 7 after chapter 37.5: *de qualitate uentorum sub caeli axe liquido et aperte inueniet si circuli similitudinem prudens lector requirat*, 'if the skilled reader investigates the likeness of the circle, he may learn clearly and openly about the quality of the winds under the axis of heaven'.[1] Diagram 7, which in other manuscripts is placed after chapter 37.4, in these MSS follows this addition at the end of chapter 37.5, with the exception of Laon 423, where it is not displaced, and Strasbourg 326, where it follows the poem *De Ventis* (see Appendix 4). With the exception of Strasbourg 326, Diagram 7 includes the KOCMOC inscription.

MSS: Avranches 109, Bern A.92/20, Bern 224, Cambrai 937, Cologne 83(II), Laon 423, NY Plimpton 251, Paris 6400G, Paris NA 448, St Gall 240, and Strasbourg 326.

One manuscript adds an abbreviated variant of the formula: *de qualitate autem uentorum positionibus sub caeli axe circuli similitudinem*, but, unlike the above MSS, places it within Diagram 7 and inserts Diagram 7 in the first sentence of chapter 37.4 after ... *ab occidente interiore flat*.

MS: Laon 422.

(b) *Quorum ordinibus*. Several MSS add this formula of introduction (with minor variations) for Diagram 7 after chapter 37.4: *quorum ordinibus et institucionibus per nomina iuxta grecos et latinos duodecim uentorum subiecta demonstrat figura*, 'the figure appended below displays these matters in their right orders and arrangements by the names in Greek and Latin of the twelve winds'.

MSS: Bamberg Msc. Nat. 1, Besançon 184, Vatican City Pal. 834, and Verdun 26.

---

1 According to Fontaine, 'La Diffusion carolingienne', 120 and n. 26, Milan H 150 inf. belongs in this group, but he does not provide details.

(c) *Rotatim.* Three MSS (all medium recension?) add this formula of introduction for Diagram 7 (with minor variations) before chapter 37.1: *quorum nomina et elementa in pictura inuenies, id est, primum incipies a septentrionali et unumquemque cardinalem cum bino ramusculo inuenies dextro et sinistro. Hos inter rotam circuli, si a septentrione uersos rotatim legas, in gyrum inuenies ordines eorum* (cf. Teyssèdre, 'Les Illustrations', p. 34 n. 10). Diagram 7 is placed after chapter 37.5.

MSS: Florence 22 dex.12, Lisbon 446, Oxford Ashmole 393 (*Si a septembrione ... ordines eorundem*, only).

One MS (medium recension) adds this formula after chapter 37.4 and *precedes* it with the second paragraph described below.

MS: Paris 15171 (also Du Breul, ed.).

Following Diagram 7 comes a second paragraph (with variations): *Inde flatus uentorum pares inuicem sibi cursus emittunt, id est, facie ad faciem, hic ordo sequitur. Septentrio cum austro. Subsolanus cum fauonio. Circius cum euro austro. Vulturnus et affricus. Aquilo et austro affricus. Eurus et chorus. hii in flatibus suis lineas rectas secuntur. Omnes enim uenti nuncupantur qui de thesauro egrediuntur et unus quisque eorum secundum ordinem suum nubibus imperantur in hoc mundo ueniunt sicut psalmographum dicit. Producit uentos de thesauris suis. Quorum uentorum quattuor apostoli tenent flatum. Iudas aquilonem. Iohannes orientalem. Eurus austrum. Iacobus occiduum.*

MSS: Florence 22 dex.12, Lisbon 446 (Oxford Ashmole 393 omits). Paris 15171 (also Du Breul, ed.) places this paragraph [up to *lineas rectas secuntur*] before the *rotatim* formula.

(d) *Hoc ordine.* One MS (medium recension?) adds this formula for Diagram 7 before chapter 37.1: *Ventus est aer commotus. Qui hoc ordine in hac spera figuratur.* Diagram 7 is placed after chapter 36.3.

MS: Escorial E.IV.13.

(e) *Quo ordine.* Two manuscripts introduce Diagram 7 with excerpts from the text of chapter 37 inside it with this formula: *Hic quoque potest uideri quo ordine spirant uenti.*

MSS: Berlin, Phillipps 1830, Trier 2500.

3 EXTRACTS FROM CHAPTER 37 ARRANGED WITHIN THE DIAGRAM OF THE WINDS

In some MSS an abbreviated text of chapter 37.1–4, with several additions from *Etym.* 13.11, is inserted within the 12 segments of a diagram of the winds (= Diagram 7) with a T-O map in its centre; the text has been rearranged to fit into the appropriate points of the compass. The diagram is oriented with East at the top. Thus the 12 segments in clockwise order begin with East **(1)** and end with East-Northeast **(12)**. Text transcribed from Karlsruhe 106, fol. 52v:

1 (E): *Secundus cardinalis Subsolanus qui et Apheliotes. Hic ab ortu intonat et est temperatus [DNR 37.2]. Subsolanus quia ab ortu solis* [cf. *Etym.* 13.11.4].

2 (ESE): *Eurus ex sinistro latere ueniens Subolani, orientem nubibus inrigat [DNR 37.2]. Eurus eo quod ab eo flat idem ab oriente [Etym.* 13.11.4].

3 (SSE): *Euroauster calidus a dextris intonat Austri [DNR 37.3]. Euroauster dictus eo quod una parte abeat Eurum et altera Austrum [Etym.* 13.11.6].

4 (S): *Tertius cardinalis uenter* [sic] *Auster, qui et Nothus, plage meridiane humidus atque fulmineus, generans largos imbres et pluuias laetissimas et soluit florem [DNR 37.3].*

5 (SSW). *Euronothus uentus temperatus et calidus a sinistra Austri spirat [DNR 37.3].*

6 (WSW). *Africus qui dicitur Lipis, ex Zephyri dextera intonans, generat tempestates et pluuias et nubium conlisiones et tonitura et fulgora et fulminum inpulsus [DNR 37.4].*

7 (W). *Quartus cardinalis Zephyrus, qui et Fabonius, ab occidente interiore flat; hic hieme frigora relaxat, florem producit [DNR 37.4].*

8 (WNW). *Chorus, qui et Arcestes, ex sinistra parte Fauonii aspirans; eo flante, in oriente nebula sunt, India serena* [*DNR* 37.4].

9 (NNW). *Circius, qui et Thrascias; hic a dextris Septentrionis intonat et facit nubes et grandinum coagulationes* [*DNR* 37.1]. *Circius eo quod Euro sit iunctus* [*Etym.* 13.11.12].

10 (N). *Ventorum primus cardinalis Septentrio, qui et Aparchias, frigidus et niualis; flat rectus ab axie* [sic] *et facit arida et frigora et siccas nubes* [*DNR* 37.1].

11 (NNE). *Aquilo uentus, qui et Boreas, ex alto flans, gelidus atque siccus et sine pluuia, q*[*?*] *non discutet sed constringit* [*DNR* 37.1].

12 (ENE). *Vulturnus, qui et Chaecias uocatur, dexterior Subsolani; hic dissoluit cuncta atque desiccat* [*DNR* 37.2].

This diagram of the winds sometimes accompanies the poem *De Ventis* (see Appendix 4) independently of the text of *DNR*. Bede's copy of Isidore's *DNR* may have included this variety of Diagram 7. The diagram always reads *qui et Aparctias* rather than *hic et Aparctias* (see segment 10).[1] *Qui et Aparctias* is the wording which Bede employs in ch. 27 of his *On the Nature of Things*.[2]

MSS: Berlin 1830; Bern 212/1; Florence 29.39 (the MS contains *DNR* but the wind diagram appears separately); Karlsruhe 106; Strasbourg 326 (ch. 37 + the poem *De Ventis* [see Appendix 4] + the wind diagram); Trier 1084/115 (puts an abbreviated excerpt of Bede's *DNR* 27 beneath the wind diagram); Trier 2500.

---

1 *Aparctias* is variably spelled both in Isidore and in Bede.

2 The suggestion (see von Büren, 'Le *DNR* de Winithar', 393 and n. 22) that Bede could have taken the name *Aparctias* from Vegetius, *Epitoma Rei Militaris* 4.38.12, ed. M.D. Reeve (Oxford: Clarendon Press, 2004), 151 is unnecessary, since Bede followed Isidore's list of names, including *Aparctias*, exactly as they appear in Isidore's medium and long recensions (the names of several of the winds in Vegetius do not correspond to those in Isidore and Bede).

4 THE POEM OF THE WINDS[1]

Four winds arise from the four corners of the earth.[2] Two more are joined around each of these on the right and the left, and thus they encircle the world with a twelvefold blast.[3]

The first, Aparctias (N), blows from the north: (5) Septentrio is the name given it in our language. Frigid Circius (NNW) roars around this from the hollow on its right; the Greeks call it Thrascias in their own tongue. Boreas (NNE) on its left bellows with an icy swirling gust; in our speech it is usually referred to as chill Aquilo.

(10) Next Subsolanus (E) blows straight from due east; the Greek calls it, by a fit name, Apeliotes. Beside it is Vulturnus (ENE), which arises on the right side; the Attic language calls it Caecias among the Greeks. Eurus (ESE) refreshes the left side with its cloud-bearing breeze, (15) which the Grecian tongue indicates by the same name.

Next Notus (S) emits its winds from the midday track of the sun: the Romans rightly they call it Auster, because it 'ousts' the clouds by its breezes.[4] Beside it on the right is Euronotus (SSE), which the Latins called by the hybrid name Euroauster in the Latin language. (20) Libonotus (SSW) contaminates the left with its hot breezes; this is burning Austroafricus with its tremendous heat.

The flowery trumpet of Zephyrus (W) inhabits the setting of the sun; to it the name of Favonius is given in the language of Italy. The wind called Lips (WSW) in the Attic tongues touches its right side; (25) this, coming from its own region, is called Africus. But you, Corus (WNW), you rage

---

1 Our translation is based on the text in *Anth. Lat.* 1.2:6–8 (no. 484). The incipit is '*Quattuor a quadro consurgunt limite uenti*'; see D. Schaller and E. Könsgen, *Initia Carminum Latinorum saeculo undecimo antiquorum* (Göttingen: Vandenhoeck & Ruprecht, 1977), no. 13113 for bibliography and a list of manuscripts.

2 Literally, 'from a square boundary'.

3 The winds (personified) are imagined in a circle around the earth facing inwards. Therefore, a wind to the 'right' of north would be somewhere to the northwest, etc.

4 Isidore, *Etym.* 13.11.6 gives the same etymology (*Auster < haurire*, to exhaust), though in a different sense from the way we take it here.

from the left side of Zephyrus; the Greeks are accustomed to call you by your old name Argestes.

COMMENTARY

The Poem of the Winds, De Ventis (Quattuor a quadro consurgunt limite uenti), consists of 27 lines of Latin hexameter verse. It is very closely related to the diagram of the winds that Isidore or one of his assistants inserted into *DNR* as Diagram 7. It has been attributed to Isidore's dedicatee, King Sisebut,[5] but in our view it is more than likely Italian and even possibly the work of a sixth-century monk in Cassiodorus' monastery of Vivarium. *De Ventis* is found in more than twenty manuscripts. The earliest, Bern 611, dates to the end of the eighth or beginning of the ninth century. Six of these manuscripts contain Isidore's *DNR* in whole or in part. *De Ventis* must not be confused with a similar poem of 63 lines (*Quattuor a quadris uenti flant partibus orbis*), which Gustav Becker published in his edition of Isidore's *DNR*, under the mistaken impression that it was the work of Suetonius.[6] *Quattuor a quadris uenti* is in common leonine hexameters, a verse form which dates it to not earlier than the Carolingian period and more probably the eleventh or twelfth century.[7]

In its developed form in late Antiquity the wind-rose was divided into twelve points. The chief function of the poem is to provide the Greek and Latin names for the twelve winds, together with a few rather vague meteorological phenomena associated with each. The poet, if he can be dignified by that name, might have been a student of Latin metrics in a monastic school, demonstrating his accomplishments.

Pliny's *Natural History* is a convenient repository of the Graeco-Roman scientific tradition that would filter down in attenuated form into Late Antiquity and the early Middle Ages.[8] His list of the names of the winds

---

5 Martyn, *King Sisebut*, 111–20.

6 Ed. Becker, xviiii–xxi. The poems are published together in *C. Suetoni Tranquilli praeter Caesarum libros reliquiae* (Leipzig, 1860), ed. Augustus Reifferscheid, 304–08, nos. 151a and 151b.

7 For the structure and use of leonine verse in the Middle Ages, see Calvin B. Kendall, *The Allegory of the Church: Romanesque Portals and their Verse Inscriptions* (Toronto: University of Toronto Press, 1998), 71–72. 'Common' leonines display monosyllabic internal rhyme, as opposed to 'full' leonines with disyllabic internal rhyme. *Quattuor a quadris uenti* cannot be older than *De Ventis*, contrary to what Obrist, 'Wind Diagrams', 71 states.

8 Seneca's list of winds in his *Natural Questions* 5.16 is similar to Pliny's. He gives the

is similar to that of *De Ventis*, with two major exceptions. His sole name for the south-south-east wind is *Phoenix* instead of *Euronotus/Euroauster*. And he gives *Vulturnus* (*Volturnus*) as the Latin name for the east-south-east wind also known as *Eurus* (Gr). This was in accordance with classical usage. But before Isidore's time *Vulturnus* unaccountably shifted in the main current of the Italian tradition to a wind between north and east, where it lingered for some centuries. The misplacement of *Vulturnus* is already found on the second- or third-century marble Roman wind-vane known as the Vatican table.[9] The Vatican table also gives *Euroauster, Austroafricus,* and *Circius* as the Latin terms for *Euronotus, Libonotus,* and *Thrascias.* Apart from various misspellings of names both in Latin and Greek, the only substantive variant between the Vatican table and *The Poem of the Winds* is the Greek name *Iapyx* in place of *Argestes* in the Vatican table. The poem is clearly heir to the Italian tradition.

As printed by Fontaine, the names in Diagram 7 are: Septentrio/ Aparctias (N), Aquilo/Boreas (NNE), Vulturnus/Caecias (ENE), Subsolanus/Apeliotes (E), Eurus/Eurus (ESE), Euroauster/Euronotus (SSE), Auster/Notus (S), Austroafricus/Libonotus (SSW), Africus/Lips (WSW), Fauonius/Zephyrus (W), Corus/Argestes (WNW), and Circius/ Thrascias (NNW).

These are precisely the names of the winds as they are catalogued in *The Poem of the Winds.* However, *Caecias* (ENE) is Fontaine's normalization of the manuscript readings *calcias/cilcias/circias*, etc. Munich 14300 reads *cilcias*. In the text of *DNR*, ch. 37.2, Fontaine also normalizes to *Caecias* from manuscript readings *calcias, calcius, cilitias.*[10] But in the poem the name of the wind is given correctly as *Caecias* (in the Greek accusative form *Caecian*).[11] The author of *De Ventis* appears to have taken his inspiration from a diagram of the winds very similar to, but more accurate than, Diagram 7.

R.A.B. Mynors calls attention to a group of manuscripts of

name *Euronotus* instead of *Phoenix* to the south-southeast wind, and in that respect is closer than Pliny to *The Poem of the Winds.* But he omits the Greek equivalents.

9 Taub, *Ancient Meteorology,* 108, fig 3.3; P.G. Lais, 'Monumento greco-latino di una rosa classica dodecimale in Vaticano', *Pubblicazioni della Specola Vaticana* 4 (1894), xi–xv and plate 1.

10 Bede's MS of Isidore's *DNR* probably read *calcias* (cf. Bede, *DNR* 27, ed. Charles W. Jones, *De natura rerum,* CCSL 123A (Turnhout: Brepols, 1975), 218.

11 The MSS of *De Ventis* consulted by Riese, *Anth. Lat.* 1.2:7 (*apparatus*), support *Caecias.*

Cassiodorus' *Institutions* that also include *The Poem of the Winds*. These manuscripts descend from a common archetype, which contained five items in the following order (hereafter 'Cassiodorus/Boethius'): (1) Cassiodorus, *Institutions* 2; (2) *Excerptum de quattuor elementis*; (3) excerpts from works of Augustine; (4) *The Poem of the Winds* followed by a diagram of the winds; (5) an excerpt from Boethius.[12] Mynors believes that this grouping probably had its origin in the eighth century.[13] However that may be, the fact that Cassiodorus' *Institutions* 2 was an important inspiration as well as source for Isidore, and that both the diagram of the winds and the poem preserve the Italian tradition of late Antiquity that lies back of Isidore's chapter 37 and the diagram which he or one of his followers selected for inclusion in *DNR*, makes it virtually certain that these materials, irrespective of whether they arrived in the grouping Cassiodorus/Boethius, came to Spain from Italy in the sixth or early seventh century.

Manuscripts of *The Poem of the Winds* (*De Ventis*). The symbol (*) = lost or destroyed; (†) = manuscript containing *DNR*, chapter 37.1–4, inscribed within the diagram of the winds [for which, see Appendix 3]); (††) = manuscript containing the text or part of the text of *DNR*:

1 †BERLIN 1830, fol. 3v. See above, Manuscripts of Isidore's *DNR*, p. 69.

2 †BERN 212(I), fol. 109r. Cassiodorus/Boethius, with (4) *De Ventis* and the diagram of the winds. See above, Manuscripts of Isidore's *DNR*, p. 70.

3 BERN, Burgerbibliothek 611, fol. 42r, s. viii[1/2], eastern France. *Quattuor a quadro consurgunt limite uenti ... Argestem quem graia suo uocat ore ecamena.* Corresponding diagram of the winds on fol. 93v. In the diagram the Greek names of the winds are in Greek lettering. The centre of the diagram's circle is inscribed with a T-shaped figure which appears to be the preparation for a T-O map which was never filled in with the names of the three continents.

Hagen, *Catalogus*, 480; Homburger, *Die Illustrierten Handschriften*, 21–23 and ill. 13.

---

12 *Institutiones*, ed. Mynors, xxx–xxxix.
13 *Institutiones*, ed. Mynors, xxxix.

4 *CHARTRES, Bibliothèque municipale 102, fols. 1r–61v, s. x, Chapter Library (destroyed?). Cassiodorus/Boethius, with (4) *De Ventis*.
Mynors, *Cassiodori Institutiones*, xxx–xxxii.

5 ††DIJON 448, fol. 75r. See above, Manuscripts of Isidore's *DNR*, p. 75.

6 ††ERLANGEN 186, fol. 246r. See above, Manuscripts of Isidore's *DNR*, p. 76.

7 ESCORIAL, Real Biblioteca de San Lorenzo d.I.1, fol. 11v, s. x. Diagram of the winds and *De Ventis* (3 lines only?): *Quattuor a quadro consurgunt limite uenti / Hos circum gemini dextera lebaque iunguntur / atque ita bisseno circundant flamine mundum.*
Antolín, *Catálogo* 1:323.

8 ESCORIAL, Real Biblioteca de San Lorenzo d.I.2, fol. 14v, s. x. Diagram of the winds and *De Ventis*: *Quattuor a quadro consurgunt limite uenti*, etc.
Antolín, *Catálogo* 1:375.

9 ††ESCORIAL R.II.18, fol. 34v. See above, Manuscripts of Isidore's *DNR*, p. 77.

10 ††FLORENCE 16.39, fol. 99r–v. See above, Manuscripts of Isidore's *DNR*, pp. 78–79.

11 †KARLSRUHE 106, fol. 52r. Cassiodorus/Boethius, with (4) *De Ventis*. *De Ventis* is headed by a three-pronged diagram above a playful inscription ('either of singers or of flute players [= *tibicinantium*?] or of lute players'):

AVT CAN	AVT TIBI	AVT CITHARI
TANTIVM	ZANTIVM	ZANTIVM

See above, Manuscripts of Isidore's *DNR*, pp. 80–81.

12 KARLSRUHE, Badische Landesbibliothek, Karlsruhe 442 (Durlach 36), fol. 18v, s. x–xi. *De Ventis*.
    Anth. Lat. 1.2:6; Baehrens, *Poetae Latini minores* 5:383; *Die Handschriften der Grossherzoglich Badischen Hof* 4, 78–79.

13 LEIDEN, Bibliotheek der Rijksuniversiteit, Vossianus Latinus Q 33, fol. 134, s. x. *De Ventis.*
Baehrens, *Poetae Latini minores* 5:383.

14.OSLO and LONDON, The Schøyen Collection 1800 (Phillipps 16278 [Libri 229]), s. ix². Cassiodorus/Boethius (lacking item [5]), with (4) *De Ventis.* Whether the manuscript preserves the diagram of the winds is unclear.
Mynors, *Cassiodori Institutiones,* xxx–xxxiii.

15 ††PARIS 5239, fols. 138v–139r. See above, Manuscripts of Isidore's *DNR,* p. 88.

16.††PARIS 5543, fol. 140r. See above, Manuscripts of Isidore's *DNR,* p. 88.

17 PARIS, Bibliothèque nationale lat. 12963 (St Germaine 782), Corbie, s. x¹. Cassiodorus/Boethius (lacking item [5]), with (4) *De Ventis.* Whether the manuscript preserves the diagram of the winds is unclear.
Mynors, *Cassiodori Institutiones,* xxx–xxxiii.

18 ††PARIS 15171, fols. 216v–217r. See above, Manuscripts of Isidore's *DNR,* p. 90.

19 ST GALL, Stiftsbibliothek 270, p. 51, s. ix², Carolingian minuscule with insular influence. *De Ventis* is headed by a diagram with inscription nearly identical to one in Karlsruhe 106. There is no wind diagram or trace of *DNR* 37.
Mynors, *Cassiodori Institutiones,* xxxivxxxv. Online: *e-codices.*

20 ††STRASBOURG 326, 119v–120r. See above, Manuscripts of Isidore's *DNR,* p. 93.

21 †TRIER 2500, fol. 20r. See above, Manuscripts of Isidore's *DNR,* p. 93.

22 VALENCIENNES, Bibliothèque municipale 195(?) (Mangeart BM 164), s. ix. Cassiodorus/Boethius (lacking item [5]), with (4) *De Ventis. De Ventis* is headed by a diagram with inscription nearly identical to one in Karlsruhe 106. There is no wind diagram or trace of *DNR* 37.

Mangeart, *Catalogue descriptif,* 150–51; Mynors, *Cassiodori Institutiones,* p. xxx–xxxiv. Mynors's shelf-mark for this MS is BM 195, but the MS now labelled BM 195 features material from Alcuin and does not contain Cassiodorus/Boethius.

23 ††VATICAN CITY, BAV, Reg. lat. 1260, fol. 38. See above, Manuscripts of Isidore's *DNR*, p. 95.

24 VATICAN CITY, BAV, Regin. lat. 1263, fol. 78r, s. xi[in], St-Mesmin de Micy. *De Ventis* (lines 1–3) inscribed above a diagram of the winds.

Obrist, 'Le Diagramme isidorien des saisons', 128 and fig. 6; 'Wind Diagrams', p. 66 and fig. 26.

25 VIENNA, Österreischische Nationalbibliothek 378, fol. 1v, s. xiii[in]. *De Ventis* inscribed within a diagram of the winds.

Obrist, 'Wind Diagrams', 83 and fig. 34.

5 TEXTUAL INSERTIONS IN CHAPTER 48
AND T-O MAP

(a) Vegetius and *De Trinitate* interpolations

Vegetius interpolation, ch. 48.3: [*milium stadiorum aestimauerunt*] *CCCC autem in longitudinem CC in latitudinem. In historiis catholicis hoc dicitur: *omnes nationes quae uicinae sunt soli, nimio calore perustas amplius quidem sapere, sed minus habere sanguinis dicunt. Propterea constantiam ac fiduciam cominus non habere pugnandi, quia metuunt uulnera, quia exiguum sanguinis se habere nouerunt; at contra septentrionalis populus, a solis ardore remotus, in quo insultores quidem, sed tamen largo sanguine redundantes, sunt ad bella promptissimi*, 'In universal histories we find this statement: *they say that all nations which are located under the sun, burned by excessive heat, have indeed more prudence but less blood; for that reason they do not have the firmness and courage for fighting hand to hand, since they fear wounds because they know they have very little blood; but on the contrary northern people, far from the heat of the sun, being indeed more rash, but nevertheless overflowing with an abundance of blood, are very ready for war*'. The portion set within asterisks (*) is from Vegetius, *Epitoma* 1.2 (ed. Reeve, pp. xiii and 6–7). Text based on Zofingen Pa 32 as transcribed by Fontaine, 'La Diffusion de l'œuvre d'Isidore', 327.

De Trinitate insertion, chapter 48(47): 'The Trinity is so named because from a certain three is made one whole, as it were a "Tri-unity" – just like memory, intelligence, and will, in which the mind has in itself a certain image of the divine Trinity. Indeed, while they are three, they are one, because while they persist in themselves as individual components, they are all in all. Therefore, the Father, Son, and Holy Spirit are a trinity and a unity, for they are both one and three. They are one in nature, three in person. One because of their shared majesty, three because of the individuality of the persons. For the Father is one person, the Son another, the Holy Spirit another – but another person, not another thing, because they are equally and jointly a single thing, immutable, good, and coeternal. Only the Father is not derived from another; therefore, he is called Unbegotten. Only the Son is born of the Father; therefore, he is called Begotten. Only the Holy Spirit proceeds from the Father and the Son; therefore, it alone is referred to as "the Spirit of both the others."

'For this Trinity some names are appellative, and some are proper. The proper ones name the essence, such as God, Lord, Almighty, Immutable, Immortal. These are proper because they signify the very substance by which the three are one. But appellative names are Father and Son and Holy Spirit, Unbegotten and Begotten and Proceeding. These same are also relational because they have reference to one another. When one says "God", that is the essence, because he is being named with respect to himself. But when one says Father and Son and Holy Spirit, these names are spoken relationally, because they have reference to one another. For we say "Father" not with respect to himself, but with respect to his relation to the Son, because he has a son; likewise we speak of "Son" relationally, because he has a father; and so "Holy Spirit", because it is the spirit of the Father and the Son. This relationship is signified by these "appellative terms", because they have reference to one another, but the substance itself, in which the three are one, is not thus signified.

'Hence the Trinity exists in the relational names of the persons. Deity is not tripled, but exists in singleness, for if it were tripled we would introduce a plurality of gods. For that reason, the name of "gods" in the plural is said with regard to angels and holy people, because they are not his equal in merit. Concerning these is the Psalm, "I have said: You are gods" (Ps. 81:6). But for the Father and Son and Holy Spirit, because of their one and equal divinity, the name is observed to be not "gods" but "God", as the Apostle says: "Yet to us there is but one God" (1 Cor. 8:6), or, as we hear from the divine voice, "Hear, O Israel: the Lord thy God is one God" (Mark 12:29), namely inasmuch as he is both the Trinity and the one Lord God.

'This tenet of faith concerning the Trinity is put in this way in Greek: "one ousia", as if one were to say "one nature" or "one essence"; "three hypostaseis", which in Latin means "three persons" or "three substances". Now Latin does not speak of God properly except as "essence"; people say "substance", indeed, but metaphorically, for in Greek the term "substance" actually is understood as a person of God, not as his nature.' *Etym.* 7.4; trans. Barney, et al., *Etymologies*, 159–60.

See von Büren, 'Isidore, Végèce et Titanus'.

MSS (Vegetius and *De Trinitate* passages, with minor variants): Paris 10616, Zofingen Pa 32, and Einsiedeln 167 (*De Trinitate* unconfirmed).

MSS (Vegetius passage, with minor variants): Einsiedeln 360, London Harley 2660, London Harley 3035.

MSS (first portion of Vegetius only, *CCCC autem ... hoc dicitur*, again with variants): Berlin Ham. 689, Cambrai 937 (e), Florence S. Marco 582, Florence Ricc. 379/4, and Vatican City Urb. 100.

(b) T-O map, legends: *ASIA. Post confusionem linguarum et gentes dispersae fuerant per totum mundum. Habitauerunt filii Sem in Asia, de cuius posteritate descendunt gentes XXVII, et est dicta Asia ab Asia regina. Quae est tertia mundi pars. REGIO ORIENTALIS.*

Europa dicta ab Europa filia Agenoris regis Lybiae uxoris Iouis. Vbi filii Iapheth uisi sunt terram tenere, de cuius originis gentes XV. Et habet ciuitates CXX. REGIO SEPTEMTRIONALIS. EVROPA.

Africa dicta ab Afer uno de posteris Habrae, quam possederunt filii Cham, de quo sunt egressae gentes XXX. Et habent ciuitates CCCLX. REGIO AVSTRALIS. AFRICA.

'ASIA. After the confusion of languages, the nations had also been scattered throughout the whole world. The children of Shem lived in Asia, from whose posterity 27 nations descend. Asia is named after Queen Asia. It is the third part of the world. THE REGION OF THE EAST.

'Europe is named after Europa, the daughter of King Agenor of Lybia [and] the wife of Jupiter. It is where the children of Japheth are known to have possessed the land, from [the descendants] of whose lineage come 15 nations. It has 120 cities. THE REGION OF THE NORTH. EUROPE.

'Africa is named after Afer, one of the descendants of Abram. It was possessed by the children of Ham, from whom 30 nations derive. It has 360 cities. THE REGION OF THE SOUTH. AFRICA.'

Text from Becker, ed., *De natura rerum*, p. 80.

MSS: Bamberg Msc Nat 1, Bern 417, Florence 29.39, Vatican City Pal. 834, Vatican City Reg. 1573, Verdun 26.

6 THE ZOFINGEN AND ENGLISH TYPES OF THE LONG RECENSION

(a) The chapters of the Zofingen type, with their equivalents in the regular order of *DNR*, are listed in Table 4.

TABLE 4 THE CHAPTERS OF THE ZOFINGEN TYPE, WITH THEIR EQUIVALENTS IN THE REGULAR ORDER OF *DE NATURA RERUM*

Zofingen chapter	Title	*DNR* chapter
Ch. 1	Preface	Preface
Ch. 2	Days	Ch. 1
Ch. 3	Night	Ch. 2
Ch. 4	The Week	Ch. 3
Ch. 5	The Months	Ch. 4
Ch. 6	The Concordance of the Months	Ch. 5
Ch. 7	The Years	Ch. 6
Ch. 8	The Seasons	Ch. 7.1–6
Ch. 9	[Z: Recapitulation of the above]	Ch. 7.7
Ch. 10	The Solstice and the Equinox	Ch. 8
Ch. 11	The World	Ch. 9
Ch. 12	The Five Circles of the World	Ch. 10
Ch. 13	The Parts of the World	Ch. 11
Ch. 14	Heaven and Its Name	Ch. 12
Ch. 15	The Planets of Heaven	Ch. 13
Ch. 16	The Heavenly Waters	Ch. 14
Ch. 17	The Course of the Stars	Ch. 22

Zofingen chapter	Title	*DNR* chapter
Ch. 18	The Position of the Seven Wandering Stars	Ch. 23
Ch. 19	The Light of the Stars	Ch. 24
Ch. 20	The Fall of the Stars	Ch. 25
Ch. 21	The Names of the Stars	Ch. 26.1–2
Ch. 22	[Zof.: Arcturus]	Ch. 26.3–4
Ch. 23	[Zof.: Boötes]	Ch. 26.5
Ch. 24	[Zof.: The Pleiades]	Ch. 26.6–7
Ch. 25	[Zof.: Orion]	Ch. 26.8–9
Ch. 26	[Zof.: Lucifer]	Ch. 26.10–11
Ch. 27	[Zof.: Vesper]	Ch. 26.12
Ch. 28	[Zof.: The Comet]	Ch. 26.13
Ch. 29	[Zof.: Sirius]	Ch. 26.14
Ch. 30	Whether the Stars have a Soul	Ch. 27
Ch. 31	Night	Ch. 28
Ch. 32	The Parts of the Earth	Ch. 48(47)
Ch. 33	Thunder	Ch. 29
Ch. 34	Lightning	Ch. 30
Ch. 35	The Rainbow	Ch. 31
Ch. 36	Clouds	Ch. 32
Ch. 37	Rains	Ch. 33
Ch. 38	Snow	Ch. 34
Ch. 39	Hail	Ch. 35
Ch. 40	The Nature of the Winds	Ch. 36
Ch. 41	The Names of the Winds	Ch. 37
Ch. 42	Signs of Storms	Ch. 38
Ch. 43	Pestilence	Ch. 39
Ch. 44	The Ocean's Tide	Ch. 40
Ch. 45	Why the Sea Does Not Grow in Size	Ch. 41
Ch. 46	Why the Sea has Bitter Waters	Ch. 42
Ch. 47	The River Nile	Ch. 43

Zofingen chapter	Title	*DNR* chapter
Ch. 48	The Position of the Earth	Ch. 45(44)
Ch. 49	Earthquake	Ch. 46(45)
Ch. 50	Mount Etna	Ch. 47(46)
Ch. 51	The Names of the Seas and the Rivers	Ch. 44(–)
Ch. 52	The Nature of the Sun	Ch. 15
Ch. 53	The Size of the Sun and the Moon	Ch. 16
Ch. 54	The Course of the Sun	Ch. 17
Ch. 55	The Light of the Moon	Ch. 18
Ch. 56	The Course of the Moon	Ch. 19
Ch. 57	The Eclipse of the Sun	Ch. 20
Ch. 58	The Eclipse of the Moon	Ch. 21
Ch. 59	[Zof.: Recapitulation of the above]	(–)

(b) The English type. The *Epistle* of King Sisebut, which accompanies many manuscripts of the short and medium recensions, disappears entirely from the long recension. However, its absence is noted in several manuscripts of the long recension that were prepared in England. We refer to them collectively as the 'English type'. The oldest of these, Exeter 3507 and London, Cotton Domitian I, date from around 970 and both may have been written in Canterbury. Wesley Stevens has described the English type as being 'a more elaborate version than has been found anywhere else'.[1] The 'elaboration', however, does not pertain to *DNR* itself, which exhibits the same 48-chapter text with its associated diagrams that is found elsewhere. Rather it has to do with two short, but interesting insertions, one before the first chapter and the other at the end of the last chapter.

Following the *capitula* and before chapter 1, the English type inserts: *Alii autem prologum cuius initium tu fortem loculentis uaga carmina gignis in hunc locum introducunt. Alii autem isidori esse respuunt sed gilde* [= Gildae? apparently a reference to Gildas]. 'Some introduce in this place a prologue the beginning of which is, "Perhaps in the woods you are leisurely composing idle songs". Others, however, declare that this prologue is not

---

1 Stevens, 'Sidereal Time', 136. See also, Stevens, 'Scientific Instruction', 99–100 and n. 57.

by Isidore, but by Gildas'. *Tu fortem loculentis uaga carmina gignis* is the first line of the *Epistle* of Sisebut, the verse letter that frequently follows the chapter, 'On Mount Etna', in the short and medium recensions, but which is never included in the long recension. Several inferences can be drawn from the insertion. In the scriptorium where the English type originated (in the tenth century? in southern England, perhaps at Canterbury?), Sisebut's *Epistle* was known to be associated with Isidore's *DNR*, though apparently it had been transferred from its normal place to the beginning of the text (there is, as noted above, one continental text of the short recension where such a transfer can be observed: Cologne 83(II)). On the other hand, the real name of the author of the *Epistle* had been lost, and the learned faculty ('Some ... Others' suggests a community of scholars) was apparently divided between those who believed the poem was by Isidore himself and others who attributed it to Gildas (and perhaps for that reason thought it should be excluded). Three centuries earlier, Aldhelm had a copy of *DNR* that included the *Epistle* but lacked Sisebut's name.[2] Evidently, there was at least one copy similar to Aldhelm's still in the library at Canterbury, which prompted this feature of the English type of the long recension.

At the other end of *DNR* the English type adds (with minor variations) to the final sentence of chapter 48: [... *estimauerunt*] *Cuius terre expositionem In medio ociano subiecta declarat formula Finiunt expositiones numero quadraginta NOVEM Explicit liber isidori psalensis episcopi de natura rerum.* 'The figure appended below makes clear the position of the land in the middle of the Ocean. The explanations end with a 49th chapter. This concludes the book of Bishop Isidore of Seville on the nature of things'. Below is placed a T-O map.

MSS: Exeter 3507, London Cotton Domitian I, Cotton Vitellius A XII, Oxford Auct. F.2.20, Auct. F.3.14, St John's 178.

In Exeter 3507, the T-O map is preceded by a brief passage from the anonymous *Historia Brittonum* (ascribed to Nennius), composed in 829 or 830: *Tres filii noae diuiserunt orbem terrarum in .iii. partes post diluuium. Sem in asiae. Cham in affrica. Iaphet in europa*,[3] 'the three sons of Noah divided the world into three parts after the Flood. Shem in Asia. Ham in Africa. Japheth in Europe'. The three sections of the map itself are inscribed with the numbers of the provinces of, respectively, Asia, Africa,

---

2 See above, p. 56.

3 Nennius, *British History and The Welsh Annals*, ed. and trans. John Morris (London: Rowman & Littlefield, 1980): *British History* (section 17), p. 63.

and Europe. The T-O map in the contemporary MS of the English type, Cotton Domitian I, is introduced by the words: *Trimoda sic mundus constat ratione diuisus*, 'it is well known that the world is divided in this threefold way', and the three sections are simply inscribed, ASSIA // EVROPA / AFFRICA.

BIBLIOGRAPHY

Primary Works

Aldhelm. *Opera*. Ed. Rudolf Ehwald. MGH: AA 15 (1919).

Ambrose. *Hexaemeron*. Ed. Karl Schenkl. CSEL 32.1. Vienna, 1897.

Anthologia Latina sive Poesis Latinae Supplementum (ed. Franciscus Buecheler and Alexander Riese) 1.2: *Reliquorum Librorum Carmina*. Ed. Alexander Riese. 2nd edn. Leipzig: Teubner, 1906.

Aratus. *Phaenomena*. Ed. and trans. Douglas Kidd. Cambridge: Cambridge University Press, 1997.

Aristotle. *Meteorologica*. Ed. H.D.P. Lee. Cambridge, MA: Harvard University Press/London: Heinemann, 1952.

Augustine. *Contra Faustum*. Ed. Joseph Zycha. CSEL 25.1. Vienna, 1891.

—— *De ciuitate Dei*. Ed. Bernhard Dombart and Alphonse Kalb. CCSL 47–48. Turnhout: Brepols, 1955.

—— *De Genesi ad litteram*. Ed. Joseph Zycha. CSEL 28.1. Vienna, 1894.

—— *De quantitate animae*. Ed. Wolfgang Hörmann. CSEL 89. Vienna: Hölder-Pichler-Tempsky, 1986.

—— *Enarrationes in Psalmos* (Pss. 1–50). Ed. Clemens Weidmann. CSEL 93/1A. Vienna: Verlag der Österreichischen Akademie der Wissenschaften, 2003 (Pss. 1–50).

—— *Enarrationes in Psalmos* (Pss. 134–40). Ed. Franco Gori. CSEL 95/4. Vienna: Verlag der Österreichischen Akademie der Wissenschaften, 2002.

—— *Enarrationes in Psalmos* (Pss. 141–50). Ed. Franco Gori. CSEL 95/5. Vienna: Verlag der Österreichischen Akademie der Wissenschaften, 2005.

—— *Quaestiones Evangeliorum*. PL 35.

Avienus. *Rufi Festi Avieni Aratea*. Ed. Alfred Breysig. Leipzig, 1882.

Baehrens, Emil, ed. *Poetae Latini minores*. vol. 5. Leipzig, 1883.

Bede. *Bede: Commentary on Revelation*. Trans. Faith Wallis. TTH 58. Liverpool: Liverpool University Press, 2013.

—— *Bede's Ecclesiastical History of the English People*. Ed. and trans. Bertram Colgrave and R.A.B. Mynors. Oxford: Clarendon Press, 1969.

—— *De natura rerum*. Ed. Charles W. Jones. CCSL 123A. Turnhout: Brepols, 1975. Trans. Calvin B. Kendall and Faith Wallis. *Bede: On the Nature of Things and On Times*. TTH 56. Liverpool: Liverpool University Press, 2010.

— *De temporibus liber.* Ed. Charles W. Jones. *Bedae Opera De Temporibus.* Cambridge, MA: The Mediaeval Academy of America, 1943. Ed. Jones (with the addition of chs. 17–22 = *Chronica minora*). CCSL 123C. Turnhout: Brepols, 1980. Trans. Calvin B. Kendall and Faith Wallis. *Bede: On the Nature of Things and On Times.* TTH 56. Liverpool: Liverpool University Press, 2010.

— *De temporum ratione.* Ed. Charles W. Jones. *Bedae Opera De Temporibus* (ed. Charles W. Jones) (with the addition of chs. 66–71 = *Chronica maiora*). CCSL 123B. Turnhout: Brepols, 1977. Trans. Faith Wallis. *Bede: The Reckoning of Time.* TTH 29. Liverpool: Liverpool University Press, 1999.

— *Lives of the Abbots of Wearmouth and Jarrow.* In Christopher Grocock and I.N. Wood, ed. and trans., *Abbots of Wearmouth and Jarrow.* Oxford Medieval Texts. Oxford: Clarendon Press, 2013.

— *On Ezra and Nehemiah.* Trans. Scott DeGregorio. TTH 47. Liverpool: Liverpool University Press, 2006.

— *On Genesis.* Trans. Calvin B. Kendall. TTH 48. Liverpool: Liverpool University Press, 2008.

Biblia Sacra Iuxta Vulgatam Versionem. Ed. Robert Weber. 4th edn, rev. Roger Gryson. Stuttgart: Deutsche Bibelgesellschaft, 1994.

Braulio of Zaragoza. *Renotatio librorum domini Isidori.* In José Carlos Martín, ed., *Scripta de vita Isidori Hispalensis episcopi.* CCSL 113B. Turnhout: Brepols, 2006.

Calcidius. *Timaeus a Calcidio translatus commentarioque instructus.* Ed. Jan Hendrik Waszink. 2nd edn. London: Warburg Institute, 1975. Trans. Béatrice Bakhouche. *Commentaire au Timée de Platon.* Paris: J. Vrin, 2011.

Cassiodorus. *Cassiodori Senatoris Institutiones.* Ed. R.A.B. Mynors. Oxford: Oxford University Press, 1937. Trans. James W. Halporn and Mark Vessey. *Cassiodorus: Institutions of Divine and Secular Learning and On the Soul.* TTH 42. Liverpool: Liverpool University Press, 2004.

Pseudo-Clement. *Recognitiones.* See Rufinus.

Corpus Christianorum Series latina. Turnhout: Brepols, 1953– .

Corpus Scriptorum Ecclesiasticorum Latinorum. Vienna, 1866– .

De ratione conputandi. Maura Walsh and Daíbhí Ó Cróinín, *Cummian's Letter De controversia Paschali and the De ratione conputandi.* Toronto: Pontifical Institute of Mediaeval Studies, 1988.

De Ventis. In *Anth. Lat.* 1.2:6–8 (no. 484).

Dracontius. *De laudibus dei.* In *Blossii Aeminii Dracontii Carmina.* Ed. Friedrich Vollmer. MGH: AA 14:23–113 (Berlin: Weidmann, 1905).

Einhard. *Vita Karoli Magni: The Life of Charlemagne.* Ed. Evelyn Scherabon Firchow and Edwin H. Zeydel. Dudweiler: AQ-Verlag, 1985.

Ennius. In *Remains of Old Latin in Three Volumes.* vol. 1. *Ennius and Caecilius*, ed. E.H. Warmington. Cambridge, MA: Harvard University Press/London: Heinemann, 1935.

Festus. *Sexti Pompei Festi de verborum significatu quae supersunt cum Pauli epitome.* Ed. Wallace M. Lindsay. Leipzig: Teubner, 1913.

Germanicus. *Germanici Caesaris Aratea cum scholiis.* Ed. Alfred Breysig. Berlin, 1867.

Gregory. *Homiliae in Hiezechihelem prophetam.* Ed. Marcus Adriaen. CCSL 142. Turnhout: Brepols, 1971.

—— *Moralia in Iob.* Ed. Marcus Adriaen. CCSL 143, 143A, 143B. Turnhout: Brepols, 1979–1985.

Hesiod. *Works and Days.* Trans. Richard Lattimore. Ann Arbor: University of Michigan Press, 1973.

Hilary of Poitiers. *Tractatus super psalmos.* Ed. Jean Doignon. CCSL 61B. Turnhout: Brepols, 2009.

Hippocrates. *Prognostic; Regimen in Acute Diseases; The Sacred Disease; The Art; Breaths; Law; Decorum; Physician (Ch. 1); Dentition.* Trans. and ed. W.H.S. Jones. Cambridge, MA: Harvard University Press/London: Heinemann, 1923.

Hyginus. *De astronomia.* Ed. Ghislaine Viré. Stuttgart: Teubner, 1992.

Isidore of Seville. *Allegoriae. PL* 83. A new edition is in preparation by Dominique Poirel.

—— *Chronica maiora.* Ed. Theodor Mommsen. MGH: AA 11.2 (Berlin, 1894). Ed. José Carlos Martín. CCSL 112. Turnhout: Brepols, 2003.

—— *Chronicorum epitome.* Ed. Theodor Mommsen. MGH: AA 11.2 (Berlin, 1894) = *Etym.* 5.38–39.

—— *De differentiis rerum.* Bk 1: *Diferencias Libro I*, ed. Carmen Codoñer. Paris: Belles Lettres, 1992. Bk 2: *Isidori Hispalensis episcopi Liber differentiarum (II)*, ed. Maria Adelaida Andrés Sanz, CCSL 111A. Turnhout: Brepols, 2006.

—— *De fide catholica contra Iudaeos. PL* 83.

—— *De haeresibus (S. Isidori Hispalensis Episcopi De Haeresibus liber).* Ed. Ángel Custodia Vega, in *Ciudad de Dios* 152 (1936), 5–32. Repr. PL Suppl. 4 (1940): 1815–20.

—— *De natura rerum (Traité de la nature).* Ed. with French translation by Jacques Fontaine. Paris: Institut d'études augustiniennes, 2002 (repr. of Bordeaux 1960 edn.). Trans. into Italian by Francesco Trisoglio. *La natura delle cose.* Rome: Città Nuova, 2001. For other editions, see Inventory of Manuscripts and Editions.

—— *De origine officiorum (De officiis).* Ed. Christopher Lawson. CCSL 113. Turnhout: Brepols, 1989.

—— *De ortu et obitu patrum, Vida y muerte de los santos.* Ed. with Spanish translation by César Chaparro Gómez. Paris: Belles Lettres, 1985.

—— *De uiris illustribus (El 'De uiris illustribus' de Isidoro de Sevilla).* Ed. Carmen Codoñer Merino. Salamanca: Consejo Superior de Investigaciones Científicas, 1964 (repr. 1997).

— *Etymologiae* (*Isidori Hispalensis Episcopi Etymologiarum sive Originum libri XX*). Ed. W.M. Lindsay. Oxford: Clarendon Press, 1911. 2 vols. (unpaginated); *Etym.* 3. Ed. Giovanni Gasparotto and Jean-Yves Guillaumin. Paris: Belles Lettres, 2009; *Etym.* 5. Ed. Valeriano Yarza Urquiola and Francisco Xavier Andrés Santos. Paris: Belles Lettres, 1981; *Etym.* 12. Ed. J. André. Paris: Belles Lettres, 1986. Trans. Stephen A. Barney, W. J. Lewis, J. A. Beach, and Oliver Berghof. *The Etymologies of Isidore of Seville*. Cambridge: Cambridge University Press, 2006.

— *Historia Gothorum Wandalorum Sueborum*. Ed. Theodor Mommsen. MGH: AA 11.2 (Berlin, 1894). Trans. Guido Donini and Gordon B. Ford. *History of the Kings of the Goths, Vandals, and Suevi*. Leiden: E.J. Brill, 1966. Trans. Kenneth Baxter Wolf. *Conquerors and Chroniclers of Early Medieval Spain*. 2nd edn. TTH 9. Liverpool: Liverpool University Press, 1999.

— *Liber numerorum / Le Livre des nombres*. Ed. with French translation by Jean-Yves Guillaumin. Paris: Belles Lettres, 2005.

— *Mysticorum expositiones sacramentorum* (*Quaestiones in Vetus Testamentum*). PL 83; Genesis section only: *Expositio in Vetus Testamentum Genesis*. Ed. Michael M. Gorman and Martine Dulaey. Aus der Geschichte der lateinischen Bibel 38. Freiburg im Breisgau: Herder, 2009.

— *Prooemia*. PL 83. A new edition is in preparation for CCSL by M.A. Andrés Sanz.

— *Sententiae*. Ed. Pierre Cazier. CCSL 111. Turnhout: Brepols, 1998.

— *Synonyma* (*Isidori Hispalensis episcopi Synonyma*). Ed. Jacques Elfassi. CCSL 111B. Turnhout: Brepols, 2009.

— *Versus* (*Isidori Hispalensis versus*). Ed. José María Sánchez Martín. CCSL 113A. Turnhout: Brepols, 2000.

Jerome. *Commentaria in Amos*. Ed. Marcus Adriaen. CCSL 76. Turnhout: Brepols, 1969.

— *Commentaria in Danielem*. Ed. F. Glorie. CCSL 75A. Turnhout: Brepols, 1964.

— *Commentaria in Isaiam*. Ed. Marcus Adriaen. CCSL 73, 73A. Turnhout: Brepols, 1963.

— *Commentaria in Osee*. Ed. Marcus Adriaen. CCSL 76. Turnhout: Brepols, 1969.

— *Commentaria in Zachariam*. Ed. Marcus Adriaen. CCSL 76A. Turnhout: Brepols, 1970.

— *Commentarius in Ecclesiasten*. Ed. Marcus Adriaen. CCSL 72. Turnhout: Brepols, 1959.

— *Homiliae XIV in Jeremiam*. PL 25.

Justinus, *Epitome. Iustini ex Trogi Pompeii historiis ex Ternis libri XLIII*. London, 1593; repr. 1984.

A Late Eighth-Century Latin–Anglo-Saxon Glossary Preserved in the Library of Leiden University (MS Voss. Q° Lat. N°. 69). Ed. John Henry Hessels. Cambridge: Cambridge University Press, 1906.

Macrobius. *Commentarii in Somnium Scipionis.* Ed. James Willis. Leipzig: Teubner, 1963. Trans. William Harris Stahl. *Commentary on the Dream of Scipio.* New York: Columbia University Press, 1952.

Martianus Capella. *De nuptiis Philologiae et Mercurii.* Ed. James Willis. Leipzig: Teubner, 1983.

Monumenta Germaniae Historica. Leipzig, Hanover, Berlin, 1826– .

Nennius. *British History and The Welsh Annals.* Ed. and trans. John Morris. London: Rowman & Littlefield, 1980.

Origen. *De principiis.* See Rufinus.

Patrologiae cursus completus. Series graeca. Ed. J.-P. Migne. 167 vols. Paris, 1857–1864.

Patrologiae cursus completus. Series latina. Ed. J.-P. Migne. 221 vols. Paris, 1841–1880, with vols. reissued by Garnier to 1905.

Plato. *Plato's Cosmology: The Timaeus of Plato translated with a running commentary.* Trans. Francis M. Cornford. London: K. Paul, Trench, Trubner & Co./New York: Harcourt, Brace, 1937.

Pliny. *Natural History* [*Naturalis historia*]. Ed. H. Rackham, W.E.S. Jones, and D.E. Eichholz. 10 vols. Cambridge, MA: Harvard University Press/London: Heinemann, 1938 (repr. 1961).

Plutarch. *The Complete Works of Plutarch: Essays and Miscellanies.* Trans. John Dowel. vol. 3. Boston: Little, Brown and Co., 1911.

Proclus. *Commentaire sur le Timée.* Ed. A.J. Festugière. 5 vols. Paris: J. Vrin, 1966–1968.

Rufinus. [Ps.-Clementis] *Origenis Peri archon libri quatuor interprete Rufino Aquileiensi Presbytero* (*Origen, De principiis*). PG 11.

— *Recognitiones Rufino Interprete.* Ed. Bernhard Rehm, *Die Pseudoklementinen. II. Rekognitionen in Rufins Übersetzung.* 2nd edn, ed. Georg Strecker. Berlin: Akademie Verlag, 1994.

Seneca. *Natural Questions* [*Naturales quaestiones*]. Ed. T.H. Corcoran. Cambridge, MA: Harvard University Press/London: Heinemann, 1971.

Sisebut. *Epistula Sisebuti.* Ed. Jacques Fontaine, *Traité,* 328–35.

Solinus. *Collectanea rerum memorabilium.* Ed. Theodor Mommsen. Berlin, 1895.

Suetonius. *C. Suetoni Tranquilli praeter Caesarum libros reliquiae.* Ed. August Reifferscheid. Leipzig, 1860.

Tyconius. *Liber Regularum.* Ed. F. Crawford Burkitt, *The Book of Rules of Tyconius.* Cambridge: Cambridge University Press, 1894.

Vegetius. *Epitoma Rei Militaris.* Ed. M.D. Reeve. Oxford: Clarendon Press, 2004. Trans. Nicholas P. Milner. *Epitome of Military Science.* 2nd edn. TTH 16. Liverpool: Liverpool University Press, 1996.

Secondary Works

Alberto, Paulo Farmhouse. 'King Sisebut's *Carmen de luna* in the Carolingian School'. In Paulo Farmhouse Alberto and David Paniagua, eds., *Ways of Approaching Knowledge in Late Antiquity and the Early Middle Ages: Schools and Scholarship*. Nordhausen: Traugott Bautz, 2012: 177–205.

Anderson, Jeffrey C., ed. *The Christian Topography of Kosmas Indikopleustes (Firenze, Biblioteca Medicea Laurenziana, plut. 9.28): The Map of the Universe Redrawn in the Sixth Century, with a contribution on the Slavic Recensions*. Rome: Edizioni di Storia e Letteratura, 2013.

—— 'Description of the Miniatures and Commentary'. In Jeffrey C. Anderson, ed., *The Christian Topography of Kosmas Indikopleustes*, 33–63.

Anglo-Saxon Manuscripts in Microfiche Facsimile. vol. 5. *Manuscripts with Anglo-Saxon Glosses*, with descriptions by Peter J. Lucas, A.N. Doane, and I.C. Cunningham. Tempe, AZ: Arizona Center for Medieval and Renaissance Studies, 1997; vol. 14. *Manuscripts of Durham, Ripon, and York*, with descriptions by Sarah Larratt Keefer, David Rollason, and A.N. Doane. Tempe, AZ: Arizona Center for Medieval and Renaissance Studies, 2007; vol. 22. *Exeter Manuscripts*, with descriptions by Matthew T. Hussey. Tempe, AZ: Arizona Center for Medieval and Renaissance Studies, 2014.

Antolín, P. Guillermo. *Catálogo de los Códices Latinos de la Real Biblioteca del Escorial*. 5 vols. Madrid: Imprenta Helénica, 1910–1923.

Arévalo, Faustino. *S. Isidori Hispalensis Episcopi ... Opera Omnia*. vols. 1–2. *Isidoriana*. Rome, 1797.

Auerbach, Erich. *Scenes from the Drama of European Literature*. Trans. Ralph Manheim. Minneapolis: University of Minnesota Press, 1984.

Baker, Peter S., and Michael Lapidge, eds. *Byrhtferth's Enchiridion*. Early English Text Society SS 15. Oxford: Oxford University Press, 1995.

Beeson, Charles Henry. *Isidor-Studien*. Quellen und Untersuchungen zur lateinischen Philologie des Mittelalters 4.2. Munich: C.H. Beck (O. Beck), 1913.

Bischoff, Bernhard. 'Die europäische Verbreitung der Werke Isidors von Sevilla'. In Bernhard Bischoff, *Mittelalterliche Studien: Ausgewählte Aufsätze zur Schriftkunde und Literaturgeschichte*. vol. 1. Stuttgart: Hiersemann, 1966: 171–94.

—— *Katalog der festländischen Handschriften des neunten Jahrhunderts (mit Ausnahme der wisigotischen)*. Ed. Birgit Ebersperger. part I. Aachen–Lambach. Wiesbaden: Harrassowitz, 1998; part II. Laon–Paderborn. Wiesbaden: Harrassowitz, 2004; part III. Padua–Zwickau. Wiesbaden: Harrassowitz, 2014.

Bischoff, Bernhard, and Michael Lapidge. *Biblical Commentaries from the Canterbury School of Theodore and Hadrian*. Cambridge Studies in Anglo-Saxon England 10. Cambridge: Cambridge University Press, 1994.

Black, Jonathan, and Thomas L. Amos. *The Fundo Alcobaça of the Biblioteca*

Nacional, Lisbon. vol. 3. *Manuscripts 302–456.* Collegeville, MN: Hill Monastic Manuscript Library, 1990.

Black, William Henry. *A Descriptive, Analytical, and Critical Catalogue of the Manuscripts Bequeathed unto the University of Oxford by Elias Ashmole.* Oxford: Oxford University Press, 1845.

Blackburn, Bonnie J., and Leofranc Holford-Strevens, eds. *The Oxford Companion to the Year.* Oxford: Oxford University Press, 1999.

Bober, Harry. 'An Illustrated Medieval School-Book of Bede's *De natura rerum*'. *Journal of the Walters Art Gallery* 19/20 (1956–1957): 65–97.

Boese, Helmut. *Die lateinischen Handschriften der Sammlung Hamilton zu Berlin.* Wiesbaden: Otto Harrassowitz, 1966.

Borland, Catherine R. *A Descriptive Catalogue of Western Mediaeval Manuscripts in Edinburgh University Library.* Edinburgh: T. & A. Constable for the University of Edinburgh, 1916.

Borst, Arno. *Die karolingishe Reichskalender und seine Überlieferung bis ins 12. Jahrhundert.* MGH: Libri memoriales 2. 3 vols. Hanover: Hahnsche Buchhandlung, 2001.

— ed. *Schriften zur Komputistik im Frankenreich von 721 bis 818.* MGH: Quellen zur Geistesgeschichte des Mittelalters 21. 3 vols. Hanover: Hahnsche Buchhandlung, 2006.

Callataÿ, Godefroid de. *Annus Platonicus: A Study of World Cycles in Greek, Latin and Arabic Sources.* Louvain-la-Neuve: Université catholique de Louvain, Institut orientaliste, 1996.

The Cambridge History of Science. vol. 2. *Medieval Science.* Ed. David C. Lindberg and Michael H. Shank. Cambridge: Cambridge University Press, 2013.

Catalogue général des manuscrits des bibliothèques publiques de France. Départements. Paris, 1885– .

Catalogue général des manuscrits des bibliothèques publiques des départements. 7 vols. Paris, 1846–1885.

Catalogue général des manuscrits latins. Ed. Philippe Lauer. 7 vols. Paris: Bibliothèque nationale, 1939–1988.

Catalogus Codicum Latinorum Bibliothecae Regiae Monacensis. vol. 1.1. Codices num. 1–2329. 2nd edn. Munich, 1892.

Chadwick, Henry. *Priscillian of Avila: The Occult and the Charismatic in the Early Church.* Oxford: Clarendon Press, 1976.

Chartrand, Mark R. *National Audubon Society Field Guide to the Night Sky.* New York: Alfred A. Knopf, 1991.

Chernow, Ron. *Washington: A Life.* New York: Penguin, 2010.

Chiesa, Paolo, and Lucia Castaldi, eds. *La trasmissione dei testi latini del Medioevo. Mediaeval Latin Texts and their Transmission (Te.Tra.)* 1. Florence: SISMEL/ Edizioni Galluzzo, 2004.

— eds. *La trasmissione dei testi latini del medioevo. Mediaeval Latin Texts and their Transmission (Te.Tra.)* 2. Florence: SISMEL/Edizioni Galluzzo, 2005.

Codoñer, Carmen, José Carlos Martín, and M. Adelaida Andrés Sanz. 'Isidorus Hispalensis Ep.'. In Paolo Chiesa and Lucia Castaldi, eds., *La trasmissione* (*Te. Tra.*) 2:274–417 (the single author of each of the 14 numbered sections of this article is identified in a note at the foot of p. 417).

Collins, Roger. 'Literacy in Early Medieval Spain'. In Rosamond McKitterick, ed., *The Uses of Literacy in Early Medieval Europe*. Cambridge: Cambridge University Press, 1990: 109–33.

— *Visigothic Spain 409–711*. Oxford: Blackwell, 2004.

Contreni, John J. *The Cathedral School of Laon from 850 to 930: Its Manuscripts and Masters*. Münchener Beiträge zur Mediävistik und Renaissance-Forschung 29. Munich: Arbeo-Gesellschaft, 1978.

Davidson, Norman. *Astronomy and the Imagination: A New Approach to Man's Experience of the Stars*. London: Routledge & Kegan Paul, 1985.

Decker, Elly. *Illustrating the Phaenomena: Celestial Cartography in Antiquity and the Middle Ages*. Oxford: Oxford University Press, 2013.

Delpit, Jules. *Bibliothèque municipale de Bordeaux. Catalogue des manuscrits*. vol. 1. Bordeaux, 1880.

Di Sciacca, Claudia. *Finding the Right Words: Isidore's Synonyma in Anglo-Saxon England*. Toronto: University of Toronto Press, 2008.

Díaz y Díaz, Manuel C. *Index Scriptorum Latinorum Medii Aevi Hispanorum*. Acta Salmanticensia, Filosofía y Letras 13.1–2 (continuously paginated). Salamanca: University of Salamanca, 1958–1959.

Ducos, Joëlle. 'Eau douce et eau salée'. In Danièle James-Raoul and Claude Thomasset, eds., *Dans l'eau, sous l'eau: Le monde aquatique au moyen âge*. Paris: Presses de l'Université de Paris-Sorbonne, 2002: 121–38.

Dúran, Angel Benito. 'Valor catequético de la obra "De natura rerum", de San Isidoro de Sevilla'. *Atenas* 9 (1938): 41–51.

Eastwood, Bruce S. 'Celestial Reason: The Development of Latin Planetary Astronomy to the Twelfth Century'. In Susan J. Ridyard and Robert G. Benson, eds., *Man and Nature in the Middle Ages*. Sewanee, TN: University of the South Press, 1995: 157–72.

— 'The Diagram of the Four Elements in the Oldest Manuscripts of Isidore's "De natura rerum"'. *Studi Medievali* 42 (2001): 547–64 (with 6 plates).

— 'Early Medieval Cosmology, Astronomy and Mathematics'. In David C. Lindberg and Michael H. Shank (eds.), *The Cambridge History of Science*, 2:302–22.

— *Ordering the Heavens: Roman Astronomy and Cosmology in the Carolingian Renaissance*. Leiden and Boston: Brill, 2007.

Elfassi, Jacques, D. Poiret, Carmen Codoñer, José Carlos Martín, and M. Adelaida Andrés Sanz. 'Isidorus Hispalensis Ep.'. In Paolo Chiesa and Lucia Castaldi, eds., *La trasmissione* (*Te.Tra.*) 1:196–226.

Fischer, Hans. *Katalog der Handschriften der Universitätsbibliothek*

Erlangen. vol. 1. *Die lateinischen Pergamenthandschriften.* Erlangen: Universitätsbibliothek, 1928.

Fontaine, Jacques. 'La Diffusion carolingienne du *De natura rerum* d'Isidore de Séville d'après les manuscrits conservés en Italie'. *Studi medievali* 7 (1966): 108–27.

—— 'La Diffusion de l'œuvre d'Isidore de Séville dans les scriptoria helvétiques du haut moyen âge'. *Revue suisse d'histoire* 12 (1962): 305–27.

—— 'Fins et moyens de l'enseignement ecclésiastique dans l'Espagne wisigothique'. In *La scuola nell'occidente latino dell'alto medioevo*, Settimane di studi del Centro italiano di studi sull'alto medioevo 19. Spoleto: Presso la sede del Centro, 1972. 1:145–202.

—— 'Isidore de Séville et l'astrologie'. *Revue des études latines* 31 (1953): 271–300.

—— *Isidore de Séville et la culture classique dans l'Espagne wisigothique.* 3 vols. (continuously paginated). 2nd edn. Paris: Études augustiniennes, 1983.

—— *Isidore de Séville: Genèse et originalité de la culture hispanique au temps des Wisigoths.* Turnhout: Brepols, 2000.

—— 'King Sisebut's *Vita Desiderii* and the Political Function of Visigothic Hagiography'. In Edward James, ed., *Visigothic Spain: New Approaches*, 93–129.

—— 'Qui a chassé le carthaginois Sévérianus et les siens? Observations sur l'histoire familiale d'Isidore de Séville'. In *Estudios en Homenaje a Don Claudio Sánchez Albornoz en sus 90 años, I.* Buenos Aires: Instituto de Historia de España, 1983. 349–400. Repr. in Jacques Fontaine, *Tradition et actualité chez Isidore de Séville.* London: Variorum Reprints, 1988.

Franz, Gunther, and Ulrich Lehnart, eds. *Karolingische Beda-Handschrift aus St. Maximin.* Trier: Die Stadtbibliothek, 1990.

Gameson, Richard. *The Manuscripts of Early Norman England (c.1066–1130).* Oxford: Published for the British Academy by Oxford University Press, 1999.

García Moreno, Luis A. *Historia de España Visigoda.* 3rd edn. Madrid: Cátedra, 2008.

Gasparotto, Giovanni. *Isidoro e Lucrezio. Le fonti della meteorologia isidoriana.* Verona: Libreria, 1983.

Gautier Dalché, Patrick. 'De la glose à la contemplation. Place et fonction de la carte dans les manuscrits du haut moyen âge', *Testo e immagine nell'alto medioevo*, Settimane di studi del Centro italiano di studi sull'alto medioevo 41. Spoleto: Presso la sede del Centro, 1994: 693–764.

Gil, Juan. *Miscellanea Wisigothica.* Seville: Publicaciones de la Universidad de Sevilla, 1972.

Gneuss, Helmut, and Michael Lapidge. *Anglo-Saxon Manuscripts: A Bibliographical Handlist of Manuscripts and Manuscript Fragments Written or Owned in England up to 1100.* Toronto: University of Toronto Press, 2014.

Gorman, Michael. 'The Diagrams in the Oldest Manuscripts of Isidore's 'De natura rerum' with a Note on the Manuscript Traditions of Isidore's Works'. *Studi Medievali* 42 (2001): 529–45 (with 8 plates).

Gryson, Roger, Bonifatius Fischer, Hermann Josef Frede, and Pierre Sabatier. *Répertoire général des auteurs ecclésiastiques latins de l'Antiquité et du haut moyen âge.* vol. 2. *Répertoire des auteurs I–Z.* Freiburg: Herder, 2007.

Gudiol i Cunill, Josep, and Eduard Junyent. *Catàleg dels Llibres manuscrits anteriors al segle XVIII del Museu Episcopal de Vich.* Barcelona: Imprenta de la Casa de Caritat, 1934.

Hagen, Hermann. *Catalogus Codicum Bernensium.* Bern, 1875. Repr. Hildesheim: Olms, 1974.

Hegedus, Tim. *Early Christianity and Ancient Astrology.* New York: Peter Lang, 2007.

Hen, Yitzak. *Roman Barbarians: The Royal Court and Culture in the Early Medieval West.* Basingstoke and New York: Palgrave Macmillan, 2007.

— 'A Visigothic King in Search of an Identity: "Sisebutus Gothorum gloriosissimus princeps"'. In Richard Corradini, Matthew Gillis, Rosamond McKitterick, and Irene van Renswoude, eds., *Ego Trouble: Authors and their Identities in the Early Middle Ages.* Vienna: Verlag der Österreichischen Akademie der Wissenschaften, 2010: 89–99.

Henderson, John. 'The Creation of Isidore's *Etymologies or Origins*'. In Jason König and Tim Whitmarsh, eds., *Ordering Knowledge in the Roman Empire.* Cambridge: Cambridge University Press, 2007.

— *The Medieval World of Isidore of Seville: Truth from Words.* Cambridge: Cambridge University Press, 2007.

Hermann, Marek. 'Die astronomischen Metaphern in Isidors von Sevilla *Origines* und *De natura rerum*'. *Analecta Cracoviensia* 38–39 (2006–2007): 442–53.

— 'Zwischen heidnischer und christlicher Kosmologie. Isidor von Sevilla und seine Weltanschauung'. *Analecta Cracoviensia* 34 (2002): 311–28.

Herren, Michael W. 'Classical and Secular Learning among the Irish before the Carolingian Renaissance'. *Florilegium* 3 (1981): 118–57.

— 'On the Earliest Irish Acquaintance with Isidore of Seville'. In Edward James, ed., *Visigothic Spain: New Approaches*: 243–50.

— ed., *The Hisperica Famina: I. The A-Text. A New Critical Edition with English Translation and Philological Commentary.* Toronto: Pontifical Institute of Mediaeval Studies, 1974.

— 'Scholarly Contacts Between the Irish and the Southern English in the Seventh Century.' *Peritia* 12 (1998): 24–53.

Hillgarth, J. N. 'Ireland and Spain in the Seventh Century'. *Peritia* 3 (1984): 1–16.

— 'Visigothic Spain and Early Christian Ireland'. *Proceedings of the Royal Irish Academy. Section C: Archaeology, Celtic Studies, History, Linguistics, Literature* 62 (1962): 167–94.

Homburger, Otto. *Die Illustrierten Handschriften der Burgerbibliothek Bern: Die Vorkarolingischen und Karolingischen Handschriften.* Bern: Burgerbibliothek Bern, 1962.

Horden, Peregrine, and Nicolas Purcell. *The Corrupting Sea: A Study of Mediterranean History.* Oxford: Blackwell, 2000.

Inglebert, Hervé. *Interpretatio Christiana: Les mutations des savoirs (cosmographie, géographie, ethnographie, histoire) dans l'Antiquité chrétienne, 30–630 après J.-C.* Paris: Institut d'études augustinienne, 2001.

James, Edward, ed. *Visigothic Spain: New Approaches.* Oxford: Clarendon Press, 1980.

James, Montague Rhodes. *A Descriptive Catalogue of the Manuscripts in the College Library of Magdalene College Cambridge.* Cambridge: Cambridge University Press, 1909.

Jones, A.H.M. *The Later Roman Empire, 284–602: A Social, Economic and Administrative Survey.* Oxford: Blackwell, 1964.

Jones, Charles W. *Bedae Pseudepigrapha: Scientific Writings Falsely Attributed to Bede.* Ithaca, NY and London: Cornell University Press, 1939.

Kendall, Calvin B. *The Allegory of the Church: Romanesque Portals and their Verse Inscriptions.* Toronto: University of Toronto Press, 1998.

Ker, N.R. *Medieval Manuscripts in British Libraries,* 5 vols. Oxford: Clarendon Press, 1969–2002.

King, Margot H., and Wesley M. Stevens, eds. *Saints, Scholars and Heroes: Studies in Medieval Culture in Honour of Charles W. Jones.* 2 vols. Collegeville, MN: Saint John's University, 1979.

Kline, Naomi Reed. *Maps of Medieval Thought: The Hereford Paradigm.* Woodbridge: Boydell Press, 2001.

Köhler, Franz. *Die Gudischen Handschriften.* Frankfurt am Main: Klostermann, 1966.

Kominko, Maja. 'The Science of the Flat Earth'. In Jeffrey C. Anderson, ed., *The Christian Topography of Kosmas Indikopleustes,* 67–82.

— *The World of Kosmas: The Byzantine Illustrated Codices of the Christian Topography.* Cambridge: Cambridge University Press, 2013.

Kühnel, Bianca. 'Carolingian Diagrams, Images of the Invisible'. In Giselle de Nie, Karl F. Morrison, and Marco Mostert, eds., *Seeing the Invisible in Late Antiquity and the Early Middle Ages.* Turnhout: Brepols, 2005: 359–89.

Lais, P.G. 'Monumento greco-latino di una rosa classica dodecimale in Vaticano'. *Pubblicazioni della Specola Vaticana* 4 (1894): xi–xv.

Laistner, M.L.W. 'A Fragment from an Insular Manuscript of Isidore'. *Mediaevalia et Humanistica* 2 (1944): 28–31.

— *A Hand-List of Bede Manuscripts,* with the collaboration of H.H. King. Ithaca, NY: Cornell University Press, 1943.

— *Thought and Letters in Western Europe AD 500 to 900.* rev. edn. Ithaca, NY: Cornell University Press, 1957.

—— 'The Western Church and Astrology during the Early Middle Ages'. *Harvard Theological Review* 34 (1941): 251–75.

Lapidge, Michael. *The Anglo-Saxon Library*. Oxford: Oxford University Press, 2006.

—— 'The School of Theodore and Hadrian'. *ASE* 15 (1986): 45–72.

Law, Vivien. *Wisdom, Authority and Grammar in the Seventh Century: Decoding Virgilius Maro Grammaticus*. Cambridge: Cambridge University Press, 1995.

Le Boeuffle, André. *Astronomie, astrologie: lexique Latin*. Paris: Picard, 1987.

Le Bourdellès, H. *L'Aratus Latinus. Étude sur la culture et la langue dans le nord de la France au VIIIᵉ siècle*. Lille: Université de Lille, 1985.

Lehoux, Daryn. *Astronomy, Weather, and Calendars in the Ancient World*. Cambridge: Cambridge University Press, 2007.

—— *What Did the Romans Know? An Inquiry into Science and Worldmaking*. Chicago: University of Chicago Press, 2012.

Leitschuh, Friedrich, and Hans Fischer. *Katalog der Handschriften der Königlichen Bibliothek zu Bamberg*, 3 vols. Bamberg: C.C. Buchner, 1887–1912.

Lejbowicz, Max. 'Les Antécédents de las distinction isidorienne *astrologia/ astronomia*'. In Bernard Ribémont, ed., *Observer, lire, écrire le ciel au moyen âge: actes du colloque d'Orléans 22–23 avril 1989*. Paris: Klincksieck, 1990: 175–212.

Lexikon des Mittelalters, ed. Robert Auty. 10 vols. Munich and Zurich: Artemis-Verlag, 1977–1999.

Lindsay, W.M. 'The Affatim Glossary and Others'. *Classical Quarterly* 11 (1917): 185–200.

Lof, Laurens Johan van der. 'Der Mäzen König Sisebutus und sein "De eclipsi lunae"'. *Revue des études augustiniennes* 18 (1972): 145–51.

Long, Kim. *The Moon Book: Fascinating Facts about the Magnificent, Mysterious Moon*. Boulder, CO: Johnson Books, 1988.

Louth, Andrew. 'The Byzantine Empire in the Seventh Century'. In Paul Fouracre, ed., *The New Cambridge Medieval History*. vol. 1. c.*500*–c.*700*. Cambridge: Cambridge University Press, 2005: 291–316.

Lowe, E. A. *Codices Latini Antiquiores: A Paleographical Guide to Latin Manuscripts Prior to the Ninth Century*. 11 vols. plus Supplement. Oxford: Clarendon Press, 1934–1971.

Maas, Michael. *John Lydus and the Roman Past*. London: Routledge, 1992.

McCluskey, Stephen C. *Astronomies and Cultures in Early Medieval Europe*. Cambridge: Cambridge University Press, 1998.

—— 'Natural Knowledge in the Early Middle Ages'. In David C. Lindberg and Michael H. Shank (eds.), *The Cambridge History of Science*, 2:286–301.

McCready, William D. 'Bede, Isidore, and the Epistola Cuthberti'. *Traditio* 50 (1995): 75–94.

McCulloh, John. 'Martyrologium excarpsatum: A New Text from the Early Middle

Ages'. In Margot H. King and Wesley M. Stevens, eds., *Saints, Scholars and Heroes*, 2:179–237.

McVaugh, Michael. 'The *humidum radicale* in Thirteenth-Century Medicine'. *Traditio* 30 (1974): 259–83.

Mangeart, J. *Catalogue descriptif et raisonné des manuscrits de la bibliothèque de Valenciennes*. Paris, 1860.

Manitius, Max. *Geschichte der lateinischen Literatur des Mittelalters*. vol. 1. *Von Justinian bis zur Mitte des 10. Jahrhunderts*. Munich: C.H. Beck, 1911.

Markus, R.A. *Gregory the Great and his World*. Cambridge: Cambridge University Press, 1997.

Martin, Dale B. *Inventing Superstition: From the Hippocratics to the Christians*. Cambridge MA: Harvard University Press, 2004.

Martín, José Carlos. 'Isidorus Hispalensis Ep.: 8. De rerum natura'. In Paolo Chiesa and Lucia Castaldi, eds., *La trasmissione* (*Te.Tra.*) 2:353–62. See Carmen Codoñer, José Carlos Martín, and M. Adelaida Andrés Sanz, 'Isidorus Hispalensis Ep'.

— 'Réflexions sur la tradition manuscrite de trois œuvres d'Isidore de Séville'. *Filologia mediolatina* 11 (2004): 205–63.

— 'Sisebutus Visigothorum Rex'. In Paolo Chiesa and Lucia Castaldi, eds., *La trasmissione* (*Te.Tra.*) 1:402–10.

Martyn, John R.C. *King Sisebut and the Culture of Visigothic Spain, with Translations of the Lives of Saint Desiderius of Vienne and Saint Masona of Mérida*. Lewiston, NY: Edwin Mellen Press, 2008.

Mateu Ibars, Josefina, and María Dolores Mateu Ibars. *Colectánea paleográfica de la Corona de Aragón: siglos IX–XVIII*. vol. 1. Barcelona: Universidad de Barcelona, 1991.

Menéndez-Pidal, G. 'Mozárabes y asturianos en la cultura de la Alta Edad Media en relación especial con la historia de los conocimientos geográficos'. *Boletin de la Real Academia de la Historia* 134 (1952): 137–291 (and plates).

Merrills, Andy. 'Isidore's *Etymologies*: On Words and Things'. In Jason König and Greg Woolf, eds, *Encyclopaedism from Antiquity to the Renaissance*. Cambridge: Cambridge University Press, 2013: 301–24.

Meyer, Heinz. *Lexikon der mittelalterlichen Zahlenbedeutung*. Munich: Fink, 1987.

— *Die Zahlenallegorese im Mittelalter. Methode und Gebrauch*. Munich: Fink, 1975.

Montagu, Ashley, and John C. Lilly. *The Dolphin in History. Papers Presented ... at a Symposium at the Clark Library, 13 October 1962*. Los Angeles: William Andrews Clark Memorial Library, 1983.

Mostert, Marco. *The Library of Fleury: A Provisional List of Manuscripts*. Hilversum: Verloren, 1989.

Murdoch, John E. *Album of Science: Antiquity and the Middle Ages*. New York: Scribner, 1984.

Nutton, Vivian. 'The Seeds of Disease: An Explanation of Contagion and Infection from the Greeks to the Renaissance'. *Medical History* 27 (1983): 1–34.

Ó Cróinín, Dáibhí. 'A Seventh-Century Irish Computus from the Circle of Cummian'. In Dáibhí Ó Cróinín, *Early Irish History and Chronology.* Dublin: Four Courts Press, 2003: 99–132.

O'Loughlin, Thomas. '*Aquae super caelos* (Gen. 1:6–7): The First Faith–Science Debate?'. *Milltown Studies* 29 (1992): 92–114.

— 'The Waters above the Heavens: Isidore and the Latin Tradition'. *Milltown Studies* 36 (1995): 104–17.

Obrist, Barbara. 'The Astronomical Sundial in Saint Willibrord's Calendar and its Early Medieval Context'. *Archives d'histoire doctrinale et littéraire du moyen âge* 67 (2000): 71–118.

— *La Cosmologie médiévale: textes and images.* vol. 1. *Les fondements antiques,* Micrologus' Library 11. Florence: SISMEL/Edizioni del Galluzzo, 2004.

— 'Le Diagramme isidorien de l'année et des saisons: Son contenu physique et les représentations figuratives'. *Mélanges de l'École française de Rome: Moyen âge* 108/1 (1996): 95–164.

— 'Wind Diagrams and Medieval Cosmology'. *Speculum* 72 (1997): 33–84.

Palmer, James T. 'The Ends and Futures of Bede's *De temporum ratione*'. In Peter Darby and Faith Wallis, eds., *Bede and the Future.* Farnham: Ashgate, 2014: 139–60.

Porter, David W. 'Isidore's *Etymologiae* at the School of Canterbury'. *ASE* 43 (2014): 7–44.

Reallexikon für Antike und Christentum. Ed. Theodor Klauser, Georg Schöllgen, and Franz Joseph Dölger. Multiple vols. Stuttgart: A. Hiersemann, 1950– .

Recchia, Vincenzo. *Sisebut di Toledo, il 'Carmen de luna',* Quaderni di 'Vetera christianorum' 3 (Bari, 1971).

— 'Sul Carmen de luna di Sisebuto di Toledo'. *Invigilata Lucernis* 20 (1998): 201–19. Reprinted in his *Lettera e profezia nell'esegesi di Gregorio Magno,* Quaderni di Invigilata Lucernis 20 (Bari, 2003): 137–55.

Ribémont, Bernard. *Les Origins des encyclopédies médiévales d'Isidore de Séville aux Carolingiens.* Paris: Champion, 2002.

Riché, Pierre. *Education and Culture in the Barbarian West from the Sixth through the Eighth Century.* Trans. John J. Contreni. Columbia: University of South Carolina Press, 1976.

Rotili, Mario. *La miniatura nella Badia di Cava.* 2 vols. Naples: Di Mauro, 1976–1978.

Rouse, Richard H., and Mary A. Rouse. 'Donatist Aids to Bible Study: North African Literary Production in the Fifth Century'. In Richard H. Rouse and Mary A. Rouse, eds., *Bound Fast with Letters: Medieval Writers, Readers, and Texts.* Notre Dame, IN: University of Notre Dame Press, 2013: 24–49.

Schove, D.J., with Alan Fletcher. *The Chronology of Eclipses and Comets, AD 1–1000.* Woodbridge: Boydell, 1984.

Scott, Alan. *Origen and the Life of the Stars: A History of an Idea.* Oxford: Clarendon Press, 1991.

Simek, Rudolf. 'The Shape of the Earth in the Middle Ages and Medieval Mappaemundi'. In P.D.A. Harvey, ed., *The Hereford World Map: Medieval World Maps and their Context.* London: British Library, 2006: 293–303.

Smyth, Marina. *Understanding the Universe in Seventh-Century Ireland.* Woodbridge: Boydell, 1996.

Speroni, Charles. 'The Folklore of Dante's Dolphins'. *Italica* 25 (1948): 1–5.

Stahl, William H. *Roman Science: Origins, Development, and Influence to the Later Middle Ages.* Madison: University of Wisconsin Press, 1962.

Stevens, Wesley M. 'Compotistica et Astronomica in the Fulda School'. In Margot H. King and Wesley M. Stevens, eds., *Saints, Scholars and Heroes,* 2:27–63.

—— 'The Figure of the Earth in Isidore's *De natura rerum*'. *Isis* 71 (1980): 268–77.

—— 'Scientific Instruction in Early Insular Schools'. In Michael W. Herren, ed., *Insular Latin Studies: Papers on Latin Texts and Manuscripts of the British Isles, 550–1066.* Toronto: Pontifical Institute of Mediaeval Studies, 1981: 83–111.

—— 'Sidereal Time in Anglo-Saxon England'. In Calvin B. Kendall and Peter S. Wells, eds., *Voyage to the Other World: The Legacy of Sutton Hoo.* Minneapolis: University of Minnesota Press, 1992: 125–52.

Sukale-Redlefsen, Gude. *Die Handschriften des 12. Jahrhunderts.* Katalog der illuminierten Handschriften der Staatsbibliothek Bamberg, 2. Wiesbaden: Harassowitz, 1995.

Tabulae codicum manu scriptorum praeter graecos et orientales in Bibliotheca Palatina Vindobonensi asservatorum. 10 vols. Vienna, 1864–1899.

Taub, Liba. *Aetna and the Moon: Explaining Nature in Ancient Greece and Rome.* Corvallis: Oregon State University Press, 2008.

—— *Ancient Meteorology.* London and New York: Routledge, 2003.

Teyssèdre, Bernard. 'Les Illustrations du *De natura Rerum* d'Isidore'. *Gazette des Beaux-Arts* 56 (1960): 19–34.

Thulin, C. *Die Handschriften des Corpus Agrimensorum Romanorum.* Abhandlungen der Königlich Preussischen Akademie der Wissenschaften, Philosophisch-Historische Class. Fasc. 2. Berlin: Verlag der Königlichen Akademie der Wissenschaften, 1911: 1–102.

von Büren, Veronika. 'Le "De natura rerum" de Winithar'. In Carmen Codoñer and Paulo Farmhouse Alberto, eds., *Wisigothica After M.C. Díaz y Díaz.* Florence: SISMEL/Edizioni Galluzzo, 2014: 387–404.

—— 'Isidore, Végèce et Titanus au VIIIe siècle'. In Pol Defosse, ed., *Hommages à Carl Deroux,* vol. 5. *Christianisme et moyen âge: Néo-latin et survivance de la latinité.* Brussels: Latomus, 2003: 39–49.

Vossen, P. 'Über die Elementen-Syzygien'. In Bernhard Bischoff and Suso Brechter, eds., *Liber Floridus: mittellateinische Studien Paul Lehmann gewidmet.* St. Ottilien: Eos Verlag der Erzabtei, 1950: 33–46.

Wallis, Faith. 'What a Medieval Diagram Shows: A Case Study of *Computus*'. *Studies in Iconography* 35 (2015): 1–40.

Walsh, Maura, and Dáibhí Ó Cróinín. *Cummian's Letter De controversia Paschali and the De ratione conputandi*. Toronto: Pontifical Institute of Mediaeval Studies, 1988.

Warntjes, Immo. *The Munich Computus: Text and Translation; Irish Computistics between Isidore of Seville and the Venerable Bede and its Reception in Carolingian Times*. Stuttgart: Franz Steiner, 2010.

— Review of Calvin B. Kendall and Faith Wallis, *Bede: On the Nature of Things and On Times*. In *The Medieval Review* 12.08.01 (2012).

Wilmart, André, ed. *Codices Reginenses Latini*. 2 vols. Rome: Bibiotheca Vaticana, 1937–1945.

Winsbury, Rex. *The Roman Book: Books, Publishing and Performance in Classical Rome*. London: Duckworth, 2009.

Wood, Ian. 'The Problem of Late Merovingian Culture'. In Gerald Schwedler and Raphael Schwitter, eds., *Exzerpieren–Kompilieren–Tradieren. Transformationen des Wissens zwischen Spätantike und Frühmittelalter*. Forthcoming.

Woodward, David. 'Medieval *Mappaemundi*'. In J.B. Harley and David Woodward, eds., *The History of Cartography*. vol. 1. *Cartography in Prehistoric, Ancient and Medieval Europe and the Mediterranean*. Chicago: University of Chicago Press, 1987: 286–370.

Wright, M.R. *Cosmology in Antiquity*. London and New York: Routledge, 1995.

Zaluska, Yolanta. *Manuscrits enluminés de Dijon*. Paris: Éditions du Centre national de la recherche scientifique, 1991.

Digitized Manuscript Repositories and Catalogues

Belgica. Bibliothèque numérique de la Bibliothèque royale de Belgique. http:// belgica.kbr.be/.

Biblioteca Apostolica Vaticana (BAV). *Catalogo Manoscritti della Biblioteca Apostolica Vaticana*. http://193.43.102.72/gui/html/index.jsp.

Biblioteca Medicea Laurenziana. *Biblioteca Firenze. Teca digitale*. http://teca. bmlonline.it/TecaRicerca/index.jsp.

Biblioteca Pinacoteca Accademia Ambrosiana. *Biblioteca digitale*. http:// ambrosiana.comperio.it/biblioteca-digitale/.

Bibliotheca Laureshamensis digital. *Virtual Monastic Library of Lorsch*. http:// bibliotheca-laureshamensis-digital.de/.

Bibliothèque de Verdun. *Communauté de Communes de Verdun. Manuscrits de Verdun*. [Manuscrits numérisés]. www1.arkhenum.fr/bm_verdun_ms/_app/.

BLB: Badische Landesbibliothek. *Digitale Sammlungen*. http://digital. blb-karlsruhe.de/blbhs/Handschriften.

BL-Digitised Manuscripts. British Library. *Digitised Manuscripts.* http://bl.uk/ manuscripts/.

BNP. Biblioteca Nacional de Portugal. *Biblioteca digital.* http://purl.pt/index/geral/ PT/index.html.

BVMM: Bibliothèque virtuelle des manuscrits médiévaux. http://bvmm.irht.cnrs. fr/.

Carolingian Culture at Reichenau and St Gall. *The Libraries of Reichenau and St Gall.* www.stgallplan.org/stgallmss/.

CEEC: Codices Electronici Ecclesiae Coloniensis. www.ceec.uni-koeln.de/.

Columbia University Archive. http://archive.org/.

The Digital Walters. www.thedigitalwalters.org/01_ACCESS_WALTERS_ MANUSCRIPTS.html.

e-codices: Virtual Manuscript Library of Switzerland. www.e-codices.unifr.ch/en.

Gallica. *Bibliothèque numérique.* http://gallica.bnf.fr/.

HathiTrust *Digital Library.* www.hathitrust.org/.

HMML (Hill Museum and Manuscript Library). *The OLIVER database.* www. hmml.org/oliver.html.

Internet culturale. *Cataloghi e collezioni digitale delle biblioteche italiane.* www. internetculturale.it/opencms/opencms/it/.

Jordanus. An International Catalogue of Mediaeval Scientific Manuscripts. http:// archimedes.mpiwg-berlin.mpg.de/iccmsm.html.

Det Kongelige Bibliotek. *E-manuskripter.* www.kb.dk/da/nb/materialer/ haandskrifter/HA/e-mss/e_mss.html.

Manus online. Censimento dei manoscritti delle biblioteche italiane. http://manus. iccu.sbn.it/.

MDZ: Münchener Digitalisierungszentrum. *Digitale Bibliothek.* www.digitale-sammlungen.de/.

ODL: *Oxford Digital Library.* www.odl.ox.ac.uk/.

Parker Library, on the Web. http://parkerweb.stanford.edu/parker/actions/page. do?forward=home.

— *The Production and Use of English Manuscripts 1066 to 1220.* University of Leicester. http://www.le.ac.uk/english/em1060to1220/.

Ville de Laon. *Bibliothèque municipal.* http://manuscrit.ville-laon.fr/_app/index. php.

Wallis, Faith. 'The Calendar and the Cloister'. Digital facsimile of St John's College Oxford MS 17, with commentary. http://digital.library.mcgill.ca/ms-17.

INDEX OF SOURCES AND PARALLEL PASSAGES

1. Biblical citations

Gen. 1:5   111
  1:7–8   136
  1:14   123
  8:5   119
  9:13–15   159

Exod. 23:15   126
  25:30   112
  32:19   112
  34:18   126

Num. 14:33–36   113

2 Kings/2 Sam. 11:1   114

3 Kings/1 Kings 8:27   135

4 Kings/2 Kings 25:1–7   112
  25:8–9   113
  25:25   113

2 Paral/2 Chron. 2:6   135
  6:18   135

Job 9:9   151
  26:7   172
  37:21   160
  38:31–32   151
  38:32   154

Ps. 17:8 (18:8)   174
  18:2 (19:1)   133
  80:4 (81:4)   112
  102:2 (103:2)   138
  103:4 (104:4)   163
  131:7 (132:7)   174
  135:6 (136:6)   172
  146:8   150
  148:4   135

Eccles. 1:6   155
  1:7   169

Isa. 11:2–3   143
  14:12   154
  14:13–19   154
  51:6   133
  59:9   140
  61:1–2   121
  61:2 [+58:5?]   120

Jer. 39:1–8   112
  41:2–3   113
  52:12–14   113

Dan. 9:24   115
  12:7   123

Hos. 4:5–6   113

Amos 5:8   160
    9:6   160

Nahum 1:3   165

Zech. 8:19   112, 113
    14:4   174

Mal. 4:2   137, 140

Wisd. of Sol. 5:6   140
    7:17   107

Ecclus. 16:18   135
    18:2   172
    18:3   173

Matt. 16:2–3   166
    25:41   175

Luke 4:19   121
    12:55   166

John 1:10   127

Rom. 1:20   151

1 Cor. 7:31   127
    15:41   150

2 Cor. 4:6   111
    6:2   120
    12:2   136

2 Thess. 2:8   173

Apoc. 2:26, 28   154
    20:9–10   175
    20:14–15   159
    22:16   154

2. Classical, patristic and medieval writers

Ambrose *Hexaemeron* 1.1.4   134
    1.3.10   134
    1.6.21   122
    1.6.22   172
    1.7.25   172
    2.2.5–6   135
    2.2.6   135
    2.3.9   136
    2.3.12   124, 134, 137
    2.3.13   137
    2.3.14   137, 170
    2.4.15   133
    2.4.16   129, 157, 161
    3.2.8   169
    3.4.18   131, 132
    4.3.9   137
    4.3.11   156

    4.5.21   123, 124
    4.5.23   124
    4.5.24   121, 144
    4.6.25   138
    4.6.26   138
    4.6.27   150
    4.7.29   142, 144
    4.7.30   168
    4.8.32   142
    4.11   114
    6.2.7   176
    6.2.8   156

Aratus *Phaenomena*   166

Atta, Titus Quinctius   369

Augustine *Contra Faustum* 12.22
    159
    18.5   117, 118
De civitate Dei 4.10   117
    5.6   144
    8.2   121
    12.16   123
    15.14   119
    16.17   175
    18.19   118
De Genesi ad litteram 2.5   136
    2.18   155
    4.34.54   168
De quantitate animae 4.6   162
Enarrationes in Psalmos 10.3   141,
    142
    10.4   146
    138.16   113
    147.2.3   162
Epistolae 55.7   142
Quaestiones Evangeliorum 1.7   111

Augustus 171

Bede *De natura rerum* 4   130
De temporibus 1   126
De temporum ratione 3   126
    7   156
    11   120, 126
    39   122
    41   122

Calcidius *Timaeus a Calcidio*
        translatus commentarioque
        instructus 69   146

Cassiodorus *Institutiones* 2.7.2   134,
    138

Cicero *Timaeus* 9.29   148

ps.-Clement *Recognitiones*, tr. Rufinus
    8.21.2   123, 134

8.23.1–2   162
8.24.2   169
8.42.3–7   159
8.45.2   139
8.45.3   140, 167
8.45.5   140
8.45.6

Dracontius *De laudibus Dei*
    1.733–37   142

Ennius *Annales* 557   133

Gregory *Homeliae in Hiezechihelem*
        prophetam 1.8.29   159
    1.10.28   174
Moralia in Iob 9.11.13   152
    9.11.14   153
    17.26.36   160
    18.19.31   165
    18.20.32   163
    20.2.5   161
    29.20.37   162
    29.31.67   153
    29.31.68   153
    29.31.73   152
    32.15.25   133
    34.14.25   142

Hilary of Poitiers *Tractatus super*
        Psalmos 135.9   136
    135.10   136

Horace *Odes* 2.10.11–12   158

Hyginus *De astronomia* preface   146,
    147
    1.4   134
    1.8   129, 130
    1.9   175
    2.21   152, 153
    2.35   155
    2.42–43   116

4.8   146
4.9   156
4.13   139, 150
4.14   138, 142, 143, 144, 145, 146,
      147
4.19   111, 114, 117

Isidore of Seville *Differentiae* 1   113
  Etymologiae 3.23.2   127
  3.36   133
  3.43.1   134
  3.44.3   128
  3.44.4   128
  3.45   134
  3.49   137
  3.50   138
  3.61   150
  3.66.3   147
  3.67   147
  3.71.4–5   113
  3.71.6–9   152
  3.71.8   148
  3.71.10–11   153
  3.71.14   155
  3.71.15   155
  3.71.16   154
  3.71.19   154
  5.29.1   126
  5.30.1   111
  5.30.5   111, 116
  5.30.6   116
  5.30.7   116
  5.30.8   116
  5.31.1   114, 125
  5.31.4   115
  5.31.6–7   115
  5.31.8   115
  5.31.9   115
  5.31.11   115
  5.31.12   115
  5.33.13   119
  5.33.14   119
  5.36.3   121

5.36.4   122
5.37.1   122
5.37.3   122
6.17.21–22   121
6.17.27   122
6.17.28   114
6.18.4   112
6.18.11   112
9.7.10   160
12.6.11   165
13.3.2   130
13.6.5   128
13.7.1   160
13.7.2   160
13.11.16   165
13.14.2   169
13.15.1   168
13.16.7   171
13.17.1   171
13.18.1   171
13.18.2   171
13.18.6   171
13.21.2   172
14.1.2   173
14.8.41   171
14.8.43   172
Liber numerorum 8.45–46   117

Jerome *Commentaria in Amos*
      2.5.8   160
  Commentaria in Danielem
      12.7   123
  Commentaria in Isaiam
      6.13.10   150
      17.61   121
  Commentaria in Osee 3.12.10   112
  Commentaria in Zachariam
      1.1   112
      2.8.18–19   113, 118
  Commentarius in Ecclesiasten
      1.5   139
      1.6   139, 156
  Homeliae IV in Jeremiam 5   157

Justinus *Epitome* 174

Leiden Glosses

Lucan *Civil War* 1.529 154
  1.538–9 146
  1.643 147
  2.269–71 158
  3.252 152
  4.106–07 128
  5.561–64 151
  5.623–26 169
  7.160 158
  10.199–200 135
  10.201–03 147

Lucretius *De natura rerum*
  2.381–87 158
  6.164–72 157
  6.225–27 158
  6.685 162

Macrobius *Commentarii in Somnium Scipionis* 1.12.2–8 118
  12.20–21 118

Martianus Capella *De nuptiis Philologiae et Mercurii*
  8.883 148

Naevius, Gnaeus 171

Nigidius 166

Origen *De principiis* tr. Rufinus
  2.3.6 168

Pacuvius, Marcus 171, 172

Plato *Timaeus* 34A 135
  38D 148
  43B 135

Pliny *Historia naturalis* 2.6.39 148
  18.87.361 165

Seneca *Naturales quaestiones* 7.4.1 155

Solinus *Collectanea rerum memorabilium* 1.93–94 127
  23.20–21 168

Statius *Thebaid* 1.354 157

Suetonius *Prata* 152 165

Tyconius *Liber regularum* 7 140

Varro Atacinus 128, 332

Vegetius Epitoma rei militaris

Vergil *Georgics* 1.233 128
  1.235–36 127
  1.237–38 130
  1.240–43 127
  1.336 148
  1.365–67 151
  1.432 166
  1.441–43 166
  1.475 173
  1.479 173
  1.487–88 158
  1.488 154
  Aeneid 1.47 117
  1.301 172
  2.698 158
  3.138–39 167
  5.20 160

GENERAL INDEX

Aetius 217n87
Aetna (Latin poem) 250–51
Africa *see* continents
Agamemnon 243
Alcobaça 44
Alcuin 271
Aldhelm 47, 53, 56, 278
allegory 19–21, 23, 47, 50, 55, 62,
    192, 215, 218, 227, 231,
    233–34, 250
Ambrose 9, 12, 15–16, 21–22, 42,
    123, 131, 133, 135–37, 156,
    161, 170, 172, 189–90,
    202–04, 209–13, 215, 217,
    227, 231, 234, 246–47, 249,
    251–52
Amos 160
Amyclae 171
Anaximander 217n87
Anglo-Saxon England 33, 47, 60–61
Anio (river) 172
Aparctias interpolation 31, 34,
    46, 49–50, 58–59, 65,
    163n321
Apollo 243–44
Aratus of Soli 139, 166, 195, 226,
    240–42
    Aratus latinus 240
Arévalo, Faustino 64–65
Aristotle 190, 197, 202–03, 223, 230,
    232–36, 239–40, 246–47,
    249–50
Asia *see* continents
astral determinism 23, 188

astrology 21, 182, 226
    Priscillianist 23n33
astronomy 12, 226
Asturia 257n2
atomic theory 231–32, 246
Atta 48, 171
Athens 246
Augsburg 64
augury, pagan 231n122
Augustine 5, 12, 14, 16, 22, 25, 42,
    54, 57, 141, 155, 175, 177,
    187–88, 210–11, 215, 227–28,
    235, 247–48, 251
Augustus 48, 117, 171, 183
Aulus Gellius 180, 236
authority, biblical vs. classical 22
Avienus (Rufius Festus Avienus)
    240

Babylon 113, 180
Baetica 4
Basil of Caesarea 202, 210, 228
Becker, Gustav 65, 266
Bede 27, 47, 56–58, 62–64, 118n64,
    126n101, 159n299, 190–92
    On the Nature of Things (DNR) 53,
    56–57, 62–63, 67, 74–75,
    83–84, 87–88, 91, 93–95, 99,
    130n125, 202n54, 212–13,
    214n81, 229n116, 233n126,
    243, 264
    On Times (DT) 57, 62–63, 75,
    83–84, 87–88, 91, 93–95, 99,
    183, 229n116

The Reckoning of Time (DTR)
62–63, 67, 69, 71, 73–75, 81,
84, 87–88, 93–94, 99, 120n69,
122n82, 156n280, 182–84,
191–92, 202n54, 216, 224,
243
Benedict Biscop 56, 58
Berlin 65
Bischoff, Bernhard 55
bissextile day 122n84, 123
Bobbio 57–58
Boethius 268–71
Boniface 61
Braulio, bishop of Zaragoza 7–8, 16,
27, 54
Byzantine (Roman) Empire 4–5
in Spain 3

Cadiz 4
Calcidius 15, 64, 197, 199–200, 219,
223
calendar, Egyptian 114n26, 182–84
Julian 114n26, 120n69, 122n84, 191
Jewish 182
medieval 185
Roman 180, 182–83
Cantabria 257n2
Canterbury 55–58, 61, 278
Cartagena 3
Cassiodorus 46, 118n59, 200, 266,
268
Institutions 46, 67, 92, 200, 214n81,
268–71
Censorinus 187
chapter 44(–), as addition 30, 32, 34,
47–49, 56, 63, 65
chapter 48(47), as addition 30, 34,
42–45, 56–57, 63, 65
textual insertions in 272–74
Charlemagne 204
Chelles 49, 60
'Chronicle of 754' 260
Chryses 243

Cicero 9, 15, 135, 222, 226, 240, 242
Clemens Scottus 53n8
Clement I (pope) 232
Clement of Alexandria 188
(pseudo-)Clement 15, 22, 139, 159,
162, 168–69, 205, 232, 235,
247
clouds 159–61, 233
mystical sense of 160–61
comets 154–55, 227
compilatio, technique of 25–26
computus 55, 59, 63–64, 186–88, 191,
195–96, 223
Constantinople 4, 33
constellations 219–20, 226
continents, three (Asia, Europe,
Africa) 175, 251–52,
278–79
Cordoba 43
Cosmas Indicopleustes 13–14, 21,
218n88
cosmography 14
Christian 10
cosmology 11, 14, 64, 179
councils held at Toledo 6, 9
court days 113
Creon 243–44

Dalché, Patrick Gautier 252
Daniel 115, 123, 188
David 135, 138, 244
day 111–14, 179–81
mystical sense of 111–13
de qualitate formula 35, 42, 44, 50,
60n40, 261
De Trinitate interpolation 272–73
De Ventis see 'Poem of the Winds'
diagrams (*rotae*) 28–30, 219
Diagram 1 (wheel of the months)
28–29, 184–85
Diagram 1A (seasons) 29n10, 81
Diagram 2 (wheel of the year) 28,
189–90

Diagram 3 (wheel of the circles of the world) 29, 195–96
Diagram 4 (cube of the elements), 29, 198–202
Diagram 5 (wheel of the microcosm) 29, 202–03
Diagram 5A (phases of the moon) 29, 216
Diagram 6 (wheel of the planets) 29, 223
Diagram 6A (four colours of the rainbow) 29n10, 73, 76, 78, 83, 86, 94
Diagram 7 (wheel of the winds) 29–30, 32, 34–35, 42–46, 50, 235–39, 261–64, 267–68
  KOCMOC inscription in 35, 42–44, 236–37, 261, 266
Diagram 7A (mountains of Sicily) 29n10, 69, 78, 84, 86, 90, 96–97
T-O Map 29–30, 42–43, 62, 197, 251–52, 263, 278–79
textual insertions in 274
Don (river) 175
Donation of Constantine 9
Donatus 6–7
du Breul, Jacques 64

earth 172–76, 249–53
  circumference of 175–76, 252–53
  parts of 175–76, 251–53
  position of 172–73, 175
  size of 214
  where inhabited 251
earthquakes 173–74, 250–51
  mystical sense of 173–74
Eastwood, Bruce 200–01
Ebroin 58
Ecgbert, archbishop of York 32
eclipse 10, 17–18, 23, 51–52, 121, 150, 187, 209, 215

allegorical meaning of 23, 145–46
  at Crucifixion 218
  of moon 145–46, 217–19
  of sun 144–45, 217–19, 225
education in Visigothic Spain 6–7
Egypt 170
Einhard 204
elements, four (earth, air, fire, water) 130–32, 197
  fifth element (ether) 225
Ennius 48, 133
epacts 114, 180
Epicureans 231–32, 245
Epicurus 246
equinoctial days 114
equinox 126–27, 191–92
era 122
Eratosthenes 252
Etna, Mount 174–75, 250–51
  mystical sense of 174–75, 251
etymology 8, 11n3, 25–26, 203, 249, 265n4
Europe see continents
Eusebius of Caesarea 5, 178
Eutropius 7
evaporation 168
exhalations, theory of 230, 244n151, 246–47, 250

Fabianus 248
fast days 112–13
feast days 112
feria 115
Festus 48
flat-earth 13, 195n40, 252n161
Fleury 35, 59, 60, 63
Florentina, sister of Isidore 4, 9
Fontaine, Jacques 5, 11, 13, 17, 25, 28–32, 35, 42–44, 47–51, 55, 58, 65, 179, 184, 189–90, 195, 198–99, 206, 221, 231, 234–36, 267, 272

Fredegar 260
Fulda 17, 61
Fulgentius, bishop of Ecija, brother of
    Isidore 4, 9

Galen 246
Gedaliah 113
Gennadius 9
Germanicus Julius Caesar 240
Germany 65
Gildas 277–78
Gog and Magog 5
Gorman, Michael 30, 54–55, 58, 200
Goths 5
grammar/*grammatica* 5n8, 6–8, 10,
    19, 26
  as a principle of knowledge 7–8
Gregory the Great 15–16, 18, 227, 250
Grial, Juan 64
Grosseteste 179
Guadalquivir (river) 4
Guillaumin, Jean-Yves 11
Gundemar (Visigothic king) 259

Hadrian (abbot) 55–56, 58
hail 161–62, 234
  mystical sense of 162
Ham 278
hemerology 11, 14–16, 179, 188–89
heaven 133–35, 203–09
  rotation of 134, 209
  spiritual sense of 133, 203
  parts of 133–34
  zones of 21, 116n46
Hen, Yitzak 52
heresy, Priscillianist 23–24, 188, 228
Herren, Michael 48
Hesiod 243
hexaemera 15
hexameter verse 266
hiemisphaeria interpolation 31, 34,
    46, 49–50, 56, 58–59, 65,
    134n140

Hilary of Poitiers 16, 21, 210
Hippocrates 244–45
Hisperica Famina 48, 55
history
  Christian genre of 5
  moral function of 5
Homer 243
Horace 158
Hosea 113
hours 185–86, 192
Hrabanus Maurus 17, 27
humours, four 203
Hyginus 14, 22–23, 42, 45, 139, 175,
    196, 216, 220, 226, 229, 251

intercalary days 114
intercalary month 144n201
Ireland 55, 57
  Irish influence on Isidore 32–33,
    48–49, 50n58
Isaiah 120
Isidore, bishop of Seville
  and biblical exegesis 8
  and Church discipline 8
  education of 6–7
  life 3–4
  Northumbrian influence on 32
  On the Nature of Things 3, 5
    appeal to reason in 21–24
    as providing Christian erudition
      24–26
    as exercise in *correctio*
      rusticorum? 17–18
    authorship of additions 32–34,
      49, 49n55
    Bede's manuscript of 57
    Canterbury manuscript of 56
    composite construction of 27–50
    date of 10
    dedication of 12–13, 16
    illustrations in *see* diagrams
    in context 10–14
    in print 64–65

interpolations 46–47, 51
  see also *Aparctias*; chapter
    44(–); chapter 48(47); *De*
    Trinitate; *hiemisphaeria*;
    Maeotis; mystical addition;
    Sisebut, *Epistle*; Solinus;
    Vegetius
  manuscripts and editions of
    66–101
  need for new edition of 32n16
  occasion for 16–18
  popularity of 54
  preface of 13
  purposes of 18–21
  recensions of 27–28, 30–34,
    52–53, 57–58, 60, 63–65
    English type (long recension)
      61, 277–79
    long recension 47–50
    medium recension 42–45
    short recension, two types of
      34–42
    Zofingen 'metamorphosis'
      60–64, 275–77
  reception of 51–65
  structure of 14–16
  titles of, various 27–28
'research assistants' of 33–34, 51
scriptorium of 33–35
works, other
  Allegories 9
  Book of Numbers 9, 11–12, 19,
    232
  Chronicle 5, 54
  Differences 7–8, 18, 219, 225
  Etymologies 8, 10–11, 16, 19,
    24–26, 33, 46, 48–50, 54, 57,
    61–62, 183, 187, 196, 207,
    216, 219, 221, 233, 252
  Expositions on the Old Testament
    9
  History of the Goths 5, 52
  Isidoriana 9

Lives and Deaths of the Fathers
  9
On the Catholic Faith against the
  Jews 9, 52
On Famous Men 9
On Heresies 9
On the Origin of Offices 9, 44
Prefaces 9
Sentences 9, 52
Synonyms 8, 20
Versus 25
Italy 57–59, 268

Japheth 278
Jeremiah 113
Jerome 5, 9, 16, 50, 113, 213, 228, 231,
  233, 240
Jerusalem 112–13, 180
Jews 52
  in Spain 228
Job 22, 151, 159, 172, 226, 233
John Damascene 228
Julius Caesar 117
Jupiter (god) 117, 274
Jupiter (planet) 116, 148, 223
  see also Phaëton
Justinian 4
Justinus 174, 251

KOCMOC inscription *see* Diagram 7
Kühnel, Bianca 237

la Bigne, Margerin de 27, 64
Laistner, M.L.W. 33
Leander, bishop of Seville, eldest
  brother of Isidore 4, 6–7, 16,
  18n23
Leiden Glossary 55–56
leonine hexameter 266
Liber glossarum 61n45
Liber rotarum 27–28, 56, 61–62, 184,
  257
light, symbolism of 14

lightning 157–58, 230–32
Loire valley 59
Lucan 15, 22, 128, 135, 146–47,
    151–52, 158, 168–69, 173,
    209, 212, 218, 221, 225,
    233n126
Lucifer (Satan) 153–54
Lucifer (planet Venus) 116, 148–49,
    151, 153–54, 181
Lucretius 14–15, 22, 27, 52, 158,
    162, 167n345, 178, 231–32,
    245–46, 248–51
lunar-solar cycles 223–24
lustrum 122

Macrobius 15, 64, 195–96, 207–08,
    223
    Saturnalia 118n64, 180, 183
Máeldub, Irish monk 56
Maeotis interpolation 31, 34, 45, 65
Malachi 137
Malmesbury 56
Manilius 193, 231
manuscripts
    Amiens, Bibliothèque communale
        222 53n7
    Avranches, Bibliothèque municipale
        109 60n37, 261
    Baltimore, Walters Art Gallery 73
        63n50
    Bamberg, Staatsbibliothek, Msc.
        Nat. 1 65, 261, 274
    Bamberg, Staatsbibliothek, Patr. 61
        58, 65, 100n1
    Barcelona, Biblioteca de Catalunya
        569 46n43
    Basel, Universitätsbibliothek Lat.
        F.III.15a 61
    Basel, Universitätsbibliothek Lat.
        F.III.15f 60–61
    Berlin, Staatsbibliothek zu Berlin,
        Preussischer Kulturbesitz,
        Hamilton 689 274

    Berlin, Staatsbibliothek zu Berlin,
        Preussischer Kulturbesitz,
        Phillipps 1830 262, 264
    Berlin, Staatsbibliothek zu Berlin,
        Preussischer Kulturbesitz,
        Phillipps 1833 60n39
    Bern, Burgerbibliothek A.92/20 35,
        60n37, 261
    Bern, Burgerbibliothek 212/1 264
    Bern, Burgerbibliothek 224 60n37,
        261
    Bern, Burgerbibliothek 417 60n39,
        274
    Bern, Burgerbibliothek 610 49,
        60n38
    Bern, Burgerbibliothek 611 64,
        266
    Besançon, Bibliothèque municipale
        184 261
    Cambrai, Bibliothèque municipale
        937 42–45, 49, 261, 274
    Cambridge, Corpus Christi College
        291 63n50
    Cava de' Tirreni, Archivio della
        Badia 1 173n381
    Cava de' Tirreni, Archivio della
        Badia 3 202n54
    Cologne, Erzbischöfliche Diözesan-
        und Dombibliothek 83(II)
        35, 44, 53, 60, 261, 278
    Cologne, Erzbischöfliche Diözesan-
        und Dombibliothek 99 35
    Copenhagen, Universitetsbibliothek,
        Frag. 19.VII 61
    Einsiedeln, Stiftsbibliothek 360
        273
    Escorial, Real Biblioteca de San
        Lorenzo E.IV.14 35, 64n52,
        262
    Escorial, Real Biblioteca de San
        Lorenzo K.I.12 64n52
    Escorial, Real Biblioteca de San
        Lorenzo M.II.23 35, 64n52

Escorial, Real Biblioteca de San
  Lorenzo R.II.18 27, 30n11,
  42–46, 59, 64–65, 252
Escorial, Real Biblioteca de San
  Lorenzo &.I.3 202n52
Exeter, Cathedral Library 3507
  202, 277–78
Florence, Biblioteca Medicea
  Laurenziana, Amiatino I
  173n381
Florence, Biblioteca Medicea
  Laurenziana, S. Marco 582
  274
Florence, Biblioteca Medicea
  Laurenziana, Plut. 22 dex. 12
  35n30, 43–45, 262
Florence, Biblioteca Medicea
  Laurenziana, Plut. 27 sin.
  9 55
Florence, Biblioteca Medicea
  Laurenziana, Plut. 29.39
  264, 274
Florence, Biblioteca Riccardiana
  379/4 274
Karlsruhe, Badische
  Landesbibliothek, Augiensis
  106 263
Karlsruhe, Badische
  Landesbibliothek, Augiensis
  229 49
Laon, Bibliothèque Suzanne
  Martinet 422 60, 237,
  261
Laon, Bibliothèque Suzanne
  Martinet 423 56, 261
Lisbon, Biblioteca Nacional, Alc.
  CCIX/446 43–45, 262
London, British Library, Cotton
  Domitian I 277, 279
London, British Library, Cotton
  Vitellius A XII 278
London, British Library, Harley
  2660 273

London, British Library, Harley 3017
  60n39
London, British Library, Harley
  3035 273
Madrid, Biblioteca Nacional, Vitr.
  13–1 173n381
Milan, Biblioteca Ambrosiana H 150
  inf. 35, 60n37
Munich, Bayerische Staatsbibliothek
  CLM 14300 28, 29n10,
  184, 189–90, 196, 198, 200,
  235–36, 267
New York, Columbia University
  Library, Plimpton 251
  59n34, 261
Oxford, Bodleian Library, Ashmole
  393 44n40, 262
Oxford, Bodleian Library, Auct.
  F.2.20 202, 278
Oxford, St John's College 17
  63n50
Oxford, St John's College 178 278
Paris, Bibliothèque nationale lat.
  5239 63n50
Paris, Bibliothèque nationale lat.
  5543 60n39, 63n50
Paris, Bibliothèque nationale lat.
  6400G 35, 42, 44, 59, 59n35,
  60n37, 185, 261
Paris, Bibliothèque nationale lat.
  10457 61n46
Paris, Bibliothèque nationale lat.
  10616 61, 273
Paris, Bibliothèque nationale lat.
  15171 43, 45, 64, 262
Paris, Bibliothèque nationale, Nouv.
  acq. Lat. 448 49–50, 261
Salzburg, Stiftsarchiv St Peter a. IX.
  16 173n381
St Gall, Stiftsbibliothek 238 56,
  61–62, 216
St Gall, Stiftsbibliothek 240 49–50,
  252, 261

Strasbourg, Bibliothèque nationale et universitaire 326   63n50, 261, 264
Trier, Stadtbibliothek 1084/115 264
Trier, Stadtbibliothek 2500   63n50, 262, 264
Vatican City, BAV, Pal. lat. 834   261, 274
Vatican City, BAV, Pal. lat. 1448   56
Vatican City, BAV, Reg. lat. 25 60n37
Vatican City, BAV, Reg. lat. 123 63n50
Vatican City, BAV, Reg. lat. 255 202n52
Vatican City, BAV, Reg. lat. 309 63n50
Vatican City, BAV, Reg. lat. 1260 60n37
Vatican City, BAV, Reg. lat. 1573 45n41, 274
Vatican City, BAV, Urb. lat. 100   274
Verdun, Bibliothèque municipale 26 261, 274
Vienna, Österreichische Nationalbibliothek 387   196
Weimar, Herzogin Anna Amalia Bibliothek 414a   61
Zofingen, Stadtbibliothek Pa 32   61, 272–73
Mars (planet)   115n34, 116, 148, 181, 222–24
see also Vesper
Mars (god)   117
Martianus Capella   11, 64
Martín, José Carlos   46–47
Maurice   4n6
McCready William D.   48
Mediterranean   3, 175, 246, 248
Mercury (god)   116n43, 117
Mercury (planet)   116, 147, 148n225, 149, 221n96, 222–24

Merovingian Gaul   42, 59–60
meteorology   11, 14, 18, 230
Methodists (school of ancient medical thought)   246
Metonic cycle   187, 224
Milky Way   116n43
Monte Cassino   58–59, 100n1
month   117–20, 182–86
Egyptian   120n69
moon   141–44, 213–19, 240–43
course of   143–44, 216–17
light of   141–43, 215–16
mystical sense of   142–43
sidereal   216
size of   138, 213–14
synodic   216
Moses   112, 180
Murdoch, John E.   200
Mynors, R.A.B.   267–68
mystical addition   31–32, 34, 49–50, 56, 65, 179–80

Naevius   48, 171
Nahum   165
natural phenenomena as symbols   20–21
Nebuchadnezzar   112–13, 180
Neckam, Alexander   27
Nennius   278
Neomenia, day of   112
night   114–15, 156, 181, 228–29
mystical sense of   114
Nigidius (Publius Nigidius Figulus)   166, 240, 242
Nile   170, 175, 248–49
Noah   278
North Africa   33
Numa Pompilius   118
number symbolism   11–12, 19

Obrist, Barbara   190, 196, 199–201
ocean   3, 168
as symbol   247

Oedipus 243–44
Olympiad 122
orbis terrarum 3
Origen 178, 210, 227–28
origins, Isidore and the quest for 5–6
origo 9, 25
Ovid 231
Oviedo 43

Pacuvius 48, 171–72
Paestum 171
palimpsest 59
Passover 112
Paul (apostle) 120, 127, 135, 193
Paul the Deacon 59
Pentecost 112
Persian Empire 4
pestilence 167, 243–46
Phaëton (planet Jupiter) 148–49
Phoebe (moon) 258
Phoebus (sun) 257–58
Pillars of Hercules 175
  see also Straits of Cadiz/Hercules
planets 135–36, 209–10
  names of 223
  orbital times of 149, 223
  position of 147–48, 221–24
Plato 15, 21, 134, 197–201, 205, 208
  tradition of 227
Pliny the Elder 8, 33, 48, 181, 190,
    217, 222, 232, 236, 239–40,
    246, 248–49, 253, 266–67
'Poem of the Winds' 64, 264–71
  manuscripts of 268–71
Pompeius Trogus 251
Porphyry 197
Posidonius 168n348, 252
Presocratics 230
Priscillian 23
Priscillianists 228
Proclus 197
Pseudo-Clement *see* Clement
Pseudo-Isidorean Decretals 9

Ptolemy 191, 213–14
  Ptolemaic system of universe 204
Ptolemy III 183

Quattuor a quadris uenti (poem) 266
Quintilian 193
quo ordine formula 252
quorum ordinibus formula 251

rain 160–61, 233–34
rainbow 159, 232–33
  mystical sense of 159
Reccared II (Visigothic king) 54
'reverse engineering', technique of
    12, 26
rivers, names of 249
Rome 33, 58, 65
rotae (wheels) *see* diagrams
rotatim formula 44–45, 262

Sabboth 112
Sallust 173
Sancus, king of the Sabines 118
St Gall 55, 57, 61–62, 100n1
Saturn (planet) 116, 148–49, 211, 220,
    223
scenopegia, day of 112
science 12–15, 211
  in conflict with revelation 22
sea 3, 169–70, 247–48
  depth of 248
  lack of increase of 169, 247–48
  names of 249
  salt in 170, 247–48
  see also Mediterranean
seasons 123–26, 188–91
  allegory of 125–26
  beginnings of 125, 190–91
Seneca 230, 232, 234, 248–49, 251
Septimania 60
Servius 182
Severianus, father of Isidore 4
Seville 4, 17, 33–35, 42–43, 48, 239

Shem 278
Sicily 174
Sisebut (Visigothic king) 12–13, 16, 24, 26–27, 42, 52, 54, 177–78, 218–19, 257n2, 259–60, 266
education of 259
Epistle (poem) 16–17, 31–32, 34–35, 42–45, 51–54, 56, 65, 217, 219, 259, 277
Sisenand (Visigothic king) 54
six ages of the world 5, 62, 116n46
six days of creation 15–16, 19, 50, 181
snow 161, 234
mystical sense of 162
Solinus 14, 46–47, 58–59, 168, 246
interpolation 31, 34, 46–47, 49–50, 58–59, 65, 168n346
Solomon 155, 169, 182–83
solstice 126–27, 191–92
symbolism of 191
solstitial days 114
Sophocles 243
Spain 57–59, 268
Arab invasion of 43, 60
conceived of as empire 5–6
encomium of 6
intellectual resources of 33
Spanish era 5
Stahl, William H. 252
stars 146–56, 219–21
course of 146–47
fall of 150–51, 224–25
light of 150, 224
mystical sense of 150, 152–54
names of 151–55, 226–27
souls of 155–56, 227–28
terms for 219
see also planets
Statius 157
Stevens, Wesley 277
Stoic doctrine 134n142, 146, 205, 211–13, 218

storms and fair weather, signs of 165–66, 239–43
Straits of Cadiz/Hercules (Gibraltar) 4, 171
Straits of Sicily 171
Suetonius (C. Suetonius Tranquillus) 14, 25, 48, 164, 170, 266
Prata ('The Meadows') 165
Suinthila (Visigothic king) 5, 54
sun 137–40, 212–15
course of 139–40, 214–15
size of 138, 213–14, 258
spiritual sense of 137–38, 140, 212–13
superstition 13–15, 17, 21, 23, 177–78, 226
Switzerland 65
synods *see* councils

Taub, Liba 242
tempus, ambiguity of 123n85
Tertullian 25n36
Thales of Miletus 187
Thassos 245
Thebes 243–44
Theodore, archbishop of Canterbury 55, 58
Thomas of Cantimpré 27
Thucydides 246
thunder 157, 230–31
tides 168–69, 246–47
in relation to phases of the moon 168
time-reckoning, secular 180
Titus (emperor) 113, 180
Toledo 6, 13, 16–17, 34, 42–43, 239
T-O Map *see* diagrams
Trier 56
Troy 243
trumpets, day of 112
tullius/tollus 48
Teyssèdre, Bernard 185

universe, as sphere 205, 208

Varro 8, 25n36, 128, 166, 180, 240,
    242, 246
Vegetius 61–62, 236–37, 264n2, 272
    interpolation 272–74
Venus (planet) 116, 148, 181, 221n96,
    222–24
    see also Lucifer
Vergil 15, 22, 24, 26, 51, 127–28,
    130, 148, 151, 155–58, 160,
    166–67, 171, 194, 225, 231,
    233
Verona 61
Vesper (evening star/Mars) 115n34,
    116, 148–49, 151, 154, 181,
    222
Visigoths 6
Vitruvius 236
Vivarium 266
volcanoes 251
von Büren, Veronika 59

Washington, George 17
waters 168–72
    names of bodies of 170–72
    heavenly 136–37, 204–05, 210–12
Wearmouth and Jarrow 56
week 115–16, 182
'Wheels, Book of' see Liber rotarum
wind-rose 219, 235–37, 266
    see also Diagram 7

winds 162–65, 234–39
    Etesian 248–49
    fascination with 30
    names of 163–65, 235–39, 266–67
Winithar 61–62
Winsbury, Rex 33
Wood, Ian 59–60
Woodward, David 252
world 127–32, 192–203
    five circles of 128–30, 194–97
    mystical sense of 127

year 120–23, 186–88
    beginning of 183
    Egyptian 184
    embolismic 121, 188
    Great 187–88
    Jubilee 122
    leap 122
    lunar 121, 144, 188
    natural 121
    solstitial 121, 188

Zainer, Günther 64
Zaragoza 7
Zechariah 113
Zeus 243
zodiac 23, 117, 121, 134, 186, 188,
    190, 205–09, 214, 216, 220,
    223, 226–27
Zofingen 'metamorphosis' see Isidore,
    On the Nature of Things

Printed and bound by CPI Group (UK) Ltd, Croydon, CR0 4YY

27/10/2024

14580407-0001